Water Quality

Claude E. Boyd

Water Quality

An Introduction

Third Edition

 Springer

Claude E. Boyd
School of Fisheries, Aquaculture and Aquatic Sciences
Auburn University
Auburn, AL, USA

ISBN 978-3-030-23337-2   ISBN 978-3-030-23335-8   (eBook)
https://doi.org/10.1007/978-3-030-23335-8

This Springer imprint is published by the registered company Springer Nature Switzerland AG
The registered company address is: Gewerbestrasse 11, 6330 Cham, Switzerland

Preface

Water is commonplace; two-thirds of the earth is covered by the ocean, and nearly 4% of the global land mass is inundated permanently with water. Water exists in the hydrosphere in a continuous cycle—it is evaporated from the surface of the earth but subsequently condenses in the atmosphere and returns as liquid water. Life in all forms depends on water, and fortunately the earth is not going to run out of water; there is as much as there ever was or is ever going to be.

Despite the rosy scenario expressed above, water can be and often is in short supply, a trend that will intensify as global population increases. This results because all places on the earth's land mass are not equally watered. Some places are well watered, others have little water, and water may be deficient in well-watered locations during droughts. The quality of water also varies from place to place and time to time. Most of the earth's water is too saline for most human uses, and pollution from anthropogenic sources has degraded the quality of much freshwater and lessened its usefulness. Evaporation is a water purification process, but salts and pollutants left behind when water evaporates remain to contaminate the returning rainwater.

As the human population has grown, the necessity for producing goods and services necessary or desired by the population has increased, resulting in greater water use and water pollution. Water quality has increased in importance, because the quantity of water often cannot be assessed independently of its quality. Water quality is a critical consideration in domestic, agricultural, and industrial water supply, fisheries and aquaculture production, aquatic recreation, and the health of ecosystems. Professionals in many disciplines should understand the factors controlling concentrations of water quality variables as well as the effects of water quality on ecosystems and humans. Efficient management of water resources requires the application of knowledge about water quality.

Water quality is a complex subject, and unfortunately, the teaching of this important topic is not well organized. In many colleges and universities, water quality instruction is given mainly in certain engineering curricula. The classes emphasize specific aspects of water quality management and tend to focus on the treatment of water for municipal and domestic use and on the methods of improving effluent quality to lessen pollution loads to natural water bodies. Such classes typically have prerequisites that prohibit students from other disciplines from enrolling in them. Specific aspects of water quality are taught in certain courses in the curricula of agriculture, forestry, fisheries and aquaculture, biology, chemistry, physics, geology, environmental science, nutrition, and science education. But, the coverage of water quality in such courses does not provide an understanding of water quality as a whole.

When I began teaching water quality in the College of Agriculture at Auburn University in 1971, available texts on water quality, limnology, and water chemistry were unsuitable for the class. These books were either too descriptive to be meaningful or too complicated to be understood by most students. Students who need training in water quality often have a limited background in physics and chemistry. As a result, the background for each facet of water quality should be provided using only first year college-level chemistry, physics, and algebra. The book *Water Quality: An Introduction* was based on lectures for the class. This book is slanted toward physical and chemical aspects of water quality, but it includes discussions of interactions among physical and chemical variables and biological components in water bodies. Because the chemistry is presented at a basic level, some calculations, explanations, and solutions to problems are approximate. Nevertheless, simplification allows students to grasp salient points with relatively little "weeping, wailing, and gnashing of the teeth." Hopefully, this third edition of *Water Quality: An Introduction* is an improvement over previous editions.

The preparation of this book would not have been impossible without the excellent assistance of June Burns in typing the manuscript and proofing the tables, references, and examples.

Auburn, AL, USA   Claude E. Boyd

Introduction

The universe consists of space, time, and matter behaving in obedience to very complex physical, chemical, and mathematical rules which are incompletely understood or not understood at all by humans. We know most about the earth, and it is a very special place, the only body in the universe known to support living things including what we vainly refer to as intelligent life. One of the main reasons that earth is favorable for life is the presence of liquid water in abundance. The earth has abundant liquid water because of its special features (viz., proper temperature and atmospheric pressure at its surface) and the physical characteristics of water resulting from hydrogen bonding among its molecules.

Water is essential for life and is a major component of living things. Bacteria and other microorganisms usually contain about 90–95% water; herbaceous plants are 80–90% water; woody plants are usually 50–70% water. Humans are about two-thirds water, and most other terrestrial animals contain a similar proportion. Aquatic animals usually are about three-fourths water. Water is important physiologically. It plays an essential role in temperature control of organisms. It is an internal solvent for gases, minerals, organic nutrients, and metabolic wastes. Substances move among cells and within the bodies of organisms via fluids comprised mostly of water. Water is a reactant in biochemical reactions, the turgidity of cells depends upon water, and water is essential in excretory functions.

Water plays a major role in shaping the earth's surface through the processes of dissolution, erosion, and deposition. Large water bodies exert considerable control over air temperature of surrounding land masses. Coastal areas may have cooler climates than expected because of cold ocean current offshore and vice versa. The distribution of vegetation over the earth's surface is controlled more by the availability of water than by any other factor. Well-watered areas have abundant vegetation, while vegetation is scarce in arid regions. Water is important ecologically for it is the medium in which many organisms live.

Water is essential for the production of nearly all goods and services, but it is critical for the production of food and fiber through agriculture, processing of agricultural crops, and domestic purposes, such as food preparation, washing clothes, and sanitation.

Early human settlements developed in areas with dependable supplies of water from lakes or streams. Humans gradually learned to tap underground water supplies, store and convey water, and irrigate crops. This permitted humans to spread into previously dry and uninhabitable areas, and even today, population growth in an area depends upon water availability.

Water bodies afford a convenient means of transportation. Much of the world's commerce depends upon maritime shipping that allows relatively inexpensive transport of raw materials and industrial products among continents and countries. Inland waterways also are important in both international and domestic shipping. For example, in the United States, huge amounts of cargo are moved along routes such as the Mississippi and Ohio Rivers and the Tennessee-Tombigbee waterway.

Water bodies are important for recreational activities such as sportfishing, swimming, and boating, and they are of great aesthetic appeal to people who reflect on nature. Water bodies have become an essential aspect of landscape architecture. Water has much symbolic significance in nearly all regions and especially in the Judeo-Christian faiths.

Water Quality

At the same time, that humans were learning to exert a degree of control over the quantity of water available to them, they found different waters to vary in qualities, such as warmth, color, taste, odor, etc. They noted how these qualities influenced the suitability of water for certain purposes. Salty water was not suitable for human and livestock consumption or irrigation. Clear water was superior over turbid water for domestic use. Some waters caused illness or even death when consumed by humans or livestock. The concepts of water quantity and water quality were developed simultaneously, but throughout most of human history, there were few ways for evaluating water quality beyond sensory perception and observations of its effects on living things and water uses.

Any physical, chemical, or biological property of water that effects natural ecological systems or influences water use by humans is a water quality variable. There are literally hundreds of water quality variables, but for a particular water use, only a few variables usually are of interest. Water quality standards have been developed to serve as guidelines for selecting water supplies for various uses and for protecting water bodies from pollution. The quality of drinking water is a health consideration. Drinking water must not have excessive concentrations of minerals, it must be free of toxins, and it must not contain disease organisms. People prefer their drinking water to be clear and without bad odor or taste. Water quality standards also are established for bathing and recreational waters and for waters in which shellfish are cultured or captured. Diseases can be spread through contact with water contaminated with pathogens. Oysters and some other shellfish can accumulate pathogens or toxic compounds from water, making these organisms dangerous for human consumption. Water for livestock does not have to be of human drinking water quality, but it must not cause sickness or death in animals. Excessive concentrations of minerals in irrigation water have adverse osmotic effects on plants, and irrigation water also must be free of phytotoxic substances. Water for industry also must be of adequate quality for the purposes for which it is used. Extremely high-quality water may be needed for some processes, and even boiler feed water must not contain excessive suspended solids or a high concentration of carbonate hardness. Solids can settle in plumbing systems, and calcium carbonate can precipitate to form scale. Acidic waters and saline waters can cause severe corrosion of metal objects with which they come in contact.

Water quality effects the survival and growth of plants and animals in aquatic ecosystems. Water often deteriorates in quality as a result of human use, and much of the water used for domestic, industrial, or agricultural purposes is discharged into natural water bodies. In most countries, attempts are made to maintain the quality of natural waters within limits suitable for fish and other aquatic life. Water quality standards may be recommended for natural bodies of water, and effluents must be shown to comply with specific water quality standards to prevent pollution and adverse effects on the flora and fauna.

Aquaculture, the farming of aquatic plants and animals, now supplies nearly half of the world's fisheries production for human consumption, because capture fisheries have been exploited to their sustainable limit. Water quality is a particularly critical issue in cultivation of aquatic organisms.

Factors Controlling Water Quality

Pure water is rarely found in nature. Rainwater contains dissolved gases and traces of mineral and organic substances originating from dust, combustion products, and other substances in the atmosphere. When raindrops fall on the land, their impact dislodges soil particles, and flowing water erodes and suspends soil particles. Water also dissolves minerals and organic matter from the soil and underlying formations. There is a continuous exchange of gases between water and air, and when water stands in contact with sediment in the bottoms of water bodies, there is an exchange of substances until equilibrium is reached. Biological activity in the aquatic environment has a tremendous effect upon pH and concentrations of dissolved gases, nutrients, and organic matter. Natural bodies of water tend toward an equilibrium state with regard to water quality that depends upon climatic, hydrologic, geologic, and biologic factors.

Human activities strongly influence water quality, and they can upset the natural status quo. The most common human influence for many years was the introduction of disease organisms via disposal of human wastes into water supplies. Water-borne diseases were until the past century a leading cause of sickness and death throughout the world. The problems of water-borne diseases have been reduced through the application of better waste management and public health practices in most countries. Water-borne diseases are still an issue, but the growing population and increasing agricultural and industrial effort necessary to support mankind are discharging contaminants into surface waters and groundwaters at an increasing and alarming rate. Contaminants include suspended soil particles from erosion that cause turbidity and sedimentation in water bodies and inputs of plant nutrients, toxic metals, pesticides, industrial chemicals, and heated water from cooling of industrial processes. Water bodies have a natural capacity to assimilate contaminants, and this capacity is one of the services provided by aquatic ecosystems. However, if the input of contaminants exceeds the assimilative capacity of a water body, there will be ecological damage and loss of ecological services.

Purpose of Book

Water quality is a key issue in water supply, wastewater treatment, industry, agriculture, aquaculture, aquatic ecology, human and animal health, and many other areas. The practitioners of many different occupations need information on water quality. The principles of water quality are presented in specialized classes dealing with environmental sciences and engineering, but many students in other fields who need to be taught the principles of water quality do not receive this training. The purpose of this book is to present the basic aspects of water quality with emphasis on physical, chemical, and biological factors controlling the quality of surface freshwaters. There also will be brief discussions of groundwater and marine water quality as well as water pollution, water treatment, and water quality standards. The influence of water quality on the aesthetic and recreational value of water bodies will also be discussed.

It is impossible to provide a meaningful discussion of water quality without considerable use of chemistry and physics. Many water quality books are available in which the level of chemistry and physics is far above the ability of the average readers to understand. In this book, I have attempted to use only first-year, college-level chemistry and physics in a very basic way. Most of the discussion hopefully will be understandable even to the readers with only rudimentary formal training in chemistry and physics.

Contents

Physical Properties of Water

Abstract

The water molecule is electrostatically charged with a negative site on one side and two positive sites on the other. Attractions between oppositely-charged sites on adjacent water molecules are stronger than typical van der Waals attractions among molecules and are called hydrogen bonds. The molecules in solid (ice) and in liquid water exhibit stronger mutual attractions than do the molecules of other substances of similar molecular weight. This results in water having maximum density at 3.98 °C, high specific heat, elevated freezing and boiling points, high latent heats for phase changes, remarkable cohesive and adhesive tendencies resulting in strong surface tension and capillary action, and a high dielectric constant. Light also penetrates readily into water and is strongly absorbed. The pressure of water at a given depth is a combination of atmospheric pressure and the weight of the water column above that depth (hydrostatic pressure). Water refracts light making underwater objects appear to be at lesser depth. The physical properties of water are of intrinsic interest, but they also are critical factors in geology, hydrology, ecology, physiology and nutrition, water use, engineering, and water quality measurement.

Introduction

Pure liquid water is colorless, tasteless, and odorless and consists solely of H_2O molecules. The International Union of Pure and Applied Chemistry (IUPAC) name for water is oxidane. It also is sometimes called dihydrogen oxide or dihydrogen monoxide. The water molecule is small with a molecular weight of 18.015 g/mol, but water molecules are not as simple as their low molecular weight suggests. A water molecule is negatively charged on one end and positively charged on the other. The separation of charges results in water molecules having properties very different from those of other compounds of similar molecular weight. These differences include elevated freezing and boiling points, large latent heat requirements for changes between solid and liquid phases and between liquid and vapor phases, temperature dependent density, a large capacity to hold heat, and excellent solvent action. These unique physical properties of water allow it to abound on the earth's surface and influence many of its uses.

Water is a remarkable substance that has much to do with the uniqueness of the earth. Without it there would be no life on this planet. As the twentieth century American philosopher Loren Eiseley put it, *"there is magic on this planet, it is contained in water."*

Structure of Water Molecule

Water is formed by the unions of two hydrogen atoms with one oxygen atom. Laws of thermodynamics dictate that substances spontaneously change towards their most stable states possible under existing conditions. A stable atom has two electrons in its innermost electron shell and at least eight electrons in its outermost electron shell. Atoms of nonmetals such as oxygen and hydrogen can come together to share electrons with other atoms to fill their outermost electron shell in a process called covalent bonding. The water molecule is the result of two hydrogen atoms each sharing one electron with oxygen and forming two covalent bonds (Fig. 1.1). This arrangement allows the oxygen atom to obtain two electrons to fill its outermost electron shell, while each hydrogen atom gains one electron to fill its outer and only electron shell.

Atoms and molecules often are said to be electrically neutral to distinguish them from ions that are charged, but in reality, molecules seldom are completely neutral. Electrons of a molecule are constantly in motion and they may concentrate in one or more particular areas of a molecule. When this happens, the negative charge imparted by electrons is not completely balanced by the positive charges of nearby nuclei of the molecule's atoms. This results in oppositely-charged sites or poles, and molecules with such sites are said to be dipolar. Covalent bonds in molecules also may create dipoles. In a polar covalent bond, the electrons are unequally distributed between the atoms participating in the covalent bond resulting in a slight negative charge for the more electronegative atom and a slight positive charge on the less electronegative atom. This separation of charge causes a permanent dipole moment in which the atom on one side of the bond has a positive charge and *vice versa*. Charges on molecules are electrostatic charges, i.e., they are at rest creating an electric field rather than current flow.

Completely non-polar covalent bonds occur when two atoms of the same kind bond together (H_2, N_2, O_2, etc.), but bonds in other compounds in which the difference in electronegativity of atoms is not great also are considered non-polar. Charges associated with polar covalent bonds also may be cancelled by each other in symmetrical molecules. The carbon dioxide molecule is symmetrical and it does not have a molecular dipole in spite of its two polar O-C bonds.

The electrostatic charges created on molecules by differences in electron density are weaker than ionic charges in which each electron lost or gained by an atom becoming an ion is assigned a unit charge, i.e., the charge on the monovalent sodium ion (Na^+) is +1 and the divalent sulfate ion (SO_4^{2-}) has a charge of −2. To distinguish small electrostatic charges associated with unequal electron density in molecules from ionic charges, they often are written as δ+ or δ− instead of +1 or −1 (or greater) unit charges.

The water molecule is not symmetrical because the two hydrogen atoms are on the same side of the oxygen atom (Fig. 1.1). The angle formed by imaginary lines through the centers of the hydrogen nuclei and the oxygen nucleus is 105°. The distance between the nuclei of the hydrogen and oxygen atoms in water is 0.096 nanometer (nm). The water molecule is very small—approximately 0.25 nm (0.25×10^{-9} m) in molecular diameter (Schatzberg 1967). The water molecule is not spherical, but somewhat V-shaped. Examples of common ways of depicting water molecules are shown in Fig. 1.2. The V-shape of the apex may be drawn to point in various directions.

Repulsions and attractions among molecules caused by differences in electron density were originally described by the Dutch physicist Johannes van der Waals and are called van der Waals interactions. These attractions and repulsions are relatively weak and often essentially cancel each other. Nevertheless, many of the unique

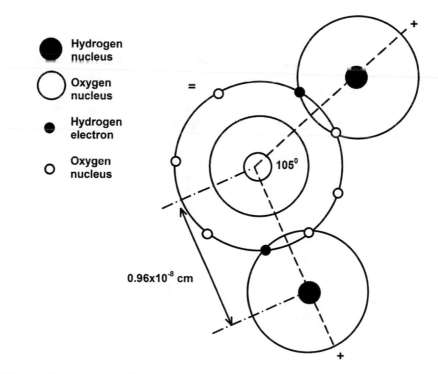

Fig. 1.1 The water molecule

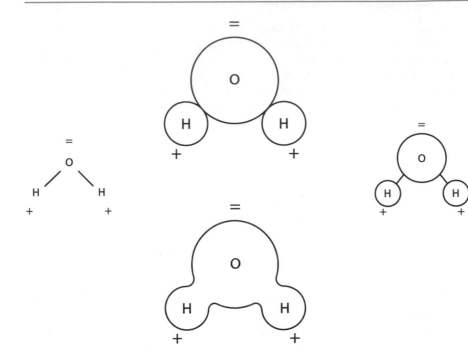

Fig. 1.2 Selected depictions of water molecules

properties of the water molecule results from the dipole formed by the two O-H covalent bonds. The oxygen nucleus has a greater positive charge than the hydrogen nucleus because it is heavier. Electrons are pulled relatively nearer to the oxygen nucleus than to the hydrogen nuclei. This imparts a small negative charge on the oxygen atom and a small positive charge on each of the hydrogen atoms. These electrostatic charges are permanent and considerably stronger than typical van der Waals attractions. Water molecules are permanent dipoles.

The dipolar nature of water leads to an attraction between oppositely-charged sites on different water molecules. These attractions are called hydrogen bonds, and they are stronger than typical van der Waals attractions but weaker than covalent or ionic bonds. Hydrogen bonding is illustrated (Fig. 1.3) in two dimensions instead of its truly three-dimensional nature. A water molecule can form hydrogen bonds with up to four other water molecules. The oxygen side of one water molecule can connect with one hydrogen in each of two other water molecules, and each hydrogen area can connect with the oxygen area of other molecules. Most molecules in liquid water participate in hydrogen bonding. Hydrogen bonding extends to all molecules in ice resulting in its regular crystalline structure. Water vapor molecules have too much energy to allow hydrogen bonds to form and vapor molecules are completely separated from each other.

Fig. 1.3 Hydrogen bonds
between water molecules

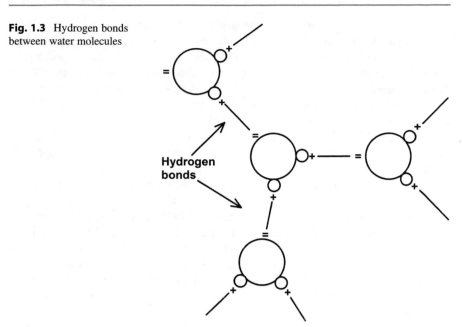

Hydrogen
bonds

Water molecules also may form electrostatic attractions with charged sites on ions
or other molecules of other substances. This phenomenon allows water to wet some
surfaces and to be repelled by other surfaces.

Thermal Characteristics and Phases

In the International System of units (SI system), the joule (J) is the basic unit for
energy; one joule is the amount of energy transferred when 1 Newton (N) acts on an
object over 1 m. Some readers may be more familiar with energy being expressed by
the calorie (1 cal = 4.184 J). The specific heat is the amount of energy that must be
applied or removed to change the temperature of 1 g of a substance by 1 °C. The
calorie definition is based on the specific heat of liquid water (1 cal will raise the
temperature of 1 g of liquid water by 1 °C), and liquid water is the standard for
comparing specific heats of substances. The specific heats of the three phases of
water are: ice at 0 °C, 2.03 J/g/°C (0.485 cal/g/°C); water at 25 °C, 4.184 J/g/°C
(1 cal/g/°C); vapor at 100 °C, 2.01 J/g/°C (0.48 cal/g/°C).

The specific heats of the three phases of water determine how much energy must
be added or removed to change the temperature of each phase by a given amount.
The ice-liquid phase change (freezing point) occurs at 0 °C at standard atmospheric
pressure, and the liquid-vapor phase change (boiling point) occurs at 100 °C at
standard atmospheric pressure. At the phase changes, the frequency of hydrogen
bonds among molecules changes abruptly when the crystalline lattice of ice forms or
collapses or when hydrogen bonds are broken to produce vapor or enough hydrogen

bonds are formed to result in condensate of vapor to liquid water. The re-arrangement of hydrogen bonds at phase changes requires a large input or removal of energy that does not cause a temperature change. The amounts of energy change necessary to cause phase changes are called the latent heat (or enthalpy) of fusion for freezing and the latent heat (or enthalpy) of vaporization for boiling. Enthalpy is a term referring to the internal energy of a substance. For water, the heat of fusion is 334 J/g (80 cal/g) and the heat of vaporization is 2260 J/g (540 cal/g).

Conversion of 1 g ice at $-10\,°C$ to water vapor at $100\,°C$ requires an energy input of 20.1 J (2.010 J/g/°C $\times$ 1 g $\times$ 10 °C) to raise the temperature to 0 °C, and 334 J more are required to convert 1 g of ice at 0 °C to liquid water at 0 °C. Raising the temperature of 1 g liquid water to its boiling pointy requires 418.4 J (4.184 J/g/° C $\times$ 1 g) of energy, and to boil it requires an additional 2260 J/g. Reversing the process described above would require removal of the same amount of energy.

Ice also can change from a solid to a vapor without passing through the liquid phase. This is the reason wet clothes suspended on a line outdoors in freezing weather may become dry. The process is called sublimation, and the latent heat of sublimation is 3012 J/g (720 cal/g). Water vapor also can change from vapor to ice without going through the liquid phase. This process is known as deposition for which the latent heat also is 3012 J/g.

Pure water is a liquid between 0 °C and 100 °C at standard (sea-level) atmospheric pressure. Other common hydrogen compounds such as methane, ammonia, and hydrogen sulfide are gases at ordinary temperatures on the earth's surface (Table 1.1). The specific heats and latent heats of fusion and vaporization also are greater for water than for the other three common hydrogen compounds (Table 1.1). The reason water differs in thermal properties from other substances of similar molecular weight results from its ability to form hydrogen bonds among its molecules. The high latent heat values for water reflect the resistance to forming and to breaking of hydrogen bonds. In the case of freezing, molecular motion must decline to the point that the crystalline lattice of ice can form, but when ice melts, enough energy must be applied to break enough hydrogen bonds to collapse to lattice of ice. The same logic applies for vaporization. Energy is necessary to free the vapor molecules from hydrogen bonding.

Table 1.1 Thermal properties of water compared with those of other hydrogen compounds of similar molecular weight

Property	Water (H_2O)	Methane (CH_4)	Ammonia (NH_3)	Hydrogen sulfide (H_2S)
Molecular weight	18.02	16.04	17.03	34.08
Specific heat (J/g/°C)	4.184	2.23	2.20	0.24
Freezing point (°C)	0	−182.5	−77.7	−85.5
Latent heat of fusion (J/g)	334	59	332	68
Boiling point (°C)	100	−162	−33.3	−60.7
Latent heat of vaporization	2260	481	1371	548

The large specific heat and latent heat requirements necessary to cause phase changes in water bodies has considerable climatic, environmental, and physiological significance. Large water bodies store heat and affect the surrounding climate. Heat loss caused by evaporation of perspiration from our skin is critical to bodily temperature control. As air masses rise the cooling rate decreases when the temperature falls low enough to allow water vapor to condense.

Vapor Pressure

Vapor pressure refers to the pressure exerted on the surface of a substance in equilibrium with its own vapor. Inside a sealed chamber initially filled with dry air, water molecules from a bowl of water will enter the air until equilibrium is reached. At equilibrium, the same number of water molecules enters the air from the water as enters the water from the air, and there is no net movement of water molecules between air and water. The pressure of water vapor acting down on the water surface in the bowl is the vapor pressure of water. Vapor pressure increases as temperature rises (Table 1.2). Bubbles that initially form in water as the temperature rises consist of atmospheric gases, because the solubility of air in water decreases with rising temperature. When the vapor pressure of pure water reaches atmospheric pressure at 100 °C, bubbles of water vapor form in the water and rise to break the water surface causing the phenomenon of boiling. Atmospheric pressure varies with altitude and weather conditions, and the boiling point of water usually is not exactly 100 °C. Water will boil at room temperature if the pressure is low enough.

An oddity called the triple point occurs when the liquid phase of a substance is simultaneously in equilibrium with its solid and vapor phases. At the triple point, the liquid boils at low pressure, and the heat lost in escaping molecules cools the liquid to allow freezing to occur. The triple point of pure water is at 0.1 °C and 611.2 Pa pressure, and can be achieved only under laboratory conditions. Although mainly an oddity, the triple point of water is sometimes used in calibrating thermometers.

Table 1.2 Vapor pressure of water in millimeters of mercury (mm Hg) at different temperatures (°C)

°C	mm Hg	°C	mm Hg	°C	mm Hg
0	4.579	35	42.175	70	233.7
5	6.543	40	55.324	75	289.1
10	9.209	45	71.88	80	355.1
15	12.788	50	92.51	85	433.6
20	17.535	55	118.04	90	525.8
25	23.756	60	149.38	95	633.9
30	31.824	65	187.54	100	760.0

Density

Molecules of ice are arranged in a regular lattice through hydrogen bonding. The regular spacing of molecules in ice creates voids absent in liquid water where molecules are closer together. Ice is less dense than liquid water (0.917 g/cm^3 versus 1 g/cm^3) allowing it to float.

The density of liquid water increases as temperature rises reaching maximum density of 1.000 g/cm^3 at $3.98\,°C$. Further warming causes the density of water to decrease (Table 1.3). Two processes influence density as water warms above $0\,°C$. Remnants of the crystalline lattice of ice break up to increase density, while bonds stretch to decrease density. From 0 to $3.98\,°C$, destruction of remnants of the lattice has the greatest influence on density, but further warming causes density to decrease through stretching of bonds caused by the greater internal energy in molecules of warmer water. The change in density resulting from temperature effects the weight of a unit volume of water. One cubic meter of pure water at $10\,°C$ weighs 999.70 kg; the same volume weighs 995.65 kg at $30\,°C$. The effect of temperature on density of water allows water bodies to undergo thermal stratification and destratification as will be explained in Chap. 3.

Densities of substances often are reported as the unitless quantity specific gravity. The basis for specific gravity is the density of water. At $25\,°C$, water has a density of 0.99705 g/cm^3 or a specific gravity of 0.99705 (Table 1.3). Another substance with a density of 1.5500 g/cm^3 at $25\,°C$ has a specific gravity of (1.5500 g/cm^3 ÷ 0.99705 g/cm^3) or 1.5546.

When an object is placed in water, it either sinks, dissolves, suspends, or floats. Which event occurs depends on several factors, but the specific gravity (density of a substance relative to the density of water) is the most important one. Particles denser than water sink—unless they are so small as to be in true or colloidal solution. Objects with specific gravity less than that of water float. Submerged portions of

Table 1.3 Density of pure water (g/m^3) at different temperatures between 0 and 40 °C

°C	g/cm^3	°C	g/cm^3	°C	g/cm^3
0	0.99984	14	0.99925	28	0.99624
1	0.99990	15	0.99910	29	0.99595
2	0.99994	16	0.99895	30	0.99565
3	0.99997	17	0.99878	31	0.99534
4	0.99998	18	0.99860	32	0.99503
5	0.99997	19	0.99841	33	0.99471
6	0.99994	20	0.99821	34	0.99437
7	0.99990	21	0.99800	35	0.99403
8	0.99985	22	0.99777	36	0.99309
9	9.99978	23	0.99754	37	0.99333
10	0.99970	24	0.99730	38	0.99297
11	0.99961	25	0.99705	39	0.99260
12	0.99950	26	0.99678	40	0.99222
13	0.99938	27	0.99652		

Table 1.4 The density of water (g/cm^3) of different salinities at selected temperatures between 0 and 40 °C

°C	Salinity (g/L)				
	0	10	20	30	40
0	0.99984	1.0080	1.0160	1.0241	1.0321
5	0.99997	1.0079	1.0158	1.0237	1.0316
10	0.99970	1.0075	1.0153	1.0231	1.0309
15	0.99910	1.0068	1.0144	1.0221	1.0298
20	0.99821	1.0058	1.0134	1.0210	1.0286
25	0.99705	1.0046	1.0121	1.0196	1.0271
30	0.99565	1.0031	1.0105	1.0180	1.0255
35	0.99403	1.0014	1.0088	1.0162	1.0237
40	0.99222	0.9996	1.0069	1.0143	1.0217

floating objects and dissolved or suspended particles displace a volume of water equal to their own volumes. For example, a 10-cm^3 marble placed in a glass of water sinks displacing 10-cm^3 of water to cause the water level in the glass to increase slightly. However, the volume of water displaced is unrelated to the density of the submerged object—two, 10-cm^3 marbles of different densities would displace equal amounts of water.

Mineral matter dissolved in natural water causes salinity. These dissolved minerals are denser than water, and 1 g of dissolved mineral matter will displace less than 1 cm^3 of water causing greater density. The relationship between salinity and density (Table 1.4) reveals that water at 20 °C with 30 g/L salinity has a density of 1.0210 g/cm^3 as compared to 0.99821 g/cm^3 for freshwater at the same temperature—this is a weight difference of 22.79 kg/m^3. Differences in density related to salinity commonly cause density stratification in estuaries where rivers run into the sea.

Surface Phenomena

The rise of water in small-bore tubes or soil pores is called capillary action. To explain capillary action, one must consider cohesion, adhesion, and surface tension. Cohesive forces are attractions between like molecules. Water molecules are cohesive because they form hydrogen bonds with each other. Adhesive forces are attractions between unlike molecules. Water adheres to a solid surface with charged sites that attract opposite charges of water molecules. Adhesive forces between water and the solid surface are greater than cohesive forces among water molecules. Such a surface is said to be hydrophilic (water loving) because it wets easily. Conversely, water will bead on a hydrophobic (water hating) surface and run off, because cohesion among water molecules is stronger than adhesion of water molecules to the surface. For example, dry, unpainted wood wets readily, but a coat of paint causes wood to shed water.

The meniscus seen at the water surface in a transparent, glass tube results from cohesion or adhesion. The meniscus in a titration buret usually is concave (Fig. 1.4) because water molecules in the titration solution are attracted more strongly to the

Fig. 1.4 The meniscus

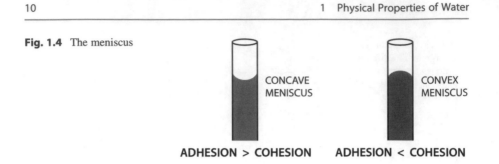

walls of the buret than to themselves. They climb up the wall of the buret causing the concave meniscus. Some fluids have stronger attractions among their molecules than with the walls of a glass tube. This causes molecules to pull away from the tube at the surface and pile up in the center causing a convex meniscus (Fig. 1.4). Such may be seen in a tube filled with mercury.

The net cohesive force on molecules within a mass of water below its surface is zero, but cohesive forces cannot act above the surface. Molecules in the surface layer of water are subjected to an inward cohesive force from molecules below. The tightly drawn surface molecules act as a skin and cause a phenomenon known as surface tension which can be thought of as the ability of the surface to resist an external force acting down upon it. Surface tension is strong enough to permit certain insects and spiders to walk over the surface of water, and to allow needles and razor blades to float when gently laid on the surface. The inward pull of water molecules also acts to make the water surface as small as possible. This is why a small amount of water on a leaf or other hydrophilic surface forms a bead—a sphere has the smallest surface area possible.

The SI unit for surface tension is the Newton/m or mN/m, and 1 mN/m = 1 dyne/cm. The surface tension of pure water (mN/m) decreases with increasing temperature: 0 °C, 75.6; 10 °C, 74.2; 20 °C, 72.8; 30 °C, 71.2; 40 °C, 69.6 mN/m). Salinity has a small effect on surface tension. According to an equation developed by Schmidt and Schneider (2011), seawater at 20 °C and 35 g/l salinity has a surface tension of 73.81 mN/m. By comparison, surface tensions of some other common liquids at 20 °C are: vinegar (acetic acid), 28.0; gasoline, 22.0; acetone, 25.2; SAE 30 motor oil, 36.0. The relatively high surface tension of water results from its ability to form hydrogen bonds among its molecules.

Surfactants are substances that when mixed with water lower its surface tension. Surfactants are relatively large molecules with both hydrophilic and hydrophobic sites. The hydrophobic sites avoid contact with water by turning on the surface towards the air. This lowers the surface tension of water, and it improves the wetting ability of water. Soaps have surfactant action—soapy water has a surface tension of about 25 mN/m as compared to 72.8 mN/m in ordinary water at 20 °C. Some commercial surfactants reduce surface tension even more than does common soap.

Water will rise in most small-diameter glass tubes when such tubes are inserted vertically into a beaker of water (Fig. 1.5). This phenomenon, called capillary action, is the combined effects of surface tension, adhesion, and cohesion. Water adheres to

Fig. 1.5 Illustration of
capillary action

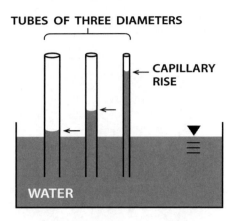

the walls of the tube and spreads upward as much as possible. Water moving up the wall is attached to the surface film, and molecules in the surface film are joined by cohesion to molecules below. As adhesion drags the surface film upward, it pulls a column of water up the tube against the force of gravity. The column of water is under tension because water pressure is less than atmospheric pressure. The height of capillary rise is inversely proportional to tube diameter.

Capillary action occurs in soils and other porous media. Soil particles do not fit together perfectly and the resulting pore space is interconnected. The pore space in soil or other porous media can function in the same manner as a thin glass tube. A good example is the rise in groundwater above the water table to cause what is known as the capillary fringe. In fine-grained soils, the capillary fringe may rise a meter or more above the top of the water table.

Viscosity

Most think of viscosity as the ease with which a fluid flows, e.g., water flows easier than table syrup because it is less viscous. All fluids have an internal resistance to flow, and viscosity (η) represents the capacity of a fluid to convert kinetic energy to heat energy and is the ratio of the shearing stress (force/area or F/A) to the velocity gradient in the fluid. Viscosity results from cohesion between fluid particles, interchange of particles between flow layers of different velocities, and friction between the fluid and the walls of the conduit. In laminar flow, water moves in layers with little exchange of molecules among layers. During laminar flow in a pipe, the water molecules in the layer in contact with the pipe often adhere to the wall and do not flow. There is friction between the pipe wall and those molecules that do flow. The influence of the pipe wall on the flow of molecules declines with greater distance from the wall. Nevertheless, there is still friction between the layers of the flowing water. When flow becomes turbulent, the molecules no longer flow in layers and the principles governing flow becomes more complex.

Table 1.5 Density (ρ) and
dynamic viscosity (μ) of
water

Temp. (°C)	Density (kg/m^3)	Dynamic viscosity ($\times 10^{-3}$ N·sec/m^2)
0	999.8	1.787
5	999.9	1.519
10	999.7	1.307
15	999.1	1.139
20	998.2	1.022
25	997.0	0.890
30	995.7	0.798
35	994.0	0.719
40	992.2	0.653

The term viscosity commonly refers to dynamic viscosity. The units for expressing dynamic viscosity can be confusing. The SI unit is the poise (P) and 1 P = 0.1 (N·sec)/m^2, but in many instances, the centipoise (cP) which equals 0.001 (N·sec)/m^2 is used. The viscosity of some common substances at 20 °C are: water, 1.00 cP; milk, 3 cP; vegetable oil, 43.2 cP; SAE 30 motor oil, 352 cP; honey, 1500 cP. Temperature has a great effect on viscosity as illustrated for water (Table 1.5). Water flow in conduits, seepage through porous media, and capillary rise in soil are favored by warmth, because viscous shear losses decrease as viscosity decreases. Viscosity also increases with density—seawater with a viscosity of 1.08 cP at 20 °C is slightly more viscous than freshwater at the same temperature.

Viscosity can also be expressed as kinematic viscosity (ν) which is the ratio of the dynamic viscosity to the density of the fluid. The kinematic viscosity usually is given in m^2/sec. Water at 20 °C has a dynamic viscosity of 1.02 cP and a kinematic viscosity of 1.00×10^{-6} m^2/sec. In many engineering applications, kinematic viscosity is used rather than dynamic viscosity.

Elasticity and Compressibility

Water, like other fluids, has little or no elasticity of form and conforms to the shape of its container. Unless completely confined, a water has a free surface that is always horizontal except at its edges. If a container of water is tilted, the water immediately forms a new horizontal surface. In other words, the old adage "water seeks its own level" is actually true for water itself. Water can travel downslope in response to very small differences in elevation because it is a liquid.

The ideal liquid is incompressible, and water often is said to be incompressible. The preceding statement is not actually correct—water is slightly compressible. Its coefficient of compressibility at 20 °C is 4.59×10^{-10} Pa^{-1}. At 4000 m depth in the ocean, water would be compressed by about 1.8% as compared to water at the surface. Of course, if a fluid is compressed, its density increases because a unit weight will represent less volume.

Water Pressure

The pressure of water at any particular depth is equal to the weight of the water column above that depth (Fig. 1.6). A water column of height h acting down on a small area ΔA has a volume of hΔA. The weight of water or force (F) is

$$F = \gamma h \Delta A \tag{1.1}$$

where γ = weight of water per unit volume.

Pressure (P) is a force acting over a unit area

$$P = \frac{F}{\Delta A}. \tag{1.2}$$

The pressure of water acting on the area ΔA (Fig. 1.5) is

$$P = \frac{\gamma h \Delta A}{\Delta A} = \gamma h. \tag{1.3}$$

This pressure is for the water only, i.e., the hydrostatic pressure, and can be calculated as shown in Ex. 1.1. To obtain absolute pressure, atmospheric pressure must be added to hydrostatic pressure (Fig. 1.7). These forces are always normal to the water surface.

Because hydrostatic pressure at a point depends mainly upon depth of water above the point, pressure often is given as water depth. The actual pressure could vary slightly among different waters of the same depth because the density of water varies with temperature and salinity.

Fig. 1.6 Pressure of water on a surface (ΔA) beneath water of depth (h)

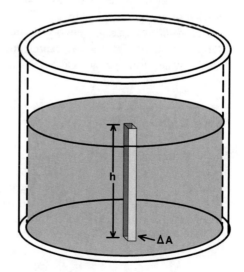

Fig. 1.7 Total pressure at a
point below a water surface

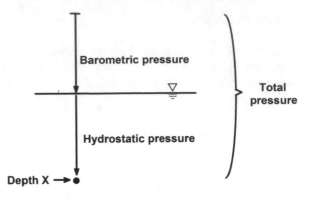

Ex. 1.1 *The hydrostatic pressure of 1-m depth of water at 20 °C will be expressed in millimeters of mercury which is a common unit for atmospheric pressure used in water quality. Note: the pascal (Pa) is the SI unit for pressure (1 mm Hg = 133.32 Pa).*

  Solution:
  The density of water at 20 °C is 0.99821 g/cm³ and the density of mercury is 13.594 g/cm³. Thus, a 1 m column of water may be converted to an equivalent depth of mercury as follows:

$$\frac{0.99821 \ g/cm^3}{13.594 \ g/cm^3} \times 1 \ m = 0.0734 \ m \ or \ 73.4 \ mm.$$

Thus, freshwater at 20 °C has a hydrostatic pressure of 73.4 mm Hg/m.
  When barometric pressure is 760 mm Hg, a point at 1 m depth in a water body has a total pressure of 760 mm Hg + 73.4 mm Hg = 833.4 mm Hg.

Pressure also can result from elevation of water above a reference plane, water velocity, or pressure applied by a pump. In hydrology and engineering applications, the term head expresses the energy of water at one point relative to another point or reference plane. Head often is expressed as depth of water. In water quality, the depth of water usually is the cause of pressure above atmospheric pressure.

Dielectric Constant

The dielectric property of a material is its ability to lessen current flow in an electric field. An electric field can be established by placing charged plates (a positively-charged anode and a negatively-charged anode) in a medium. The field can be thought of as extending from the positive plate towards the negative plate. In situations where the medium is dipolar, the positively-charged pole of molecules will tend to be displaced slightly towards the cathode while the negatively-charged

Fig. 1.8 Orientation of water molecules in an electric field

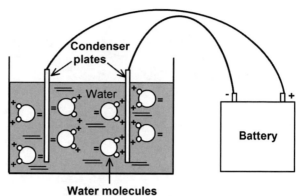

pole of molecules will be slightly displaced towards the anode (Fig. 1.8). This slight separation of charge (polarization) reduces the strength of the electric field.

The capacity of a substance to lessen an electric field is assessed from its dielectric constant (ε). The standard for the dielectric constant is a vacuum within which no polarization occurs; a vacuum is assigned a dielectric constant of unity (1.0000). The method of measuring the dielectric constant is beyond the scope of this book, but dielectric constants of some materials are: vacuum, 1.00; air, 1.0005; cotton, 1.3; polyethylene, 2.26; paper, 3.4; glass, 4–7; soil, 10–20; water, 80.2. Water has a very high dielectric constant, because its molecules have a strong dipolar moment which lessens an electric field imposed within its volume.

Water is a good solvent for compounds held together by ionic forces, e.g., sodium chloride ($Na^+ + Cl^- \rightarrow NaCl$). The strength of the electrostatic force holding NaCl and other compounds with ionic bonds together is described by Coulomb's law:

$$F \frac{(\varepsilon)(Q_1)(Q_2)}{d^2} \tag{1.4}$$

where F is the electrostatic attraction between two ions, Q_1 and Q_2 are the charges on the ions, ε = dielectric constant of the medium, and d = distance between the charges. As ε increases, electrostatic attraction decreases. Water has a large dielectric constant and insulates ions of opposite charge, lessening the electrostatic attraction between them. For example, the Na^+ and Cl^- in NaCl are strongly bonded in air ($\varepsilon = 1.0005$), but in water ($\varepsilon = 80.2$), the electrostatic attraction between them is less and NaCl dissolves readily. The same reasoning applies to other ionic substances, but all are not as soluble as NaCl. Ions that dissolve in water attract the oppositely-charged dipole of water and become surrounded by water further insulating them from ions of the opposite charge (Fig. 1.9). This process is called ion hydration.

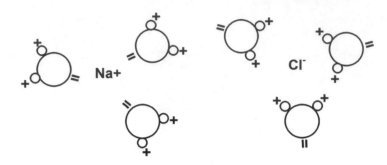

Fig. 1.9 Hydration of dissolved ions by water molecules

Conductivity

Conductivity is the ability of a substance to convey an electrical current, and electricity is conducted on unbound (free) electrons that move about in a substance. Metals such as copper have lots of free electrons and are excellent conductors. Pure water contains only small concentrations of hydrogen and hydroxyl ions resulting from its weak dissociation; it is a poor conductor. Natural waters, however, contain greater concentrations of dissolved ions than pure water and are therefore better conductors. Electrical conductivity of water increases roughly in proportion to dissolved ion concentration, and the conductivity is an important water quality variable to be discussed in Chap. 5.

Transparency

Pure water held in a clear drinking glass in sunlight appears colorless, but larger amounts of clear water tend to have a blue hue caused by the selective absorption and scattering of light. Water tends to absorb visible light at the red end of the visible spectrum more than at the blue end. In natural water, dissolved and suspended substances may affect color and transparency.

Because of the high transparency of water, much of the light that strikes a water surface is absorbed. The fraction of sunlight reflected by a surface is known as the albedo—from albus (white) in Latin. The albedo is expressed as the percentage of incoming light incident to a surface that is reflected. A completely reflective surface has an albedo of 100%; a completely absorptive surface has an albedo of 0%.

The albedo of water varies from around 1% to 100%. It is least when the surface of a very clear water body is still and the rays of sunlight are vertical, and it is greatest when the sun is below the horizon. A mirror reflects light at the same angle as the incident light as shown in Fig. 1.10. This is known as specular reflection, and in nature, the angle of incidence of the sun's rays will vary with time of day and with the progression of the seasons. Seldom will the angle of incidence be exactly

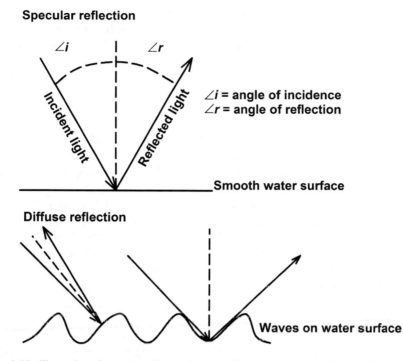

Fig. 1.10 Illustration of specular reflection from a still water surface (upper) and diffuse reflection from a water surface with waves (lower)

perpendicular with the water surface. However, water surfaces seldom are completely smooth—there usually are ripples or waves. This results in the angle of incidence varying and the angle of the reflected light also varying (Fig. 1.10). This type of reflection results in light being reflected at many different angles and is known as diffuse reflection.

In spite of the various factors affecting reflectivity, when the angle of incidence is 60° or less, e.g., the sun's rays at 30° or more above the horizon, the albedo of water usually is less than 10%. When the angle of incidence is 0%—the sun's rays are vertical—the albedo of clear water usually would be around 1 to 3%. According to Cogley (1979), depending upon the method of calculation, the average annual albedos for open water surfaces range from 4.8 to 6.5% at the equator and from 11.5 to 12.0% at 60° latitude. On a monthly basis, albedos varied from 4.5 in March and September to 5.0 in June and December at the equator. The corresponding values at 60° latitude were 7.0% in June and 54.2% in December.

Water absorbs an average of about 90% of the radiation that strikes its surface. Water, because of its high specific heat, is the major means of detaining incoming solar energy at the earth's surface. Of course, water bodies like land masses re-radiate long wave radiation continuously, and the input of solar energy is balanced by outgoing radiation.

The light that penetrates the water surface is absorbed and scattered as it passes through the water column. This phenomenon has many implications in the study of water quality, and will be discussed several places in this book—especially in Chap. 6.

Refractive Index

Light travels faster through a vacuum than through other media, because a vacuum is devoid of matter which interferes with passage of light waves. The optical density is the logarithmic ratio of incident to transmitted radiation for a material, and it usually is directly proportional to the mass density of that material (Wilson 1981). When light passes from a less optically dense to a more optically dense material, its velocity decreases. The opposite occurs when light passes from a more dense to a less dense medium. According to Snell's law, a decrease in speed causes light to refract towards the normal (Fig. 1.11), while an increase in speed causes light to refract away from the normal. The variables shown in Fig. 1.11 can be arranged into the following relationship:

$$\frac{Sin\,\theta_2}{Sin\,\theta_1} = \frac{C_2}{C_1} = \frac{n_1}{n_2} \tag{1.5}$$

where θ_1 = angle of refraction in air, θ_2 = angle of refraction in water, c_1 = speed of light in air, c_2 = speed of light in water, n_1 = refractive index of air, n_2 = refractive index of water. The refractive index can be defined as the ratio of the speed of light in

Fig. 1.11 Depiction of the Snell's law relationship

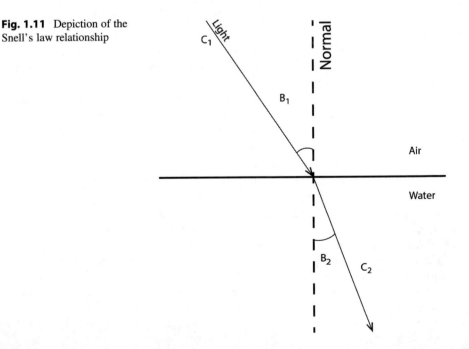

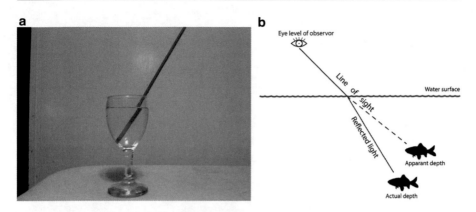

Fig. 1.12 (a) Visual evidence of the refraction of light by water. (b) Refraction makes an object appear at less depth than it is

a vacuum to the speed of light in another medium. The refractive index of water has traditionally been reported as 1.33300 (Baxter et al. 1911). However, the refractive index varies with the wavelength of measurement, increases with greater salinity and pressure, and decreases with greater temperature (Segelstein 1981).

The refraction of light by water is quite obvious when viewing from the side a drinking straw inserted into a container of clear water (Fig. 1.12a). Visible objects beneath the water's surface reflect light which is seen by the observer. The reflected light will increase in velocity and refract away from the normal upon exiting the water. As a result, a fish or other underwater object in a water body will appear to an observer standing on the shore to be at a shallower depth than it actually is (Fig. 1.12b).

Conclusions

The unique physical properties of water resulting from hydrogen bonding allow it to exist at the earth's surface. This has allowed the earth to support an abundance of living things. The physical properties of water contribute to its many beneficial uses.

References

Baxter GP, Burgess LL, Daudt HW (1911) The refractive index of water. J Am Chem Soc 33:893–901
Cogley JG (1979) The albedo of water as a function of latitude. Mon Weather Rev 107:775–781
Schatzberg P (1967) On the molecular diameter of water from solubility and diffusion measurements. J Phys Chem 71:4569–4570
Schmidt R, Schneider B (2011) The effect of surface films on the air-sea gas exchange in the Baltic Sea. Mar Chem 126:56–62
Segelstein D (1981) The complex refractive index of water. MS thesis. University of Missouri
Wilson JD (1981) Physics, 2nd edn. Health and Company, Lexington

Solar Radiation and Water Temperature

2

Abstract

Temperature is a measure of the heat content of an object which results from the energy content of that object. The main source of energy in ecological systems is solar radiation. The earth's energy budget is essentially balanced with the incoming solar radiation being balanced by the reflection of solar radiation and the re-radiation of energy absorbed by the earth as longwave radiation. The temperature of natural water bodies vary in response to diurnal and seasonal changes in solar radiation. The penetration of light into water bodies which regulates the depth to which photosynthesis occurs is strongly influenced by water clarity. Solar radiation heats the surface layers of water in lakes, reservoirs, and ponds more quickly than it warms deeper water. Water bodies often experience thermal stratification in which warmer, lighter water of the surface layer does mix with cooler, heavier deeper water. Warmth also favors greater rates of most chemical and physical process, and respiration of organisms increases with greater temperature within the temperature tolerance range of organisms.

Introduction

Temperature has a major influence on rates of physical, chemical, and biological processes. Temperature effects molecular motion, which, in turn influences physical properties of matter, abiotic and biotic reactions, and activities of organisms. Each biological species has a specific range of temperature within which it can survive and a narrower temperature range within which it functions most efficiently and does not suffer thermal stress. Rates of physical and chemical reactions influence concentrations of dissolved and suspended matter in water. Suspended particles and some colored organic compounds in water lessen light penetration and reduce aquatic plant growth which is the source of nutrition for animals and most non-photosynthetic microorganisms. The influence of temperature on activities of aquatic organisms result in water quality fluctuations.

© Springer Nature Switzerland AG 2020

C. E. Boyd, *Water Quality*, https://doi.org/10.1007/978-3-030-23335-8_2

Physical, chemical, and biological aspects of water quality are highly interrelated, and water temperature effects water quality. Water temperature depends mainly upon the amount of solar radiation received by water bodies, but it is generally considered a basic water quality variable because of its many effects. The purpose of this chapter is to explain relationships among energy, heat, and temperature, discuss solar radiation, and describe the major effects of temperature in water bodies.

Energy, Heat, and Temperature

Electrons, atoms, and molecules are in constant motion. Electrons are classically depicted by the Bohr model as moving around the nucleus of an atom in well-defined layers of orbitals with each orbital containing a fixed number of electrons. Although this depiction is extremely useful, modern quantum mechanics holds that electrons move very fast and act as if in a cloud revolving about the nucleus of an atom. The exact location of an electron at any given moment is not predictable. Electrons are negatively charged and repel each other, but their density around the nucleus is seldom uniform.

The bonds between atoms in a molecule vibrate. A molecule with two atoms vibrates when its atoms move closer together and then further apart. Molecules with three or more atoms—water has three atoms—can vibrate in more than one way. The two H-O bonds in water can go back and forth like a yo-yo in a symmetrical or in a non-symmetrical pattern, and two H-O bonds may bend towards each other and then away. The bonds, including hydrogen bonds in water, also tend to stretch as temperature increases. The rate that atoms and molecules vibrate and stretch is related to temperature. All molecular motion is said to cease (it actually becomes minimal) at a temperature of $-273.15\,^{\circ}C$ on the Celsius scale or $0\,^{\circ}$Kelvin (K) on the absolute temperature scale, and molecular vibrations increase with increasing temperature. The temperature *per se* does not cause the vibrations; the energy content of atoms and molecules is responsible for the vibrations. Temperature is merely a measure of the energy content of something, or in everyday terms, how hot or how cold something is. The temperature results from the response of the temperature sensor to the activity level of molecules in a substance according to a scale ranging from cold to hot.

Energy is a complex concept difficult to explain in a few words. It is universally recognized as what is required to warm something or to do work, and its definition relates to doing work on or transferring heat to an object. The science of thermodynamics is the study of relationships among heat, work, temperature, and energy. There are three main laws of thermodynamics, and the first law often is stated as "energy cannot be created or destroyed, but it can be changed between heat and work. The total energy content of the universe remains constant according to time's arrow." The expression time's arrow (passing of time) implies that energy, like matter, had a beginning. It was created at some time in the past, but it has not been created since. This is one of many examples of the overlaps of physics, philosophy, and theology.

Heat and temperature can be related to each other by the expression

$$Q = mc\Delta T \qquad (2.1)$$

where Q = heat (J), m = mass (kg), c = specific heat (J/kg/°K), ΔT = temperature change (°K). Of course, a Kelvin degree and a Celsius degree are exactly the same size, but the Kelvin scale has its base as absolute temperature which is 273.15 degree units lower than the zero for the Celsius scale (°K = °C + 273.15). The symbol Q often is used for heat, while energy often is indicated by E, but both are reported in joules in the SI unit system.

Ex. 2.1 *A water body of 10,000 m³ and 1 m depth has net absorption (radiation absorbed-radiation radiated) of 50 J/m²/sec. Assuming no energy loss, how much would the temperature of the water increase in 1 hr?*
Solution:
The energy input in 1 hr is

$$Q_{in} = 50\ J/m^2/sec \times 10,000\ m^2 \times 3,600\ sec/hr$$

$$Q_{in} = 1.8 \times 10^9\ J.$$

The temperature change can be calculated with Eq. 2.1

$$\Delta T = \frac{Q}{mc}$$

$$\Delta T = \frac{1.8 \times 10^9 J}{10^4 m^3 \times 10^6 g/m^3 \times 4.184\ J/g/°C}$$

$$\Delta T = \frac{1.8 \times 10^9 J}{4.184 \times 10^{10} J/°C}$$

$$\Delta T = 0.043°C.$$

The second law of thermodynamics recognizes that everything becomes more disorganized or more random in response to time's arrow. The second law often is stated as "the entropy of the universe always increases with time's arrow." In other words, the universe is becoming more random or running down. The degree of randomness or disorder is called entropy. The common way of explaining entropy involves the organization of objects or particles of matter. The top of a well-organized office desk has low entropy, but after organizing your desk, it becomes less organized as you use it—entropy increases. Molecules in a sugar cube have low entropy. When you put the cube in a cup of coffee, it disintegrates into sugar grains, molecules in the grains dissolve, and you stir the coffee to mix the molecules of sugar homogeneously. Entropy increases from cube to grains to heterogeneous coffee-sugar mixture to homogeneous coffee-sugar solution. This simple view of entropy is not useful in mathematic computations of energy relationships. It did, however, inspire the Scot-Irish physicist William Thompson (better known as Lord Kelvin) to say *"we have* (from the second law of thermodynamics) *the sober scientific certainty that the heavens and earth shall wax old as doth a garment."*

Entropy can also be defined as the energy change which occurs when energy is transferred between two objects at constant temperature. There are several equations for entropy, but a simple one is

$$\Delta S = \frac{Q}{T} \qquad (2.2)$$

where ΔS = entropy change (J/°K), Q = heat transferred (J), T = temperature (°K). The second law requires energy to flow by conduction from a hotter object to a cooler object until both objects are at the same temperature. Some heat will be lost to the surroundings and not available to do the work of raising the temperature of the second object. Entropy is the amount of energy not available for doing work. Closed systems move towards a state of equilibrium in which entropy is maximum. Entropy change in a closed system cannot decrease (be negative).

The third law acknowledges that entropy is related to temperature and establishes a zero point for entropy. This law usually is expressed as "the entropy of a perfect crystal of a substance is zero at absolute zero (°K or $-$ 273.15 °C)." In the entropy equation (Eq. 2.2), temperature must be in degrees absolute.

There is a fourth law (often called the zeroth law) indicating the rather obvious fact that "two thermodynamic systems each in thermal equilibrium with a third, are in thermal equilibrium with each other." This law is usually not of much importance in water temperature discussions.

Solar Radiation

The source of energy (and light) for earth is the sun. The sun consists mainly of hydrogen gas (73.46%) and helium gas (24.58%). The sun emits energy produced by nuclear fusion (mainly by the proton-proton chain) occurring in its core. At the high temperature (possibly 15 million °C) and enormous pressure (possibly 250 billion atmospheres) in the sun's core, hydrogen gas (H_2) is converted to electrons and naked hydrogen ions or protons (H^1). The electrons and protons are moving rapidly, and when two protons smash together, they form a deuterium nucleus (H^2) with the release of a positron (β^+) and a neutrino (v):

$$H^1 + H^1 \rightarrow H^2 + \beta^+ + v. \qquad (2.3)$$

When a deuterium nucleus collides with a proton, a helium-three nucleus (H^3) results with the release of a gamma ray (γ):

$$H^2 + H^1 = H^3 + \gamma. \qquad (2.4)$$

Two H^3 nuclei collide to make a helium-four (He^4) nucleus and two protons are released.

$$H^3 + H^3 \rightarrow He^4 + 2H^1. \qquad (2.5)$$

The mass of H^4 is slightly less than the mass of the four protons (H^1) that combined to make it. This slight loss of mass resulted from mass being converted to energy in the form of gamma rays.

The relationship of mass to energy is explained by Einstein's equation

$$E = mc^2 \qquad (2.6)$$

where E = energy (J), m = mass (kg), c = speed of light (299,792,458 m/sec or $\approx 3 \times 10^8$ m/sec). The base unit for the joule is kg·m^2/sec^2 or Newton·meter (N·m). Using Eq. 2.6, it can be seen that if 1 kg of mass is converted completely to energy, 8.99×10^{16} J of energy would result. The sun creates about 3.9×10^{26} J of energy every second by converting an estimated 600×10^6 t of hydrogen to 596×10^6 t of helium (www.astronomy.ohio-state.edu/~ryden/ast162_1/notes2.html). To put this in some kind of perspective, the daily energy consumption by the global activities of humans was only around 10^{18} J in 2018 (https://www.eia.gov/outlooks/aeo/). Of course, the earth only receives about 0.0000002% of the total amount of solar energy emitted into space by the sun.

Energy released in the sun's core as gamma rays by the proton-proton chain reaction slowly migrates through the plasma of the sun to its surface. The temperature declines from millions of degrees Celsius in the sun's core to about 5500 °C at its surface. As the gamma rays move through the plasma, they are absorbed and reradiated at lower frequencies, and the lower frequencies are likewise absorbed and reradiated at even lower frequencies.

Electromagnetic radiation from the sun (sunlight) can, according to quantum mechanics, be thought of as consisting of massless particles traveling in waves of different frequencies. The energy content of electromagnetic waves increases with decreasing wavelength. The wavelength (λ) is the distance between the crests of an electromagnetic wave. Short gamma rays have the most energy, while long radio waves have the least. A wave transports momentum through its motion, and this motion carries energy even though the particles of energy called photons of which light consists of no mass. This can be visualized as one flipping the end of a rope to cause the rope to move as a wave to the other end. The wave could cause a person holding the other end of the rope to experience a tug even though the rope carried no matter. The fact that a wave can carry energy results in an energy transfer when the wave strikes another substance.

Objects absorb electromagnetic radiation but they simultaneously emit electromagnetic radiation. Objects with temperatures above absolute zero emit radiation because electric charges on their surfaces are accelerated by thermal agitation. The wavelength of radiation is equal to the speed of light (c) divided by the frequency (f) or $\lambda = c/f$. The frequency of the radiation increases with the energy content of waves, e.g., gamma rays have more energy than visible light. The electric charges on objects are accelerated at different rates, and a spectrum of radiation is emitted.

The energy for the spectrum radiated by an object depends upon temperature. The simplified form of the Stefan-Boltzmann law describing energy radiated by objects is:

$$E = \delta T^4 \tag{2.7}$$

where E = energy emitted (J/m²/sec), δ = Stefan-Boltzmann constant $(5.67 \times 10^{-8}$ J/°K⁴/m²/sec), T = absolute temperature (°K). Of course, in reality, the total amount of energy radiated depends also upon the emissivity of objects (which ranges from 0 to 1), the temperature of the surroundings into which the object is radiating, and the area of the radiating surface. The expanded form of Eq. 2.7 (not presented here) takes all of these factors into account.

The wavelength of radiation is related to the temperature of the radiating surface, and wavelength increases at lower temperature. This phenomenon is described by Wein's law which can be used to estimate the peak of wavelength in the radiated spectrum. The equation is:

$$\lambda_{max} = \frac{2,897}{T} \tag{2.8}$$

where λ_{max} = peak wavelength (µm), T = absolute temperature (°K).

Ex. 2.2 *The amount of energy radiated from the sun and the peak wavelength of this radiation will be calculated assuming the sun's surface has a temperature of 5773°K.*
  Solution:
  Energy, using Eq. 2.7

$$E = \left(5.67 \times 10^{-8} J/°K^4/m^2/sec\right)\left(5,773°K\right)^4$$

$$E = 6.30 \times 10^7 J/m^2/sec.$$

  Wavelength, using Eq. 2.8

$$\lambda = \frac{2,897}{5,773} = 0.5\ \mu m.$$

Although the peak wavelength of sunlight calculated in Ex. 2.2 is 0.5 µm, the spectrum contains rays of many other wavelengths. Sunlight passes through space without changing in intensity of wavelength. The amount of solar radiation striking the top of the earth's atmosphere on a plane perpendicular to the sun's rays is known as the solar constant, and averaged over several years, the solar constant is

1368 J/m^2/sec (http://earthobservatory.nasa.gov/Features/SORCE/). This radiation is about 10% ultraviolet ($\lambda = 0.10$–0.40 μm), 40% visible ($\lambda = .38$–0.76 μm), and 50% infrared ($\lambda = 0.7$–5.0 μm) with small amounts of other wavelengths (Fig. 2.1). The earth's atmosphere reflects, absorbs, and reradiates solar radiation. Sunlight reaching the earth's surfaces consists of roughly 3% ultraviolet radiation, 44% visible light, and 53% infrared radiation. Radiation often is classified as shortwave (<4 μm) and longwave (>4 μm).

Sunspot activity causes the solar constant to increase slightly. Sunspots are dark regions that occur on the sun's surface as the result of magnetic activity which causes vortexes in the sun's gases near the surface. The dark region of a sunspot is actually cooler than the surrounding surface, but solar flares occur around edges of sunspots resulting in a greater release of energy than normal (https://stardate.org/astro-guide/sunspots-and-solar-flares). Sunspots are said to follow an 11 year cycle between minimum and maximum activity with a variation of only 1.4 J/m^2/sec in the solar constant. However, during the period 1645 and 1715 (often called the Maunder Minimum in astronomy), almost no sunspots were observed. This period also is sometimes referred to as the Little Ice Age because of the unusually low temperatures thought to have been the result of the absence of sunspots (Bard et al. 2000). The period 1950-present has experienced more sunspots and greater solar radiation than noted since 1600.

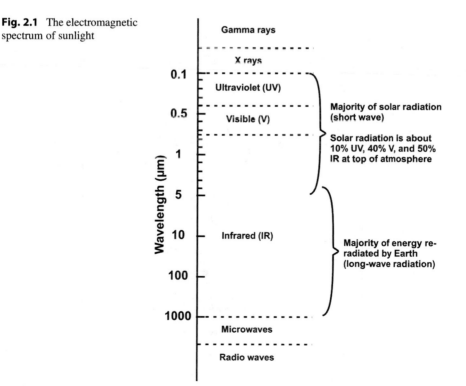

Fig. 2.1 The electromagnetic spectrum of sunlight

Earth's Energy Budget

The radiation reaching the sphere bounded by earth's atmosphere is measured on a plane perpendicular to the sun's rays. However, the earth and its atmosphere are spherical, and the radiation actually is spread over the surface of this sphere. The area of a sphere ($4\pi r^2$) is four times that of a circle (πr^2). As a result, the average amount of radiation striking the sphere of the earth's atmosphere is one-fourth of the solar constant, and it has been measured as 341.3 J/m^2/sec (Trenberth et al. 2009).

The trophosphere which extends 10–12 km above the earth's surface consists mainly of nitrogen and oxygen in a ratio of approximately 4:1. There also is a variable quantity of water vapor and small amounts of argon, carbon dioxide, and several other gases. These gases, and especially water vapor in clouds, reflect and adsorb incoming radiation. Of the incoming shortwave radiation, the atmosphere reflects 79 J/m^2/sec and absorbs 78 J/m^2/sec. Thus, about 184.3 J/m^2/sec of incoming shortwave radiation actually reach the earth's land and water surface, and 23 J/m^2/sec of this is reflected (Trenberth et al. 2009).

The earth's global solar radiation interactions are depicted (Fig. 2.2). The overall energy budget is rather simple (Table 2.1); the earth and atmosphere receive 341.3 J/m^2/sec of shortwave radiation, 102.0 J/m^2/sec of which is reflected and 238.5 J/m^2/sec of longwave radiation (340.5 J/m^2/sec is radiated back to space). The net absorption is 0.8 J/m^2/sec. The earth and the atmosphere reradiate absorbed energy

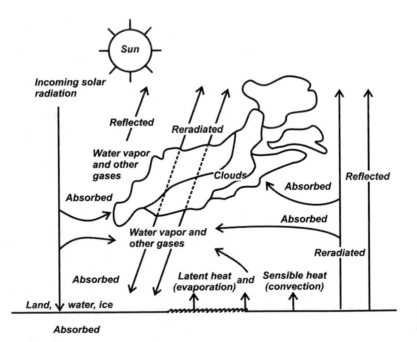

Fig. 2.2 Earth's energy budget relationships

Table 2.1 Budget for earth's incoming shortwave radiation

Radiation	Amount (J/m²/sec)
Incoming shortwave	341.3
Reflected by atmosphere	79.0
Absorbed by atmosphere	78.0
Reflected by surface	23.0
Absorbed by surface	161.3
Outgoing radiation	340.5
Reflected shortwave	102.0
Emitted longwave	238.5
Net absorbed	0.8

as longwave radiation. The shortwave radiation component is much less complex than the longwave radiation component. Equation 2.7 can be used to estimate radiation from the earth and the wavelength peak of this radiation (Ex. 2.3).

Ex. 2.3 *The average rate of re-radiation and the peak wavelength of this radiation will be calculated for an effective global average temperature of 14.74 °C (287.89° K).*

Solution:

Energy, using Eq. 2.7

$$E = \left(5.67 \times 10^{-8} J/^{\circ}K^4/m^2/sec\right)\left(287.89^{\circ}K\right)^4$$

$$E = 389.5 \, J/m^2/sec.$$

Wavelength, using Eq. 2.8

$$\lambda = \frac{2,897}{289.89} = 9.99 \, \mu m.$$

The value for E calculated in Ex. 2.3 is close to the measured value of 396 J/m²/sec presented by Trenberth et al. (2009).

The earth's surface absorbs 161.3 J/m²/sec, but of the 396 J/m²/sec of radiation emitted by the earth's surface, 333 J/m²/sec are reabsorbed as back radiation by the earth's surface. In addition, thermal convection and latent heat of evaporation contribute 80 J/m²/sec and 17 J/m²/sec of energy, respectively to the atmosphere. The actual outgoing radiation to space of 340.5 J/m²/sec consists of shortwave radiation reflected from the atmosphere and clouds (79 J/m²/sec) and the earth's surface (23 J/m²/sec) and longwave radiation emitted from clouds and the atmosphere (199 J/m²/sec) and direct emission into space (40 J/m²/sec). Longwave radiation from the earth may be absorbed and reradiated in complex patterns. Nevertheless, incoming shortwave radiation from space nearly is balanced by reflection of shortwave radiation and emission of longwave radiation.

The ocean and other water bodies absorb and hold a considerable amount of energy. According to Kirchhoff's law, there is a direct relationship between absorptivity (a) and emissivity (e), a ≈ e, and good absorbers are good radiators. The albedo of water is low (usually no more than 5–10%), and its emissivity is 0.96 on a scale of 0–1.0. The earth's land surface has an average emissivity around 0.5.

Greenhouse Effect

Nitrogen, oxygen, and argon in the atmosphere do not impede escape of longwave radiation into space. Water vapor, carbon dioxide, methane, nitrous oxide, and a few other gases absorb longwave radiation slowing its passage into space and increasing the heat content of the atmosphere. This process is known as the greenhouse effect of the earth's atmosphere; the gases responsible for this effect are called greenhouse gases. The common observation that nights are often warmer when there is cloud cover than when the sky is clear provides tangible evidence of the greenhouse effect. The greenhouse effect warms the earth, and without it, the earth would average about 30 °C cooler. Life on earth, as we know it, depends upon the natural greenhouse effect of the atmosphere.

The concentration of carbon dioxide in the atmosphere was around 280 ppm at the beginning of the industrial revolution (≈1750). The increased use of fossil fuels for energy since the 1750s has progressively increased atmospheric carbon dioxide concentration. Carbon dioxide concentration has been monitored continuously in the air at Mauna Loa Observatory in Hawaii since the late 1950s. The concentration was 320 ppm in 1960, and the average for June 2018 was 411 ppm (https://www.co2.earth). Concentrations of other greenhouse gases also have increased because of air pollution. Greater greenhouse gas concentrations are thought to be the major factor causing global warming. The annual global temperature has increased about 0.07 °C/decade since 1880, but since 1970, the increase rate averaged 0.17 °C/decade.

Light Penetration in Water

A major influence of solar radiation in aquatic ecology is the necessity of light for photosynthesis. Light for photosynthesis is almost entirely within the visible range of the electromagnetic spectrum, but the visible spectrum represents only 44% of the electromagnetic input to earth from the sun. Infrared radiation accounts for 53% of the remaining electromagnetic input. Most of the solar radiation entering water bodies is in the visible or infrared range. The focus here will be on the absorption of light energy by water, and effects of light on photosynthesis is deferred to Chap. 10. Most of the absorption of light by water is within the 0.38–4 μm wavelength range. Although Fig. 2.3 indicates strong absorption in the 0.01–0.1 μm range, relatively little of the sun's electromagnetic input falls within this range.

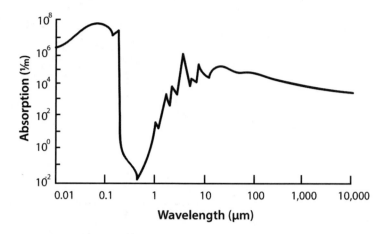

Fig. 2.3 Absorption spectrum for clear water

The absorption coefficient (k) is a function of wavelength and reported in the unit inverse of meters (1/m). The fraction of incident light (I_o) penetrating to a particular depth (I_z) may be expressed as

$$I_z/I_o = e^{-kz} \tag{2.9}$$

where e = base of natural logarithm and z = depth (m). This equation known as the Beer-Lambert law equation may be integrated to give

$$\ln I_o - \ln I_z = kz. \tag{2.10}$$

The intensity of light may be expressed as illuminance or as energy. We are more interested in energy here, so joules (J − 1 watt) will be used.

Ex. 2.4 *In clear water, about 25% of light reaches 10 m. The solar radiation on the water surface is 1000 J/m²/sec. The k for clear water will be estimated.*
Solution:
Using Eq. 2.10,

$$\ln 500 - \ln (500 \times 0.25) = k\,(10\,m)$$

$$6.215 - 4.828 = k\,(10\,m)$$

$$k = 1.387 \div 10 = 0.1387\,m^{-1}.$$

Not all of the energy in incoming radiation is absorbed by water and transformed to heat. Some of the light is scattered and reflected, but most is converted to heat. The Beer-Lambert law equation basically reveals that equal successive increments of depth in water absorb equal increments of light. As a result, the first 1-cm layer

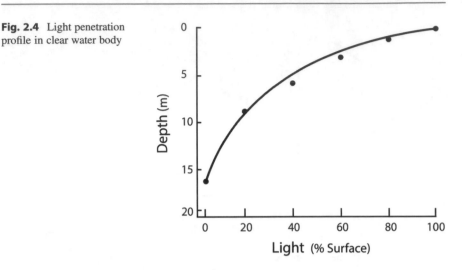

Fig. 2.4 Light penetration profile in clear water body

absorbs a certain percentage of incoming light. The second 1-cm layer absorbs the same percentage of the light reaching it, but the amount of light reaching the second, 1-cm layer is less than that reaching the first, 1-cm layer. The amount of light energy absorbed by each equal successive increment of depth decreases with greater depth. The result is a light absorption curve of the shape illustrated in Fig. 2.4.

The density of air at 20 °C is 1.204 kg/m^3, and 1 kg of air occupies 0.883 m^3. The specific heat of air at 20 °C and constant pressure is 1010 J/kg/°C. Thus, 0.83 m^3 of air would have a heat-holding capacity of 1010 J/kg/°C. By contrast, 0.83 m^3 of water would weigh 830 kg and hold 3,472,720 J of heat. Soil, depending upon its moisture content, will hold roughly 20–60% as much energy as water. Water bodies can store a lot of energy. Of course, they emit longwave radiation and tend to reach an equilibrium in which outgoing radiation equal incoming radiation.

Water Temperature

Temperature is a measure of the internal, thermal energy content of water. It is a property that can be sensed and measured directly with a thermometer. Heat content is a capacity property that must be calculated. Heat content usually is considered as the amount of energy above that held by liquid water at 0 °C. It is a function of temperature and volume. A liter of boiling water (100 °C) in a beaker has a high temperature but small heat content when compared to water at 20 °C in a reservoir of 5 million m^3 volume.

Solar radiation is usually highly correlated with air temperature, and water temperature in small lakes and in ponds closely follows air temperature (Fig. 2.5). Water temperatures usually are quite predictable by season and location. Average monthly temperatures in small bodies of water at a tropical site (Guayaquil, Ecuador at 2.1833°S; 79.8833°W) and a temperate site (Auburn, Alabama at 32.5977°N;

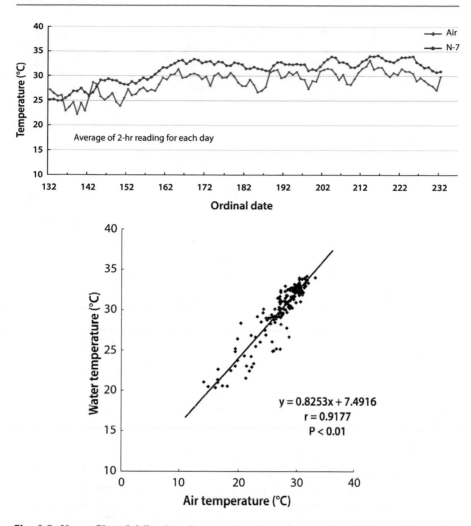

Fig. 2.5 Upper: Plot of daily air and water temperatures in a small pond. Lower: Regression between daily air and water temperatures in a small pond (Source: Prapaiwong and Boyd 2012)

85.4808°W) are provided in Fig. 2.6. Water temperatures at Auburn, Alabama, in the temperate zone change markedly with season, but water temperatures also vary to a lesser extent with season even in the tropics. Air temperature in Ecuador is higher during the warmer wet season (January through May) than during the cooler dry season. This difference in air temperature between seasons also is reflected in water temperatures. Air temperatures at a given locality may deviate from normal for a particular period causing deviation in water temperatures.

Because water can store much heat, larger bodies of water require time to warm up in the spring and to cool down in the autumn. Therefore, the temperature of large reservoirs and lakes tend to lag air temperatures. The heat budgets of lakes obviously

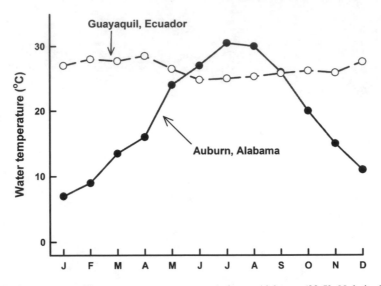

Fig. 2.6 Average monthly water temperatures at Auburn, Alabama (32.5° N latitude) and Guayaquil, Ecuador (2.1° S latitude)

are closely related to lake volumes, but deeper lakes hold more heat than shallower lakes of the same volume but greater surface area (Gorham 1964).

Stream temperature is effected by many factors such as size of drainage area, proportion of base flow to total flow, slope and turbulence, precipitation, and forest cover (Segura et al. 2014). Nevertheless, there usually is a relatively good correlation between air temperature and stream temperature (Crisp and Howson 1982)—especially for smaller streams.

The ocean is vast in extent and has currents. As a result of currents, cooler or warmer water may be delivered offshore at a particular location than would be expected from local air temperatures. For example, the Gulf Stream brings warm water along the eastern coast of North America, while the North Pacific current or California current brings cool water along the western coast of North America.

Thermal Stratification

Light energy is absorbed exponentially with depth and heat is absorbed more strongly and more quickly within the upper layer of water than in lower layers. This is particularly true in eutrophic water bodies where high concentrations of dissolved particulate organic matter greatly increase absorption of energy as compared to absorption in less turbid waters. The transfer of heat from upper to lower layers of water depends mainly upon mixing by wind.

Density of water is dependent upon water temperature (Table 1.4). Ponds and lakes may stratify thermally, because heat is absorbed more rapidly near the surface making the upper waters warmer and less dense than the deeper waters. Stratification occurs when differences in density between upper and lower strata become so great that the two strata cannot be mixed by the wind. The stratification pattern of a dimictic lake (one in which water circulates freely in spring and again in the fall) will be described. At the spring thaw, or at the end of winter in a lake or pond without ice cover, the water column has a relatively uniform temperature. Although heat is absorbed at the surface on sunny days, there is little resistance to mixing by wind and the entire volume circulates and warms. As spring progresses, the upper stratum heats more rapidly than heat is distributed from the upper stratum to the lower stratum by mixing. Waters of the upper stratum become considerably warmer than those of the lower stratum. Winds that often decrease in velocity as weather warms no longer are powerful enough to mix the two strata. The upper stratum is called the epilimnion and the lower stratum the hypolimnion (Fig. 2.7). The stratum between the epilimnion and the hypolimnion has a marked temperature differential. This layer is termed metalimnion or thermocline. In lakes, a thermocline is defined as a layer across which the temperature drops at a rate of at least 1 °C/m. The depth of the thermocline below the surface may fluctuate depending upon weather conditions, but most large lakes do not destratify until autumn when air temperatures decline and surface waters cool. The difference in density between upper and lower strata decreases until wind mixing causes the entire volume of water in a lake or pond to circulate and destratify. There are several other patterns of lake stratification and periods of free circulation in lakes (Wetzel 2001).

Small water bodies are shallower and less affected by wind than lakes. For example, the ordinary warmwater fish pond seldom has an average depth of about 2 m and a surface area of no more than a few hectares. Marked thermal stratification can develop in fish ponds because of rapid heating of surface waters on calm, sunny days. The stability of thermal stratification in a water body is related to the amount of

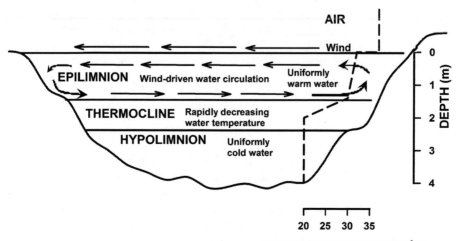

Fig. 2.7 Thermal stratification in a small lake

energy required to mix its entire volume to a uniform temperature. The greater the energy required, the more stable is stratification. Small water bodies with average depths of 1 m or less and maximum depths of 1–2 m often thermally stratify during daylight hours, but they destratify at night when the upper layers cool by conduction. Larger, deeper water bodies stratify for long periods.

Some large deep tropical lakes tend to be permanently stratified, but most occasionally destratify as the result of weather events. Smaller water bodies in the tropics typically destratify during the rainy season and stratify again in the dry season. Events that can lead to sudden destratification of water bodies—in any climate—are strong winds that supply enough energy to cause complete circulation, cold, dense rain falling on the surface that sinks through the warm epilimnion causing upwelling and destratification, and disappearance of heavy plankton blooms allowing heating to a greater depth to cause mixing.

The density of water varies with salinity (Table 1.4), so stratification may occur in areas where waters of different salt contents converge. Where rivers discharge into the ocean, freshwater will tend to float above the salt water because it is less dense. Density wedges often extend upstream for considerable distances beneath the freshwater in the final coastal reaches of rivers called estuaries.

Statification and Water Quality

The hypolimnion does not mix with upper layers of water, and light penetration into the hypolimnion is inadequate for photosynthesis. Organic particles settle into the hypolimnion, and microbial activity causes dissolved oxygen concentrations to decline and carbon dioxide concentrations to increase. In eutrophic water bodies, stratification often leads to oxygen depletion in the hypolimnion and the occurrence of reduced substances such as ferrous iron and hydrogen sulfide. In fact, lakes are classified as eutrophic if dissolved oxygen depletion occurs in the hypolimnion. Hypolimnetic water often is of low quality, and sudden destratification of lakes with mixing of hypolimnetic water with upper layers of water can lead to water quality impairment and even mortality of fish and other organisms. In reservoirs, release of hypolimnetic water may cause water quality deterioration downstream.

Ice Cover

Ice cover can have an important influence on water quality, because gases cannot be exchanged between an ice-covered water body and the atmosphere. Moreover, ice cover reduces light penetration into water, and snow cover over ice may block light penetration entirely. Photosynthetic production of dissolved oxygen is greatly reduced or prevented by ice and snow cover. Winter kills of fish because of low dissolved oxygen concentration beneath the ice can occur in eutrophic water bodies.

Temperature and Water Quality

Temperature affects most of the physical properties of water as already mentioned in Chap. 1. Rates of chemical reactions in water also are influenced by temperature. Temperature has a pronounced effect on solubility, but the effect varies among compounds. Endothermic reactions absorb heat from the environment. Thus, when mixing a solute with water causes the temperature to decrease, increasing temperature will enhance solubility as illustrated for sodium carbonate (Na_2CO_3) in Fig. 2.8. The opposite is true for compounds that release heat when they dissolved as shown for cesium sulfate [$Ce_2(SO_4)_3 \cdot 9H_2O$] in Fig. 2.8. Such reactions are called exothermic reactions. Of course, temperature has much less effect on substances such as sodium chloride (NaCl) that are not appreciably endothermic or exothermic when they dissolve (Fig. 2.8).

Most substances of interest in water quality are either endothermic or their dissolution is not strongly effected by temperature. Chemical reactions among water quality variables also are effected by temperature. Jacobus van't Hoff, a Dutch chemist, developed an equation for estimating the effect of temperature on chemical reactions. The van't Hoff equation was modified by the Swedish chemist Svante Arrhenius to show one of the most important relationships in physical chemistry. The rates of endothermic reactions usually will double or triple with a 10 °C increase in temperature while the opposite occurs with exothermic reactions.

The work of van't Hoff and Arrhenius has often been applied to physiological reactions, and there is a common opinion that biochemical processes such as

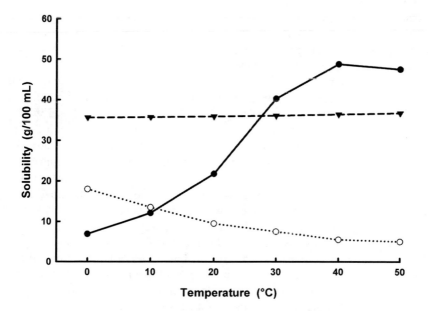

Fig. 2.8 Solubilities of sodium carbonate (*solid dots*), cesium sulfate (*open dots*), and sodium chlorine (*solid triangles*) at water temperatures of 0–50 °C

respiration or growth will double with a 10 °C increase in temperature (within the range of tolerance of the organism). However, this is not always a reliable "rule of thumb." The temperature coefficient (Q_{10}) may be calculated to determine the actual rate of change in a biological process when temperature changes by 10 °C. The Q_{10} equation is:

$$Q_{10} = (R_2 \div R_1)^{10^\circ c/(T_2 - T_1)} \qquad (2.11)$$

where R_1 and R_2 are rates for the process at temperatures T_1 and T_2 (°C). The Q_{10} of biological processes may range from <1.0 to >4.0 (Kruse et al. 2011; Peck and Moyano 2016).

Ex. 2.5 *Fish respire at 200 mg O_2/kg body weight/hr at 22 °C and at 350 mg O_2/kg/ hr at 28 °C. The Q_{10} will be calculated.*
   Solution:
   Using Eq. 2.11,

$$Q_{10} = (350 \div 200 \text{ mg/kg/hr})^{10^\circ/28-22} = (1.75)^{1.67} = 2.55.$$

Temperature effects all biological processes in aquatic ecosystems, and rates of growth of all species tend to increase with warmth. However, temperatures that are too high or too low will adversely influence living things. Each species has its particular temperature range within which it may exist within ecosystems. Aquatic animals often are classified as coldwater, warmwater, and tropical species. Coldwater species will not tolerate temperatures above 20–25 °C. Warmwater species will usually not reproduce at temperatures below 20 °C or grow at temperatures below 10–15 °C, but they will survive much lower winter temperatures. Tropical species will die at temperatures of 10–20 °C, and most do not grow well at temperatures below 25 °C.

Temperature ranges given above are very general, and each species, whether coldwater, warmwater, or tropical has its characteristic temperature requirements. The temperature effects on a tropical species of fish is illustrated in Fig. 2.9. There is a low temperature below which fish die, at slightly higher temperature, fish live, but they do not grow or grow very slowly. Above a certain temperature, growth will increase rapidly with increasing temperature until the optimum temperature is reached. As temperature rises beyond the optimum temperature, growth will slow, cease, and fish will die if the increase continues. The relationship in Fig. 2.9 is slightly different for warmwater or coldwater species which are not likely to die as a result of low temperature in natural or aquaculture waters.

The growth pattern shown in Fig. 2.9 is for animals held in the laboratory under highly controlled conditions. In natural water bodies, water temperature will fluctuate according to daily and seasonal patterns and weather patterns may result in deviations from normal trends. However, a species will do best in an environment where the water temperature remains within a non-stressful range.

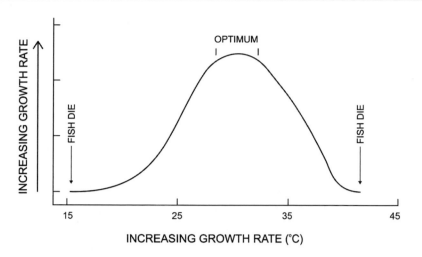

Fig. 2.9 Temperature growth curve for a species of tropical fish

Conclusions

The temperature of water depends primarily upon the amount of solar radiation which warms water and provides light necessary for photosynthesis by aquatic plants. Water temperature has a major effect on both aquatic organisms and water quality through its effect on physiological processes and physical and chemical reactions. There usually is no means of controlling water temperature in natural water bodies. The main value of knowledge of water temperature in water quality is to understand the factors controlling it and the phenomena that may be expected to occur when it changes.

References

Bard E, Reinbeck G, Yiou F, Jouzel J (2000) Solar irradiance during the last 1200 years based on cosmogenic nuclides. Tellus 52B:985–992

Crisp DT, Howson G (1982) Effect of air temperature upon mean water temperature in streams in the north Pennines and English Lake district. Fresh Bio 12:359–367

Gorham E (1964) Morphometric control of annual heat budgets in temperate lakes. Lim Ocean 9:525–529

Kruse J, Rennenberg H, Adams MA (2011) Steps towards a mechanistic understanding of respiratory temperature responses. New Phytol 189:659–677

Peck M, Moyano M (2016) Measuring respiration rates in marine fish larvae: challenges and advances. J Fish Bio 88:173–205

Prapaiwong N, Boyd CE (2012) Water temperature in inland, low-salinity shrimp ponds in Alabama. J App Aqua 24:334–341

Segura C, Caldwell P, Ge S, McNulty S, Zhang Y (2014) A model to predict stream water temperature across the conterminous USA. Hydrol Process 29:2178–2195

Trenberth KE, Fasullo JT, Kiehl J (2009) Earth's global energy budget. Bull Am Met Soc 90:311–324

Wetzel RG (2001) Limnology. Academic, New York

An Overview of Hydrology and Water Supply

3

Abstract

The familiar water cycle is driven by solar radiation, and water is continually transformed back and forth between water vapor and liquid water. In the process, water passes through several distinct compartments of the hydrosphere to include atmospheric moisture, precipitation, runoff, groundwater, standing water bodies, flowing streams, and ocean. Water budgets for water bodies allow concentrations of water quality variables to be used with inflow, outflow, and storage changes to estimate quantities of water quality variables that are contained in and pass through water bodies.

Introduction

Rainwater contains low concentrations of dissolved matter, but as water passes through the different compartments of the hydrologic cycle, it becomes more concentrated with dissolved matter in the order rainwater < overland flow < stream flow < groundwater < ocean water. A knowledge of how water moves through the hydrologic cycle allows one to appreciate why these changes occur. Water quality usually cannot be separated entirely from water quantity. One reason for this is that the amounts of nutrients entering water bodies from their watersheds depend upon both nutrient concentrations in inflows from watersheds and the amounts of inflows. Information on water quantity is needed when assessing water quality, and the student of water quality should learn the basic principles related to water quantity.

This chapter provides a brief discussion of the hydrologic cycle, the basic types of water bodies, and simple methods of water quantity measurement.

© Springer Nature Switzerland AG 2020

C. E. Boyd, *Water Quality*, https://doi.org/10.1007/978-3-030-23335-8_3

The Earth's Water

The hydrosphere consists of the earth's gaseous, liquid, and solid water and is comprised of several compartments to include the ocean, freshwater lakes and ponds, streams, wetlands, soil moisture, water vapor, etc. (Table 3.1). The ocean has a volume of around 1.3 billion km^3 and comprises 97.4% of the earth's water. The remaining water—about 36 million km^3— is contained in several freshwater compartments of the hydrosphere. The largest proportion of the earth's freshwater is bound in ice, occurs as deep groundwater, or is flood-stage stream flow, none of which can often be put to a beneficial use. The amount of freshwater available for human use at any particular time is shallow groundwater, water in lakes, non-flood-stage flow of rivers and small streams, and manmade reservoirs that capture and store river flow. Much of the available freshwater is either not accessible or not sustainable for human use.

The amount of water occurring at a given time in a compartment of the hydrosphere is not directly related to the amount of water available from that compartment for ecological or human use. The reason of this discrepancy relates to the renewal (turnover) time of water in a particular compartment of the hydrosphere (Table 3.1). The volume of water in the atmosphere is around 14,400 km^3 at any particular time, but the renewal time of water vapor in the atmosphere is only 9 days. If all water vapor could be instantly removed from the atmosphere, it would be replaced in 9 days. Water in the atmosphere can be recycled 1,500,000 times during the 37,000-year renewal time of the ocean. The amount of water cycling through the atmosphere in 37,000 years is 21,600,000,000 km^3 or about 16 times the volume of the ocean. Rainfall and other forms of precipitation directly or indirectly sustain the other components of the hydrosphere.

Table 3.1 Volumes of ocean water and different compartments of freshwater

Compartment	Volume (km^3)	Proportion of total (%)	Renewal time
Oceans	1,348,000,000	97.40	37,000 yr
Freshwater			
Polar ice, icebergs, and glaciers	27,818,000	2.01	16,000 yr
Groundwater (800 to 4000 m depth)	4,447,000	0.32	–
Groundwater (to 800 m depth)	3,551,000	0.26	300 yr
Lakes	126,000	0.009	1–100 yr
Soil moisture	61,100	0.004	280 d
Atmosphere (water vapor)	14,400	0.001	9 d
Rivers	1070	0.00008	12–20 d
Plants, animals, humans	1070	0.00008	–
Hydrated minerals	360	0.00002	–
Total freshwater	(36,020,000)	(2.60)	–

Renewal times are provided for selected compartments. (Baumgartner and Reichel 1975; Wetzel 2001)

Precipitation on the earth's land masses totals about 110,000 km^3 annually. Evapotranspiration from the land returns about 70,000 km^3 water to the atmosphere each year, and the rest—an estimated 40,000 km^3 becomes runoff and flows into the ocean (Baumgartner and Reichel 1975). Of course, if the ocean is included, evapotranspiration from land plus evaporation from the ocean must equal to precipitation onto the entire earth's surface.

Runoff consists of overland (storm) flow and base flow of groundwater into streams. Removing groundwater for water supply potentially can cause a decrease in base flow from streams, and groundwater use often is not considered sustainable. About 30% of global runoff or 12,000 km^3 is stream base flow, and another 6000 km^3 is captured in impoundments (Table 3.2). Most authorities believe about 70% of runoff is spatially available for extraction for human use, and the annual supply of water for human use is around 12,600 km^3.

The water footprint is defined as the total amount of water used for a particular purpose. Water footprints can be calculated for individuals, countries, the world, and for specific goods and services. The average water footprint for humans was an estimated 1385 m^3/cap/year (Hoekstra and Mckonnen 2012); water footprints in cubic meters per capita annually for several countries were: China, 1071; South Africa, 1255; France, 1786; Brazil, 2027; United States, 2842. Water footprints usually include rainwater that evaporates from agricultural fields. This water is called "green" water, but it would have evaporated whether or not it had fallen on cropland.

The global water footprint for agriculture given by Hoekstra and Mekonnen (2012) was 8363 km^3/year while global industrial and domestic footprints were around 400 km^3/year and 324 km^3/year, respectively. The total global water footprint is 9087 km^3/year, but this includes 6684 km^3/year of "green" water used in agriculture. A more reasonable estimate of water use is to combine water for irrigation of 3200 km^3/year (Wisser et al. 2008), animal water supply (46 km^3/year), industrial (400 km^3/year), and domestic purposes (324 km^3/year)—a total of 3970 km^3/year.

When compared to the amount of runoff that is renewable and accessible to humans, the total amount of water available for human use appears quite adequate for the future if green water is omitted. This unfortunately is not the case because of several reasons. Water is not uniformly distributed with population (Table 3.3), and many countries or areas of countries with rapidly growing populations have a natural scarcity of water. Water supply varies from year to year with respect to weather

Table 3.2 Renewable, accessible water for humankind

Hydrologic item	Volume (km^3/yr)
Rainfall onto land masses	110,000
Evapotranspiration from land masses	70,000
Renewable water: Total stream flow	40,000
Available water: Base stream flow (12,000 km^3/yr) and reservoirs (6000 km^3/yr)	18,000
Accessible water: About 70% of available	12,600

Table 3.3 Distribution of global runoff by continent and population

Continent	Total runoff (km³/yr)	Population in 2013 (millions)	Runoff (m³/capita/yr)	Runoff (% world total)	Population (% world total)
Africa	4320	1110.6	3890	10.6	15.5
Asia	14,550	4298.7	3385	35.7	60.0
Europe	3240	742.5	4364	8.0	10.4
North America (including Central America and Caribbean)	6200	565.3	10,968	15.2	7.9
Oceania (including Australia)	1970	38.3	51,436	4.8	0.5
South America	10,420	406.7	25,621	25.6	5.7
World	40,700	7162.1	5683	100.0	100.0

rainfall patterns, and places that normally have high rainfall may suffer water shortages during droughts. For example, Birmingham, Alabama, has average annual precipitation of 137 cm, but in 2007, precipitation was only 73 cm, and serious water shortages resulted. The growth of population makes water shortage more likely during drought years. Water pollution associated with growing population also degrades the quality of water supplies making them more expensive to treat for human uses. Many countries do not have adequate water supply infrastructure, suffer political instability, or are embroiled in armed conflicts. Water shortage is a frequent problem facing mankind that will become of even greater concern in the future.

The Hydrologic Cycle

The ancients were aware of the hydrologic cycle through observation. Possibly the earliest references to the water cycle are recorded in the Old Testament of the Holy Bible. Elihu, friend of Job, speaking to Job about 2000 BC said, *"he draws up the drops of water, which distill as rain to the streams; the clouds pour down their moisture and abundant showers fall on mankind"* (Job 36:27–28). Another Old Testament statement about the water cycle was made by the Teacher (probably King Solomon of Israel) about 935 BC. *"The wind blows to the south and turns to the north; round and round it goes, ever returning on its course. All streams flow into the sea, yet the sea is not full. To the place the streams come from there they return again"* (Ecclesiastes 1:6–7).

A logical place at which to begin a discussion of the hydrological cycle is evaporation. This process continually returns water to the atmosphere allowing for continued precipitation. These two processes, evaporation and precipitation, are the basis for the hydrological cycle (Fig. 3.1). Evaporation requires energy, and this energy is supplied by solar radiation. It accurately can be said that the sun drives the hydrologic cycle.

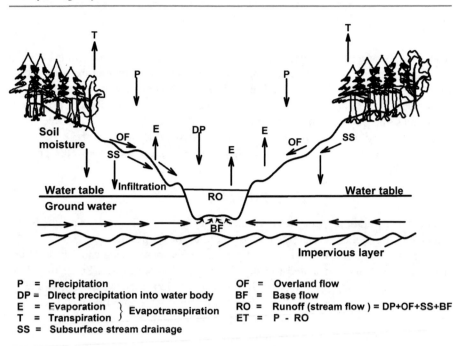

P   = Precipitation
DP = Direct precipitation into water body
E   = Evaporation }
T   = Transpiration } Evapotranspiration
SS = Subsurface stream drainage

OF = Overland flow
BF = Base flow
RO = Runoff (stream flow) = DP+OF+SS+BF
ET = P - RO

Fig. 3.1 The hydrologic cycle or water cycle

Water returns to the atmosphere as water vapor by evaporation from water bodies and moist soil, and by transpiration of plants—the combined process is called evapotranspiration. Water vapor is caught up into the general atmospheric circulation which moves from higher atmospheric pressure to lower atmospheric pressure. Where air rises, it cools and its vapor pressure declines. This results in condensation of water vapor and clouds form to favor precipitation as sleet, snow, hail, or rain.

A portion of precipitation evaporates while falling through the atmosphere, or it is intercepted by vegetation and manmade objects from which it evaporates before reaching the ground. The precipitation reaching the land surface as rainfall (or frozen water which eventually melts) continues to evaporate, but it collects, and either infiltrates the soil surface, stands in puddles, or flows over the land in response to gravity. The overland flow collects in streams, river, and lakes and other water bodies and continues to evaporate as it flows towards the ocean.

Water that infiltrates into the ground either is retained in the soil (soil moisture) or infiltrates downward until it reaches an impermeable layer and retained within a geological formation. The saturated thickness of such formations is called an aquifer, and water in underground aquifers is known as groundwater. Soil moisture can be returned to the atmosphere by evaporation from the soil directly or by transpiration of plants. Groundwater moves slowly through aquifers by infiltration in response to gravity. It will eventually seep into streams, lakes, or the ocean, and it can be removed by wells for human use.

Evaporation

The capacity of air to hold water vapor depends on its temperature (Table 1.2). The maximum amount of moisture that air can hold at a given temperature is its saturation vapor pressure. If unsaturated air is brought in contact with a water surface, molecules of water bounce from the water surface into the air until vapor pressure in air equals the pressure of water molecules escaping the surface. While water molecules continue to move back and forth across the surface, there is no net movement in either direction. The driving force for evaporation is the vapor pressure deficit (VPD)

$$\text{VPD} = e_s - e_a \qquad (3.1)$$

where e_s = saturation vapor pressure and e_a = actual vapor pressure. The greater the vapor pressure deficit, the greater is the potential for evaporation.

Energy is required to raise the temperature of a mass of water to 100 °C and additional heat must be added to cause boiling and vaporization. Some may have difficulty reconciling this fact with the observation that water evaporates at temperatures below 100 °C. Molecules in a mass of water are in constant motion, and molecular motion increases with greater temperature. However, molecules in a mass of water of a given temperature are moving at different speeds. The faster-moving molecules contain more thermal energy than do the slower-moving molecules, and some of the fastest-moving molecules escape the surface to evaporate. Nearly all evaporation occurring in nature occurs at temperatures far below the boiling point of water. Nevertheless, greater temperature favors evaporation, because more molecules gain sufficient energy to escape the water surface. Some readers may recognize the explanation above as an extension of Maxwell's "demon concept" (Klein 1970). James Maxwell, an early Scottish physicist of thermodynamic fame, envisioned a demon capable of allowing only the fast-moving molecules in air to enter a room and thereby raising the temperature in the room.

Relative humidity is a measure of the percentage saturation of air with water vapor. It is an indicator of how much more moisture can evaporate into a given mass of air. It is calculated as follows:

$$\text{RH} = \frac{e_a}{e_s} \times 100 \qquad (3.2)$$

where RH = relative humidity (%), e_a = actual vapor pressure, and e_s = saturation vapor pressure. Relative humidity usually is measured with a sling hygrometer (also called a sling psychrometer) which consists of a wet bulb and a dry bulb thermometer that can be rapidly rotated in the air. Evaporation causes the wet bulb temperature to decline to a lower temperature than measured by the dry bulb thermometer unless the air is saturated with water vapor (100% RH) and both thermometers read the same temperature. The ratio of the wet to dry bulb temperatures is used to determine relative humidity. The moisture content of air also may be expressed by the dew point or the temperature at which a parcel of air would begin to condense water vapor. The higher the dew point, the more moisture there is in the air. Saturation vapor pressure (Table 1.2) increases with warmth, so warm air has the capacity to hold more water vapor than does cool air. Evaporation rate obviously is favored by low relative humidity.

Wind velocity accelerates evaporation, because air movement usually replaces humid air with drier air. When there is no air movement over a water surface, evaporation quickly saturates the layer of air above the water surface and evaporation ceases or diminishes greatly. The influence of wind on evaporation is especially pronounced in windy, arid regions where air normally contains little moisture.

Dissolved salts decrease the vapor pressure of water. The evaporation rate under the same conditions is about 5% greater from freshwater than from ocean water. Dissolved salt concentrations do not vary enough among freshwater bodies to influence evaporation rates appreciably. Turbid waters heat faster than clear waters, and as a result, greater turbidity, and especially turbidity caused by phytoplankton, tends to enhance evaporation (Idso and Foster 1974). Changes in atmospheric pressure effect evaporation slightly.

Temperature usually has the greatest influence on evaporation, and the air and water temperatures at a locality are closely related to the amount of incoming solar radiation. The temperature difference between air and water affects evaporation. A cold water surface cannot generate a high vapor pressure, and cold air does not hold much water vapor. Cold air over cold water, cold air over warm water, and warm air over cold water are not as favorable for evaporation as is warm air over warm water. Evaporation rates tend to increase from cooler regions to warmer regions. Relative humidity must be considered, because cold, dry air may take on more moisture than warm, moist air if the moisture-holding capacity of the warm air has already been partially or completely filled.

Evaporation occurs from lakes and streams, but most terrestrial evaporation is from the leaves of plants by a process called transpiration. The combined water loss by evaporation and transpiration is given the name evapotranspiration. Unlike evaporation from the surface of a water body, evapotranspiration of terrestrial ecosystems often is limited by a lack of moisture.

Measurement of Evaporation

The water balance in a water body is complex. Water is removed from water bodies through evaporation, transpiration, seepage, consumptive use, and outflow, and water also flows into water bodies by runoff or through precipitation. The change in depth over a period of time in a water body does not result solely from evaporation, and the change in moisture content of soil over time is not the sole result of transpiration by plants and evaporation from the soil surface. Water in soil may infiltrate downward and precipitation delivers water to the soil.

One way of determining the potential for evaporation from water bodies is to measure evaporation from the surface of an otherwise "water-tight" container. Three types of evaporation instruments were recommended by the World Meteorological

Organization (Hounam 1973): the 3000-cm^2 sunken tank; the 20-m^2 sunken tank; and the Class A evaporation pan. The Class A pan (Fig. 3.2) is widely used in the United States for monitoring evaporation rate. The Class A pan is made of stainless steel, and it is 120 cm in diameter by 25 cm deep. The pan is mounted on a 10-cm-high wooden platform positioned over mowed grass. The pan is filled to within 5 cm of its rim with clear water. A stilling well in the evaporation pan allows water depth to be measured using a hook gauge or an electronic water-level monitor. A stilling well is a piece of pipe with a small hole in its side mounted vertically in the pan. The hole allows the water level to equilibrate with the water level in the pan, and it provides a smooth surface within the stilling well. The top of the stilling well supports the hook gauge that has an upward pointed hook that can be moved up and down with a micrometer. With the point of the hook below the water surface, the micrometer is turned to move the hook upward until its point pulls a pimple in the surface film of the water without breaking the surface tension. The micrometer can be read to the nearest tenth of a millimeter. Water level measurements typically are made at 24-hr intervals, and the difference in micrometer readings on successive days is the water loss by evaporation. A rain gauge must be positioned beside the evaporation pan to allow correction for rain falling into the pan.

The evaporation rate from a pan is not the same as the evaporation rate from an adjacent body of water. There are several reasons for this discrepancy, but mainly, the heat budget for a lake is much different from that of a small pan of water, and air flow over the pan is different from air flow over larger water bodies. A coefficient is used to adjust pan evaporation data to expected evaporation from a larger water body. Pan coefficients were developed by accurately measuring evaporation rates

Fig. 3.2 Class A evaporation pan and a hook gauge

from lakes by elaborate mass transfer, energy budget, or water budget techniques and relating these evaporation estimates to evaporation estimates determined simultaneously in nearby evaporation pans. The ratio of measured evaporation from a water body to the pan evaporation provides a coefficient for estimating evaporation in water bodies from pan evaporation data:

$$C_p = \frac{E_L}{E_p}$$

and

$$E_L = (C_p)(E_p) \tag{3.3}$$

where E_L = evaporation from lake, E_p = evaporation from pan, and C_p = pan coefficient. Coefficients for Class A pans range from 0.6 to 0.8, and a factor of 0.7 is most commonly recommended for general use in estimating lake evaporation. Boyd (1985) found that a factor of 0.8 was a more reliable general factor for pond evaporation.

Potential evapotranspiration (PET) can be measured in lysimeters—small, isolated soil-water systems in which PET can be assessed by the water budget method. Empirical methods such as the Thornthwaite equation or the Penman equation may be used to estimate PET based on air temperature and other variables (Yoo and Boyd 1994). The annual PET of a watershed can be calculated from annual precipitation and runoff:

$$ET = Precipitation - Runoff. \tag{3.4}$$

The equation above is valid on an annual basis because there is little change in soil water and groundwater storage from a particular month in one year to the same month in the next year, and any water falling on the watershed that did not become stream flow was evaporated and transpired.

Atmospheric Circulation

Water vapor remains in the atmosphere until it condenses and returns to the earth's surface as frozen or liquid precipitation. The general circulation pattern of the atmosphere transports large amounts of water vapor. Earth's two hemispheres each are divided into three more or less independent zones of atmospheric circulation (Fig. 3.3). Strong insolation at the equator heats air causing it to rise. This air moves poleward in both directions, and a part of it descends at about 30°N and 30°S latitude. Atmospheric pressure tends to be low in equatorial regions with rising air, but high-pressure belts dominate in subtropical latitudes with falling air. Part of the air that descends in the subtropics flows toward the low-pressure equatorial region to replace the rising air there, and this causes the trade winds. The other part of the descending air in subtropical high-pressure belts flows poleward toward

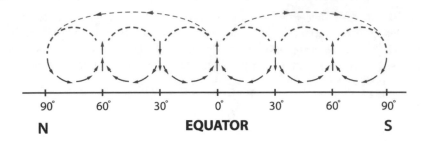

90°   60°   30°   0°   30°   60°   90°
N   **EQUATOR**   **S**

Fig. 3.3 General pattern of global atmospheric circulation

low-pressure belts at about 60°N and 60°S latitude. Cold, high-pressure air over the poles flows toward the low-pressure belts at 60°N and 60°S latitude. Cold and warm air masses converge at about 60°N and 60°S latitude resulting in rising air. The rising air divides, with a part flowing poleward and a part flowing toward the subtropics.

The general pattern described above is altered by local conditions which create smaller atmospheric movements operating within the general, global circulation pattern. Zones of maximum insolation migrate northward and then southward as the seasons change, and the cells of general atmospheric circulation also tend to migrate. The Asiatic monsoon results from the transfer of air masses across the equator in response to the migration of the sun's rays, resulting in reversal of wind directions on a seasonal basis.

The direction of winds in the atmospheric circulation is affected by the Coriolis force. To understand this force, consider a helicopter hovering above the equator. This aircraft has no north-south velocity, but it is moving toward the east at the same speed as the earth's surface ($\approx$1600 km/hr). Now, if this helicopter begins to travel northward, it will maintain its eastward velocity caused by the earth's rotation. As the helicopter moves northward, the rotational velocity of the earth's surface decreases because the diameter of the earth decreases in proportion to its distance from the equator. This causes the helicopter to deflect to the right relative to the earth's surface. If we consider another helicopter hovering some distance north of the equator, it also will deflect toward the right as it flies southward. This occurs because the rotational speed of points on the earth's surface increase in velocity as the aircraft progresses southward. In the Southern Hemisphere, the same logic prevails, but with the opposite direction of wind deflection. Moving air deflects to the right in the Northern Hemisphere and to the left in the Southern Hemisphere.

Combining the cells of atmospheric motion with the Coriolis force, the wind belts of the world can be generalized (Fig. 3.4). These wind and pressure belts tend to shift with season, because the land is cooler than the ocean in winter, resulting in cold, heavy air over landmasses, whereas in summer, the land is warmer than the ocean and cooler, heavier air builds up over the ocean.

Ocean currents are caused by wind. Where winds pile up water against a landmass, the water must then flow along the landmass in a direction effected by the Coriolis force. The trade winds cause water to pile up in the Gulf of Mexico along the eastern coast of Central America, and along the northern coast of South

Fig. 3.4 General wind belts of earth

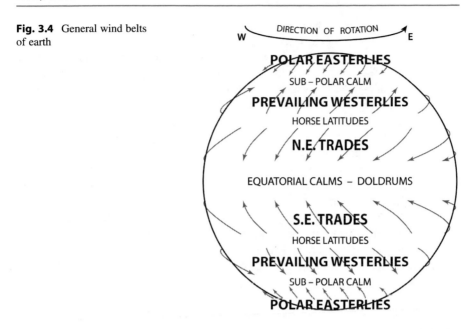

America. This water, the Gulf Stream, then flows northward along the eastern coast of North America and makes a clockwise circulation pattern in the Atlantic Ocean. Ocean currents flowing from tropical and subtropical areas are warm, whereas those coming from polar and subpolar regions are cold. Ocean currents can deliver warm water to cold regions, and *vice versa*, and greatly modify climates of coastal area.

Precipitation

Rising air is necessary for precipitation. As a parcel of air rises, the pressure around it decreases, because atmospheric pressure declines with increasing elevation, and the air parcel expands. This expansion is considered to be adiabatic implying that no energy is lost or gained by the rising air mass. The temperature of an air parcel results from the thermal energy of its molecules. The air parcel contains the same number of molecules after expansion as it did before, but after expansion, its molecules occupy a larger volume causing its temperature to decrease.

The rate of cooling is called the adiabatic lapse rate and occurs at 1 °C/100 m for rising air at less than 100% relative humidity. If air continues to rise, it will cool until it reaches 100% RH and moisture will begin to condense as water droplets. The elevation at which air begins to condense moisture (the dew point) is known as the lifting condensation level. After rising air begins to condense moisture, it is heated by the latent heat released by condensation of water vapor. If air that has reached 100% RH continues to rise, heat from condensation counteracts some of the further cooling from expansion. The adiabatic lapse rate for rising air at 100% RH (wet adiabatic lapse rate) is 0.6 °C/100 m.

Clouds form when air begins to condense moisture and the moisture collects on hygroscopic particles of dust, salts, and acids to form tiny water droplets or ice particles. Droplets grow because they bump together and coalesce when air temperature is above freezing. At freezing temperatures, both super-cooled water and ice crystals exist in clouds. The vapor pressure of super-cooled water is greater than that of ice, and ice particles grow at the expense of water droplets because of their lower vapor pressure. Precipitation occurs when water droplets or ice particles grow too large to remain suspended by air turbulence. The size of droplets or particles at the initiation of precipitation depends on the degree of turbulence. In the tops of some thunderheads, turbulence and freezing temperature can lead to ice particles several centimeters in diameter known as hail. Precipitation may begin its descent as sleet, snow, or hail, but it may melt while falling through warm air. Sometimes ice particles may be so large that they reach the ground as hail even in summer.

The major factors causing air to rise are flow over elevated topography, warm air masses rising above heavier, cold air masses, and convectional heating of air masses. The basic rules for the amounts of precipitation at different locations are summarized as follows:

- Precipitation is normally greater where air tends to rise than where air tends to fall.
- Amounts of precipitation generally decrease with increasing latitude.
- The amount of precipitation usually declines from the coast to the interior of a large land mass.
- A warm ocean offshore favors high precipitation while a cold ocean favors low precipitation.

Amounts of annual precipitation vary greatly with location; some arid regions receive <5 cm/year and some humid regions have >200 cm/year. The world average annual precipitation on the land masses is around 70 cm/year. Precipitation totals vary greatly from year to year, and at a given location some months typically have greater precipitation than others.

Measurement of Precipitation

The amount of precipitation is determined by capturing precipitation in a container—melting it if it is frozen—and measuring water depth. The standard US National Weather Service rain gauge (Fig. 3.5) consists of a brass bucket or overflow can into which a brass collector tube is placed. A removable, brass collector funnel is mounted on top of the overflow can to direct water into the collector tube. The rain gauge is mounted in a support attached to a wooden or concrete base. Rain is concentrated 10 times in the collector tube, because the area of the collector funnel is 10 times the area of the collector tube. Concentration of the rainfall facilitates measurement with a calibrated dipstick on which the wetted distance indicates rainfall depth. If rainfall exceeds the capacity of the collector tube, it overflows

Fig. 3.5 Standard, dip-stick
type rain gauge

into the overflow bucket. Water from the overflow can is poured into the collector
tube for measurement. Several other types of rain gauges including recording rain
gauges are also in common use.

Soil Water

Some of the precipitation falling on the land infiltrates through the land surface and
enters pore spaces of soil or is adsorbed onto soil particles. A completely saturated
soil is said to be at its maximum retentive capacity. Water will drain from saturated
soil by gravity, and the remaining water held against gravity is the soil moisture, and
a soil at 100% moisture content is said to be at field capacity. Soil moisture
evaporates from the soil surface or lost to the air through transpiration by plants.
When the water content of a soil drops so low that plants wilt during the day and
remain wilted at night, the soil moisture has fallen to its permanent wilting percentage. Water remaining in soil at the permanent wilting percentage is held so tightly
within soil pores and on soil particles that it is not biologically available. The
moisture status at soil moisture concentrations below the permanent wilting percentage is referred to as the hygroscopic coefficient.

The maximum retentive capacity of soil depends on the volume of pore space
alone. Fine-textured soils have higher maximum retentive capacities than do coarse-

textured soils. The proportion of biologically available water decreases in fine-textured soils, but they contain more available water than do coarse-textured ones, because they have greater pore space.

Groundwater

At some depth beneath most places on land surfaces, geological formations are saturated with water that has percolated down from above. The saturated thickness of such a formation is called an aquifer, and the top of the aquifer is known as the water table. Aquifers can be confined or unconfined (Fig. 3.6). The top of an unconfined aquifer is open to the atmosphere through voids in the geological formations above it, and unconfined aquifers are known as water table aquifers. Water in an un-pumped well in an unconfined aquifer stands at the level of the water table.

Water may be trapped between two impervious geological layers or confining strata to form what is known as a confined aquifer. The land area between outcrops of the two confining strata of a confined aquifer is its recharge area, because precipitation falling in this area can infiltrate into the confined aquifer. In a confined aquifer, water is under pressure as a result of water standing above it (Fig. 3.6). Water pressure is a function of elevation difference between a point in the aquifer and the level of the water table within the confining strata beneath the recharge area. The water in a confined aquifer is compressed slightly. When a well is drilled into the aquifer from a place on the land surface outside the recharge area, it releases pressure

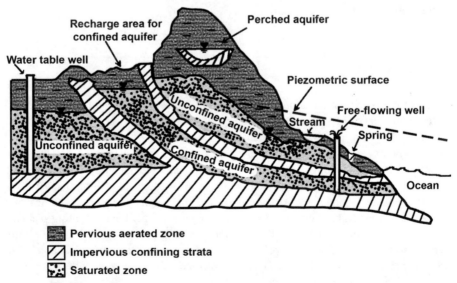

Fig. 3.6 A cross-sectional area of a place on the earth's surface illustrating different kinds of aquifers

on the aquifer allowing water in the aquifer to expand. This is a case where the slight compressibility of water plays an important role. The water level in a well casing drilled into a confined aquifer will usually rise above the top of the aquifer and sometimes reach the land surface.

The level at which water stands in an unpumped well in a confined aquifer is the piezometric level. Piezometric levels mapped over the entire aquifer form a piezometric surface. Water rises above the land surface from a well drilled into a confined aquifer when the piezometric level is above the land surface. Confined aquifers often are called artesian aquifers, and free-flowing wells frequently are said to exhibit artesian flow. The piezometric surface slopes downward, and water in a well drilled into the aquifer does not rise to the same elevation as that of the water table in the recharge area between the two confining layers (Fig. 3.6). This results primarily from reduction in hydraulic head caused by friction as water infiltrates through the aquifer. A well drilled from the recharge area into a confined aquifer will behave like a water table well.

As can be seen in Fig. 3.6, more than one water table may occur beneath a particular place on the land surface. Small perched aquifers are common on hills where water stands above a small hardpan or impervious layer. Water may leak from an artesian aquifer into aquifers above or below. One or more artesian aquifers may occur below the same point, or artesian aquifers may be absent.

Water in aquifers moves in response to decreasing hydraulic head. Groundwater may seep into streams where stream bottoms cut below the water table—this is called base flow. Aquifers may also flow onto the earth's surface from springs, or seep into the ocean.

Ocean water may seep into aquifers, and the salt water will interface with freshwater. The position of the interface depends on the hydraulic head of freshwater relative to salt water. As the hydraulic head of an aquifer decreases because of excessive pumping, salt water moves further into the aquifer. Salt water intrusion in wells of coastal regions can result in salinization of groundwater. This phenomenon usually results from over-pumping by wells resulting in a decrease in the water table elevation in coastal areas.

Aquifers vary greatly in size and depth beneath the land. Some water table aquifers may be only 2–3-m thick and a few hectares in extent. Others may be many meters thick and cover hundreds or thousands of square kilometers. Artesian aquifers normally are rather large, at least 5–10-m thick, a few kilometers wide, and several kilometers long. Recharge areas may be many kilometers away from a particular well drilled into an artesian aquifer. The depth of aquifers below the land varies from a few to several hundred meters. For obvious reasons, good aquifers nearer the ground surface are more useful for wells. The water table is not level but tends to follow the surface terrain, being higher in recharge areas in hills than in discharge areas in valleys. The vertical distance from the land surface to the water table is often less under valleys than under hills. Water table depth also changes in response to rainfall; it declines during prolonged periods of dry weather and rises following heavy rains.

Runoff and Streams

Runoff from watersheds or catchments consisting of multiple watersheds forms streams that convey both storm flow (overland flow) and subsurface inflow. Overland flow is the portion of the precipitation that flows over the land surface and enters streams. Overland flow begins after the capacity of watershed surfaces to detain water by absorption or in depressions has been filled, and the rate of rainfall exceeds the rate of infiltration of water on the surface into the soil. Features of the watershed, duration and intensity of rainfall, season, and climate influence the amount of runoff generated by a watershed. Factors favoring large amounts of overland flow are intense rainfall, prolonged rainfall, impervious soil, frozen or moist soil, low air temperature, high proportion of paved surface, steep slope, little surface storage capacity on land surface, sparse vegetative cover, soils with a low moisture-holding capacity, and a shallow water table. Overland flow moves downslope in response to gravity, it erodes the soil creating channels for water flow and ultimately forms a channel that becomes an intermittent or permanent stream.

An ephemeral stream only flows after rains and does not have a well-defined channel. An intermittent stream flows only during the wet season and after heavy rains but usually creates a channel. A perennial stream normally flows year-round and has a well-defined channel. Perennial streams receive groundwater inflow when their bottoms are below the water table and this base flow sustains stream flow in dry weather.

Streams in an area form a drainage pattern discernable on a topographic map. A tree-shaped pattern, called a dendritic pattern, forms where the land erodes uniformly and streams randomly branch and advance upslope. Where faulting is prevalent, streams follow the faults to form a rectangular pattern. Where the land surface is folded or is a broad, gently sloping plain, streams form a trellis or lattice pattern.

A stream without tributaries is a first-order stream, and its basin is a first-order basin. Two, first-order streams combine to form a second-order stream. A stream with a second-order branch is a third-order stream, and so on. The stream system that contributes to the discharge at a specified point in a higher-order stream is called a drainage network. The number of streams decreases and stream length increases as stream order increases (Leopold 1974, 1997).

Stream discharge is the amount of water flowing through the stream's cross section in a given time and at a particular place. A stream hydrograph is a plot of discharge versus time. The hydrograph in Fig. 3.7 shows the discharge of a small stream before, during, and after a storm. During dry weather, the hydrograph represents only groundwater intrusion into the stream (base flow). During the initial phase of a rainstorm, only rain falling directly into the channel (channel precipitation) contributes to the hydrograph. Channel precipitation seldom is great enough to cause an appreciable increase in discharge. During a fairly large rain, stream discharge rises sharply as storm flow enters causing the rising limb of the hydrograph. Peak discharge is reached, the crest segment of the hydrograph occurs and then declines forming the falling limb of the hydrograph. Subsurface storm drainage—water flowing downslope through the soil pores—often is blocked from

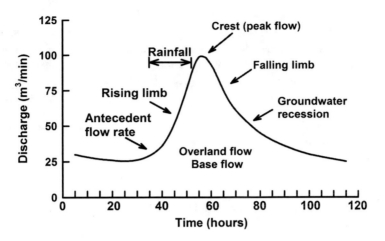

Fig. 3.7 A stream hydrograph

entering streams by high water levels in the stream until the overland flow component has passed downstream. Subsurface storm drainage cannot be separated from base flow on the hydrograph. After a heavy rain, the water table usually rises because of infiltration, and base flow increases. Because of subsurface storm drainage and greater base flow, it may be days or weeks before discharge declines to the pre-storm rate after a period of greater than normal rainfall. The plot of discharge during this time is the groundwater recession segment of the hydrograph.

Watershed characteristics influence hydrograph shape. Watershed features that favor high rates of storm flow result in steep, triangular-shaped hydrographs. Streams on such watersheds are often said to be "flashy." Basins with permeable soils, good vegetative cover, or appreciable storage capacity tend to have trapezoidal hydrographs.

Streams are classified as young, mature, or old. Young streams flow rapidly and continually cut their channels (Hunt 1974). Their sediment loads are transported with no deposition. In mature streams, slopes are less and there is no down cutting of channels. Flows are adequate to transport most of the sediment load. Old streams have gentle slopes and sluggish flows. They have broad floodplains, and their channels meander. Sediment deposition near the area where streams discharge into large water bodies leads to delta formation. Larger streams usually may be classified as young near their sources, mature along middle reaches, and old near their mouths.

Lakes, Reservoirs, Ponds, and Wetlands

Many natural and manmade basins fill with water and maintain a water surface permanently. Standing, permanent water bodies usually have both inflow and outflow, and their water levels fluctuate with season and weather conditions. A lake or reservoir is larger than a pond, but there are not widely accepted criteria for distinguishing between them—one person's large pond is another person's small lake and *vice versa*. Water resides for a longer period in lakes, reservoirs, ponds, and wetlands than in streams. Greater hydraulic retention time favors changes in water quality by physical, chemical, and biological processes.

Water Measurements

Runoff

Stream flow represents total runoff from an area. The flow rate of a stream can be expressed as

$$Q = VA \tag{3.5}$$

where Q = discharge (m^3/sec), V = average velocity (m/sec), and A = cross-sectional area (m^2). Stream discharge varies with water surface elevation (gauge height or stream stage) which changes with rainfall conditions. For any given water level, or gauge height, there is a corresponding and unique cross-sectional area and average stream velocity. Streams may be gauged by measuring and plotting discharge at different stage heights. Such a plot is a rating curve, and many streams are gauged and fitted with a water level recorder so that discharge can be estimated. The discharge of small streams often is determined by estimating velocity by timing the travel of a floating object and measuring the average stream depth by soundings.

Weirs can be installed in small streams or other open channels to estimate flow. A weir consists of a barrier plate that constricts the flow of an open channel and directs it through a fixed-shape opening. Common shapes of weirs are rectangular, trapezoidal, and triangular. The profile of a sharp-crested rectangular weir is shown in Fig. 3.8. The bottom edge of the barrier plate is the weir crest, and flow depth over the crest is measured upstream from the crest and is the effective head. The overflowing stream of water is the nappe. The effective head and the shape and size of a weir crest determine the flow rate. The flow through a weir must maintain a free discharge for an accurate measurement, and water level on the downstream side must be low enough to maintain a free-flow. This requires some amount of head loss and limits the use of weirs in channels with very little slope. The weir head may be substituted into an equation for each specific type of weir to estimate weir discharge (Yoo and Boyd 1994).

Annual runoff also can be estimated from annual precipitation using Eq. 3.4 in areas where the annual evapotranspiration is known. In humid regions, potential evapotranspiration is similar to actual evaporation from water surfaces. For example, in an analysis of stream gauge data, catchment area and annual rainfall for 73 Alabama streams revealed average runoff of 54.1 cm/year. Using annual potential evapotranspiration and rainfall data, the estimated runoff for Alabama streams was 49.7 cm/year. Of course, this method is not reliable in arid climates (Boyd et al. 2009).

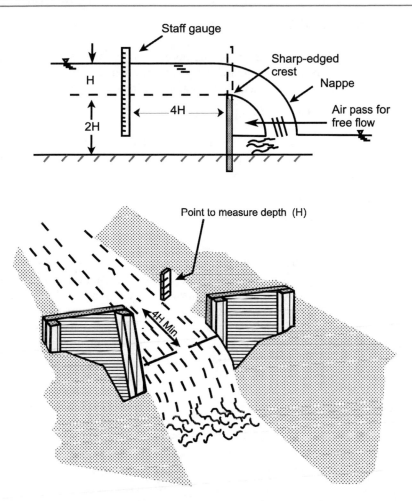

Fig. 3.8 An example with an end-contracted rectangular weir

Volumes of Standing Water Bodies

The volumes of standing water bodies can be estimated as average depth times area

$$V = Ad \tag{3.6}$$

where V = volume, A = surface area, and d = average depth. For small bodies of water, the average depth can be estimated from random soundings. Areas have traditionally been determined from aerial photographs and maps or by surveying techniques. Today, it is possible to locate standing water bodies on satellite imagery such as Google Earth Pro and use a tool provided with the software to estimate area.

Water Budgets and Mass Balance

Water budgets often are made for the purposes of describing the hydrology of catchments or watersheds, natural water bodies, or water supply systems. Water budgets can be made by the general hydrologic equation

$$\text{Inflow} = \text{Outflow} \pm \text{Change in storage}. \tag{3.7}$$

This equation appears simple, but it often becomes complex when expanded to take into account all inflows and outflows. For example, the expanded equation for a small lake made by damming a watershed might be

$$\text{Rainfall} + \text{Seepage in} + \text{Runoff} = (\text{Evaporation} + \text{Seepage out} + \text{Overflow} + \text{Consumptive use}) \pm \text{Change in storage}. \tag{3.8}$$

In Eq. 3.8, rainfall and change in storage can be measured easily. Evaporation can be estimated by applying a coefficient to Class A pan evaporation data. A weir or other flow measuring device would need to be installed in the pond to obtain overflow values. Runoff would have to be estimated by a procedure that takes into account rainfall amount and runoff-producing characteristics of the watershed. Consumptive use measurements could be made, but one would have to know the activities that withdraw water and quantify them. Seepage in and seepage out probably could not be measured independently, but if all other terms in Eq. 3.8 were known, net seepage could be estimated by difference.

The hydrologic equation allows the estimation of net seepage in a small pond as shown in Ex. 3.1.

Ex. 3.1 *A small pond has a water level of 3.61 m on day 1, and after 5 days, the water level is 3.50 m. There is no rainfall or consumptive water use over the 5-day period, and Class A pan evaporation is 6.0 cm. Net seepage will be estimated.*
    Solution:

$$Pond\ evaporation = Class\ A\ pan\ evaporation \times 0.8$$

$$Pond\ evaporation = 6\ cm \times 0.8 = 4.8\ cm.$$

The water budget equation is

$$Inflow = Outflow \pm \Delta\ Storage.$$

There was no inflow, outflows were seepage and evaporation, and storage change was 3.61 m − 3.50 m = 0.11 m or 11 cm. The seepage was outward—lost from the pond. Thus,

$$0.0 \ cm = (seepage + 4.8 \ cm) - 11 \ cm$$

$$(Seepage + 4.8 \ cm) - 11 \ cm = 0$$

$$Seepage + 4.8 \ cm = 11 \ cm$$

$$Seepage = 11 - 4.8 \ cm = 6.2 \ cm = (1.24 \ cm/d).$$

In studies of water quality, it often is necessary to calculate the amounts of substances contained in inputs, outputs, transport, or storage for specific masses of water. For example, it may be desired to calculate the amount of total suspended solids transported by a stream or the amount of phosphorus retained within a wetland. The hydrologic equation (Eq. 3.7) can be modified to make mass balance calculations of dissolved substances:

$$\text{Inputs (volume} \times \text{concentration)} = \text{Outputs (volume} \times \text{concentration)}$$
$$\pm \text{ Storage (volume} \times \text{concentration).} \quad (3.9)$$

Ex. 3.2 *The discharge of streams average 35 cm/year in an area. The stream flowing from a watershed in this area has an average total nitrogen concentration of 2 mg/L. The annual nitrogen loss per hectare will be estimated.*
Solution:

$$Runoff \ volume = 0.35 \ m/yr \times 10,000 \ m^2/ha = 3,500 \ m^3/ha/yr$$

$$Nitrogen \ loss = 3,500 \ m^3/ha/yr \times 2 \ g \ N/m^3$$
$$= 7,000 \ g \ N/ha/yr \ or \ 7 \ kg \ N/ha/yr.$$

Ex. 3.3 *A plastic lined pond (2000 m³) in an arid region is filled and supplied with city water containing 500 mg/L TDS. Over a year, 1000 m³ of water must be added to the pond to replace evaporation. What would be the approximate concentration of TDS after 3 years?*
Solution:
Quantities of dissolved solids added to the pond are

$$Filling \quad 2,000 \ m^3 \times 500 \ g/m^3 = 1,000,000 \ g$$

$$Maintenance \quad 1,000 \ m^3/yr \times 3 \ yr \times 500 \ g/m^3 = 1,500,000 \ g$$

$$Total = 2,500,000 \ g$$

$$Volume \ after \ 3 \ years \ is \ still \ 2,000 \ m^3.$$

$$Concentration \ of \ TDS \ after \ 3 \ yr = \frac{2,500,000 \ g}{2,000 \ m^3} = 1,250 \ mg/L.$$

Estimates of volumes, flow rates, or both frequently are needed in water quality studies. Most water quantity variables are more difficult to measure than are the concentrations of water quality variables.

Oceans and Estuaries

Most of the world's water is contained in the ocean. The ocean system is much larger than all other water bodies combined and contains salt water instead of freshwater. Water movements in the ocean are complex because of the large expanses of open water, great depths, wind driven currents, coriolis effect, tides, and other factors, but predictable and well-defined currents exist.

Estuaries are the entrances of streams into the ocean. Estuaries contain brackish water because the inflow of freshwater from streams dilutes ocean water. Because of tidal influence there are continuous changes in salinity at any given location in an estuary. When the tide recedes, the freshwater influence will extend farther into the estuary than when the tide advances. Salt or brackishwater is heavier than freshwater and a salt water wedge often extends into rivers beneath their freshwater flow. This can result in stratification of salinity at any given point in an estuary. Inflow of freshwater from streams increases after heavy rains and reduces the salinity of estuaries. Estuaries with poor connection to the ocean may become hypersaline in arid areas or during the dry season. The tide flushes estuaries, and the combination of freshwater inflow, size of the connection with the ocean, and tidal amplitude determines the water retention time in estuaries. Water quality in estuaries is highly dependent upon water residence time. Many estuaries are greatly influenced by human activities, and pollution problems are much more likely in estuaries where water exchange with the ocean is incomplete and slow.

Conclusions

The world's water is partitioned into several components, but nearly all of it is contained in the ocean, ice formations, or deep groundwater that is either not suitable or accessible for most human uses. The water available for human use is mainly a portion of the water continually moving through the hydrologic cycle. The three most important components of the hydrologic cycle effecting the human water supply are precipitation, evapotranspiration, and runoff. A basic knowledge of the factors affecting the quantity of water available at a particular place and time is useful to those studying water quality. The quality of water is highly dependent upon the rates at which dissolved and suspended substances enter and exit water bodies via inflow and outflow.

References

Baumgartner A, Reichel E (1975) The world water balance. Elsevier, Amsterdam

Boyd CE (1985) Pond evaporation. Trans Am Fish Soc 114:299–303

Boyd CE, Soongsawang S, Shell EW, Fowler S (2009) Small impoundment complexes as a possible method to increase water supply in Alabama. In: Proceedings 2009 Georgia water resources conference, April 27–29. University of Georgia, Athens

Hoekstra AY, Mekonnen MM (2012) The water footprint of humanity. Proc Nat Acad Sci 109:3232–3237

Hounam CE (1973) Comparisons between pan and lake evaporation. Technical Note 126, World Meteorological Organization, Geneva

Hunt CB (1974) Natural regions of the United States and Canada. WH Freeman, San Francisco

Idso SB, Foster JM (1974) Light and temperature relations in a small desert pond as influenced by phytoplanktonic density variations. Water Res Res 10:129–132

Klein MJ (1970) Maxwell, his demon and the second law of thermodynamics. Am Sci 58:84–97

Leopold LB (1974) Water a primer. WH Freeman, San Francisco

Leopold LB (1997) Water, rivers, and creeks. University Science Books, Sausalito

Wetzel RG (2001) Limnology, 3rd edn. Academic, New York

Wisser D, Frolking S, Douglas EM, Fekete BM, Vöösmarty CJ, Schumann AH (2008) Global irrigation water demand: variability and uncertainties arising from agricultural and climate data sets. Geophys Res Let 35(24):L24408. https://doi.org/10.1029/2008GL035296

Yoo KH, Boyd CE (1994) Hydrology and water supply for pond aquaculture. Chapman and Hall, New York

Solubility and Chemical Equilibrium

4

Abstract

Solubility, chemical equilibrium, and equilibrium constants are closely related to Gibbs free energy of reaction. Electrostatic interactions among ions decrease the activities of ions, and these interactions increase with greater ionic strength. As a result, for precise calculations related to solubility and equilibrium, ionic activities instead of measured molar concentrations must be used. For practical illustrations of principles, molar concentrations will be used in this book.

Introduction

A solution consists of a liquid called the solvent into which another substance the solute is uniformly mixed. Natural waters are solutions comprised of various solutes. These solutes may be solids, liquids, or gases, but in water quality, solids and gases are the most important solutes. The principles regulating the dissolution of gases and solids differ, and only the solubility of solids will be considered here. The major component of dissolved solids in natural waters is mineral matter resulting from the dissolution of minerals in surface soils and underground geological formations. Minerals have different solubilities, and their solubilities are also effected by temperature, pH, oxidation-reduction potential, presence of other dissolved substances, and the principles of chemical equilibrium.

Dissolutions

A simple dissolution occurs when a solute separates into individual molecules as illustrated below for table sugar (sucrose):

© Springer Nature Switzerland AG 2020

C. E. Boyd, *Water Quality*, https://doi.org/10.1007/978-3-030-23335-8_4

$$C_{12}H_{22}O_{11}(s) \overset{\text{Water}}{\rightarrow} C_{12}H_{22}O_{11}(aq). \tag{4.1}$$

In Eq. 4.1, (s) designates a solid and (aq) indicates the molecule is dissolved in water. A (g) beside a molecule indicates that it is in gaseous form. As it usually is implicit which forms are present, the designations are introduced here but seldom used elsewhere in this book. The sucrose molecules are in crystalline form; dissolution results in sucrose molecules in water. The sugar crystal is not held together strongly, and it has hydroxyl groups in which the bond between hydrogen and oxygen gives a slight negative charge to oxygen and a slight positive charge to hydrogen. Water is dipolar and attracted to the charges forcing the sucrose molecules from the crystal and holding them in aqueous solution. This type of solution is a purely physical process.

Dissolution may result from dissociation of a substance into its ions. This is the common way in which salts dissolve. Gypsum is an ionically bound salt that is hydrated with the formula $CaSO_4 \cdot 2H_2O$. When put in water, the hydrating water is immediately lost into the solvent, and the salt dissolves as follows:

$$CaSO_4(s) \rightleftharpoons Ca^{2+}(aq) + SO_4^{2-}(aq). \tag{4.2}$$

Water molecules, being dipolar, arrange themselves around the ions insulating and preventing recombination of the ions. This type of dissociation also is basically a physical process.

Chemical weathering is a chemical dissolution caused by reactions of water and minerals in the earth's crust. Carbon dioxide reacts with limestone to dissolve it as illustrated in the following equation in which calcium carbonate represents limestone

$$CaCO_3(s) + CO_2(aq) + H_2O \rightleftharpoons Ca^{2+}(aq) + 2HCO_3^-(aq). \tag{4.3}$$

Carbon dioxide also accelerates dissolution of feldspars common in soils and other geological formations. There are many other reactions causing chemical weathering, and nearly all of these reactions are equilibrium reactions in which only a small portion of the mineral form dissolves.

Most compounds whether inorganic or organic have a degree of water solubility; but water, because it is polar, is particularly effective in dissolving polar solutes. Hence, the old adage—"like dissolves like." Nevertheless, two polar compounds (ionically-bound compounds) can have drastically different solubilities in water; the water solubility at 25 °C of sodium chloride is 357 g/L while that of silver chloride is only 1.8 mg/L. Compounds of large molecular weight (large molecules) tend to be less soluble than compounds consisting of smaller molecules, because it is more difficult for water molecules to surround larger molecules. The chemical composition of inorganic compounds is especially important in governing solubility. This has resulted in general guidelines of solubility (Table 4.1) which have even been used as the lyrics of *"The Solubility Song"* many renditions of which have been posted online.

Table 4.1 Rules of solubility

Soluble compounds
Salts of sodium, potassium, and ammonium
Salts of chloride and other halides (Exceptions: halides of Ag^+, Hg^{2+}, and Pb^{2+})
Most fluoride compounds
Salts of nitrate, chlorate, perchlorate, and acetate
Sulfate salts (Exceptions: sulfates of Sr^{2+}, Ba^{2+}, Pb^{2+})
Insoluble compounds
Carbonates, phosphates, oxalates, chromates, and sulfides (Exceptions: combination with Na^+, K^+, and other alkali metals plus NH_4^+
Most metal hydroxides and oxides

Factors Affecting Solubility

Temperature has a pronounced effect on solubility. Endothermic reactions absorb heat from the environment, and increasing temperature will enhance solubility. The opposite is true for compounds that release heat when they dissolve; such reactions are called exothermic reactions. Temperature does not greatly affect the solubility of some compounds such as sodium chloride that are not appreciably endothermic or exothermic when they dissolve.

Dissolution occurs at the surface of minerals and other substances, and dissolution is accelerated by increasing surface area. The more surface area per unit of mass, the quicker a substance will dissolve. Stirring a solution effectively increases the surface area between solute surfaces and water thereby accelerating dissolution.

The solubilities of substances in distilled water often are reported, but natural waters already contain various solutes. The solutes already in water can affect the solubility of a substance. For example, if calcium chloride is dissolving in water, the equation is

$$CaCl_2 \rightleftharpoons Ca^{2+} + 2Cl^-. \tag{4.4}$$

Chloride already in the water is chemically indistinguishable from chloride resulting from the dissolution of calcium chloride. Chloride already dissolved in water lessens the rate of the forward reaction in Eq. 4.4 and diminishes the solubility of calcium chloride. This phenomenon is the common ion effect.

Ions in a solution different from those resulting from dissolution of a substance can affect solubility. This is the diverse or uncommon ion effect, and it sometimes is known as the salt effect. Increasing the total ionic concentration in water results in greater interionic attraction which tends to reduce the effective charges on ions making them less reactive. As the result of the diverse ion effect, a greater concentration of ions is necessary to reach equilibrium, and the solubility of a substance tends to increase as the total concentration of ions dissolved in a water increase.

The pH has a major influence on the solubility of many substances. The pH is defined as the negative logarithm of the hydrogen ion (H^+) activity, and it will be discussed in detail later (Chaps. 9 and 11). For present purposes, it will be enough to state that pH 7 is neutral, and water is acidic at lower pH, because there is a greater concentration of hydrogen ion than of hydroxyl ion. At higher pH than 7, water is basic because hydroxyl ion concentration exceeds that of hydrogen ion. An example of the influence of pH on dissolution of a compound is the reaction of ferric hydroxide [$Fe(OH)_3$] in acidic water:

$$Fe(OH)_3 + 3H^+ \rightleftharpoons Fe^{3+} + 3H_2O. \tag{4.5}$$

Increasing H^+ concentration favors the reaction towards the right (forward reaction) in Eq. 4.5.

Low redox potential also increases the solubility of many substances; insoluble ferric iron compounds tend to dissolve at low redox potential when ferric iron (Fe^{3+}) is reduced to more soluble ferrous iron (Fe^{2+}). Low dissolved oxygen concentration is associated with low redox potential (see Chap. 8). Pressure seldom has an influence on the solubility of solids in natural waters, but pressure has a tremendous effect on gas solubility (see Chap. 7).

Chemical Equilibrium

More of a substance added to the substance's saturated solution will settle to the bottom of the container—no more will dissolve. The saturated solution of a compound results from an equilibrium between the solid compound and the dissolved ions, e.g., in a calcium sulfate solution in pure water, there will be an equilibrium among $CaSO_4$, Ca^{2+}, and SO_4^{2-}.

The equilibrium state achieved in a saturated solution conforms to the Law of Mass Action—also known as Le Chatelier's Principle or as the Equilibrium Law. This principle holds that if a chemical reaction is at equilibrium and conditions of concentration, temperature, volume, or pressure are changed, the reaction will adjust itself to restore the original equilibrium. This idea often is expressed in a general manner as follows:

$$aA + bB \rightleftharpoons cC + dD. \tag{4.6}$$

This equation shows that reactants A and B combine in specific proportions to form products C and D also in specific proportions. There is a back reaction, however, in which C and D combine to form A and B. At equilibrium, the forward and reverse reactions are occurring at equal rates and there is no change in concentrations of reactants and products. At equilibrium, a mathematical relationship exists between the concentrations of the substances on each side of the equation and the equilibrium constant (K)

$$\frac{(C)^c(D)^d}{(A)^a(B)^b} = K \tag{4.7}$$

where the concentrations ideally should be expressed as molar activities. The difference in molar activities and molar concentrations will be explained later in this chapter; but for our purposes, it will be adequate to use molar concentrations. The equilibrium constant K sometimes is replaced by its negative logarithm, pK ($K = 10^{-9}$ and $pK = 9$), but K will be used in this book in all but one instance which is related to pH buffering.

Until equilibrium is reached, the left-hand side of Eq. 4.7 is called the reaction quotient (Q), but when equilibrium is attained, $Q = K$. Once equilibrium occurs, addition or removal of any one of the reactants (A and B) or products (C and D) in Eq. 4.6 will disrupt the equilibrium. The reactants and products will undergo rearrangement in concentrations establishing a new equilibrium in which $Q = K$. The reactants and products in reactions such as illustrated in Eq. 4.6 will always react in a manner to reduce stress in the reaction and maintain a state of equilibrium as depicted in Eq. 4.7.

Some reactions tend to progress almost entirely to the right and there will be very little or none of the reactants remaining. When a strong acid such as hydrochloric acid is placed in water, it dissociates almost entirely into H^+ and Cl^- with little HCl remaining. Treatment of calcium carbonate with excess hydrochloric acid results in all of the $CaCO_3$ being converted to Ca^{2+}, CO_2, and H_2O.

The K values for reactions that progress to completion are much larger than those for reaction in which a mixture of products and reactions remain resulting in an equilibrium. The smaller the value of the equilibrium constant the greater will be the concentration of the reactants relative to the concentrations of the products.

Relationship of Chemical Equilibrium to Thermodynamics

The force causing a chemical reaction to proceed is the difference in the energy of its products relative to the energy of its reactants. A reaction will progress in the direction of lower energy, and at equilibrium the energy of products equals the energy of the reactants. In Eq. 4.6, when A and B are brought together, they contain a certain amount of energy. They react to form C and D, and the energy of the "left-hand side" of the equation declines as C and D are formed on the "right-hand side." When the energy on both sides of the equation is equal, a state of equilibrium exists.

In terms of the laws of thermodynamics, the first law states that the amount of energy in the universe is constant, i.e., energy can neither be created nor destroyed. It can, however, be transformed back and forth between heat and work. In a system, heat can be used to do work, and work can be changed to heat, but at each transformation, some heat will be lost from the system and no longer useful for doing work in the system.

In a chemical reaction within a fixed volume, the output of heat equals the change in internal energy when reactants form products. This output of heat is known as the change in enthalpy (enthalpy is a measure of the internal heat content of a substance), and it is symbolized as ΔH. Thus, ΔH reaction = H products − H reactants. In an exothermic reaction, ΔH will have a negative sign because the enthalpy of the reactants is greater than the enthalpy of the products and heat is released to the environment, i.e., ΣH reactants $= \Sigma H$ products $+$ heat. In an endothermic reaction, heat must be absorbed from the environment because the enthalpy of the products is greater than that of the reactants (positive ΔH), and heat is required to cause the reactants to form products, i.e., ΣH reactants $+$ heat $= \Sigma H$ products.

The second law of thermodynamics holds that the universe continuously moves towards greater entropy (S), or put another way, the universe is becoming more random. Spontaneous chemical reactions will proceed from lower entropy to higher entropy, and the change in entropy is $\Delta S = \Sigma S$ products $- \Sigma S$ reactants. The American physicist and mathematician, J. W. Gibbs, in about 1875 proposed a way to assess chemical reactions based on free energy change in which the energy change was estimated from enthalpy, temperature, and entropy. The free energy change is the amount of the internal energy that can be used to do "work" in the reaction and is not lost as entropy (heat in a chemical raction), i.e., free energy $= \Delta H - T\Delta S$ where T is the absolute temperature. This concept of energy in chemical reactions was referred to as free energy (Garrels and Christ 1965), but today, it is usually called Gibbs free energy (G). Enthalpy and entropy can be measured experimentally and used to estimate the energy of formation of substances. The energy required to form 1 mol of a substance from its basic elements at standard conditions of 1 atmosphere pressure and 25 °C is called the standard Gibbs free energy of formation $\left(\Delta G_f^o\right)$. Lists of ΔG_f^o values are available, and selected values are provided in Table 4.2.

The Gibbs standard-state free energy (ΔG^o) is calculated with the following equation:

$$\Delta G^o = \Delta G_f^o \text{ products} - \Delta G_f^o \text{ reactants}. \tag{4.8}$$

The calculation of standard free energy of reaction is illustrated in Ex. 4.1.

Ex. 4.1 *The value of ΔG^o will be calculated for the reactions $HCO_3^- = H^+ + CO_3^{2-}$ and $2H_2O_2 = 2H_2O + O_2$.*
   Solution:
   Using ΔG_f^o from Table 4.2, the ΔG^o for the two reactions are:

$$\Delta G^o = \Delta G_f^o CO_3^{2-} + \Delta G_f^o H^+ - \Delta G_f^o HCO_3^-$$

$$= (-528.1 \, kJ/mol) + (0) - (-587.1 \, kJ/mol)$$

$$= 59 \, kJ/mol$$

and

$$\Delta G^o = 2\left(\Delta G_f^o H_2O\right) + \Delta G_f^o O_2 - 2\left(\Delta G_f^o H_2O_2\right)$$

$$= 2(-237.2 \, kJ/mol) + 0 - 2(-120.42 \, kJ/mol)$$

$$= -233.56 \, kJ/mol.$$

Table 4.2 Standard Gibbs free energies of formation ΔG_f^o of selected substances

Substance	State[a]	ΔG_f^o (kJ/mol)	Substance	State[a]	ΔG_f^o (kJ/mol)
Al^{3+}	aaq	−485.34	FeS_2	s	−166.9
$Al(OH)_3$	s	−1305.8	$FeCO_3$	s	−666.7
Ba^{2+}	aq	−560.7	Fe_2O_3	s	−742.2
$BaCO_3$	s	−1137.6	I_2	s	0.0
Ca^{2+}	aq	−553.54	I^-	aq	−51.67
$CaCO_3$	s	−1128.8	Mn^{2+}	aq	−227.6
$CaSO_4 \cdot 2H_2O$	s	−1795.9	MnO	s	−363.2
$Ca_3(PO_4)_2$	s	−3875.6	Mn_2O_3	s	−881.2
$CaHPO_4 \cdot 2H_2O$	s	−2153.3	MnO_2	s	−464.8
$Ca(H_2PO_4)_2 \cdot H_2O$	s	−3058.4	MnO_4^{2-}	aq	−500.8
CO_2	g	−394.4	$Mn(OH)_4$	s	−615.0
CO_2	aq	−386.2	Mg^{2+}	aq	−454.8
H_2CO_3	aq	−623.4	$Mg(OH)_2$	s	−833.7
HCO_3^-	aq	−587.1	$MgCO_3$	s	−1012.1
CO_3^{2-}	aq	−528.1	NO_2^-	aq	−37.2
Cl^-	aq	−131.2	NO_3^-	aq	−111.3
Cl_2	g	0.0	NH_3	g	−16.5
Cl_2	aq	6.90	NH_4^+	aq	−79.50
HCl	aq	−131.17	O_2	g	0.0
Cu^+	aq	49.99	OH^-	aq	−157.3
Cu^{2+}	aq	65.52	H_2O	l	−237.2
CuO	s	129.7	H_2O_2	l	−120.42
Cu_2S	s	−86.19	PO_4^{3-}	aq	−1094.1
$CuSO_4 \cdot 5H_2O$	s	−661.9	HPO_4^{2-}	aq	−1135.1
H^+	aq	0.0	$H_2PO_4^-$	aq	−1130.4
H_2	g	0.0	H_3PO_4	aq	−1111.7
H_2O_2	aq	−120.42	SO_4^{2-}	aq	−743.0
Fe^{2+}	aq	−78.86	H_2S	aq	−27.4
Fe^{3+}	aq	−4.60	HS^-	aq	12.61
Fe_3O_4	s	−1015.5	S^{2-}	aq	85.8
$Fe(OH)_3$	s	−694.5	H_2SO_4	l	−690.1

The elemental forms (i.e., Ca, Mg, Al, O_2, H_2, etc.) have $\Delta G_f^o = 0$ kJ/mole
[b]*aq* aqueous, *g* gas, *l* liquid, *s* solid

At equilibrium, ΔG^o of a reaction is 0.0. A negative ΔG^o means that a reaction will progress spontaneously from left to right (reactants → products), and heat will be released. Such a reaction is said to be exergonic. A positive ΔG^o indicates that reaction will not progress from reactants to products without an input of external energy. Such a reaction is known as endergonic. In Ex. 4.1, bicarbonate will not spontaneously dissociate in hydrogen and carbonate ion under standard conditions ($\Delta G^o = 59$ kJ/mol). Of course, when the pH of a solution rises above 8.3, dissociation of bicarbonate will occur. The calculation of ΔG^o for decomposition of hydrogen peroxide (Ex. 4.1) reveals that this compound will spontaneously and rapidly decompose into molecular oxygen and water ($\Delta G° = -233.56$ kJ/mol).

The $\Delta G°$ is for standard conditions, but the free energy of reaction (ΔG) for non-equilibrium conditions can be calculated as follows:

$$\Delta G = \Delta G° + RT \ln Q \tag{4.9}$$

where R = a form of the universal gas law constant (0.008314 kJ/mol/°A), T = absolute temperature (°A), ln = natural logarithm, and Q = the reaction quotient. At equilibrium, $\Delta G = 0$ and $Q = K$, and by substitution into Eq. 4.9, the standard-state free energy equation becomes

$$0 = \Delta G° + RT\ln K$$

$$\Delta G° = -RT\ln K. \tag{4.10}$$

The term, $-RT \ln$, has the value -5.709 at 25 °C [$- (0.008314$ kJ/°A) (298.15°A) (2.303) $= -5.709$], and

$$\Delta G° = -5.709 \log K. \tag{4.11}$$

Equation 4.11 provides a convenient technique for calculating the equilibrium constant of any reaction for which we know $\Delta G°$ as illustrated in Exs. 4.2 and 4.3.

Ex. 4.2 *The $\Delta G°$ will be used to estimate K for the reaction $HCO_3^- = CO_3^{2-} + H^+$.*
Solution:
The $\Delta G°$ for the equation was found to be 59 kJ/mol in Ex. 4.1 and we may write

$$\Delta G° = -5.709\log K.$$

$$\log K = \frac{\Delta G°}{-5.709} = \frac{59}{-5.709} = -10.33 \qquad \left(K = 10^{-10.33}\right).$$

This is the K value that is commonly reported for the dissociation of HCO_3^-. Note, that for the other reaction in Ex. 4.1 ($2H_2O_2 = 2H_2O + O_2$) the log K value would be 40.9 ($K = 10^{40.9}$). The very large K reveals that hydrogen peroxide decomposes strongly and rapidly.

Ex. 4.3 *The K for the reaction $NH_3 + H_2O = NH_4^+ + OH^-$ will be estimated from $\Delta G°$.*
Solution:
The $\Delta G°$ expression is

$$\Delta G° = \Delta G_f°NH_4^+ + \Delta G_f°OH^- - \Delta G_f°NH_3 - \Delta G_f°H_2O$$

Using data from Table 4.2

$$\Delta G° = -79.50 + (-157.30) - (-26.6) - (-237.2)$$

$$\Delta G^o = 27.0 \, kJ/mol.$$

From Eq. 4.11

$$27.0 = -5.709 \log K$$

$$\log K = \frac{27.0}{-5.709} = -4.73$$

$$K = 10^{-4.73}.$$

This is the K value normally reported for the reaction of gaseous ammonia with water.

Temperature affects the equilibrium constant. For temperatures other than 25 °C, Eq. 4.10 instead of Eq. 4.11 may be used to calculate K. The K values for the reaction $HCO_3^- = H^+ + CO_3^{2-}$ at temperatures of 0, 10, 20, 25, 30, and 40 °C are calculated in Ex. 4.4.

Ex. 4.4 *Values of K for the reaction of ammonia in water will be calculated for several temperatures.*
   Solution:
   The K at 25 °C is $10^{-4.73}$, but K may be calculated for other temperatures by adjusting Eq. 4.11 for temperature before solving for K.

$$\Delta G^o = -RT \ln K$$

$$and \qquad \left(\ln \times 2.303 = log_{10} \right)$$

At 0 °C the expression becomes:

$$-RT \ln K = -\left(0.008314 \, kJ \, A\right)\left(273.15^\circ / A\right)(2.303) = -5.230$$

$$and \qquad \Delta G^o = -5.230 \log K$$

From Ex. 4.3, $\Delta G^\circ = 27$ kJ/mol

$$\log K = \frac{\Delta G^o}{-5.230} = \frac{27}{-5.230} - 5.16 = 10^{-5.16}.$$

Repeating the above calculations for other temperatures gives

°C	ΔG°	K	°C	ΔG°	K
0	−5.230 log K	$10^{-5.16}$	20	−5.613 log K	$10^{-4.81}$
5	−5.326 log K	$10^{-5.07}$	25	−5.709 log K	$10^{-4.73}$
10	−5.422 log K	$10^{-4.98}$	30	−5.804 log K	$10^{-4.65}$
15	−5.517 log K	$10^{-4.89}$	35	−5.900 log K	$10^{-4.58}$

Solubility Product

The concept of equilibrium applies to all kinds of chemical reactions, but the solubility product is a special form of Le Chatelier's principle that may be generalized as

$$AB \rightleftharpoons A + B. \tag{4.12}$$

The concept assumes that there is a large excess of solid phase AB so that the amount that dissolves does not influence its concentration, i.e., the solid phase may be considered unity. This allows the equilibrium constant—also called the solubility product constant (K_{sp})—to be expressed as simply the product of the molar concentrations of A and B in the solution state

$$(B)(A) = K_{sp}. \tag{4.13}$$

In a case of a more complex molecules we could have

$$A_2B_3 \rightleftharpoons 2A + 3B. \tag{4.14}$$

In this case, the expression for the solubility product constant is

$$(A)^2(B)^3 = K_{sp}. \tag{4.15}$$

Selected K_{sp} values for compounds of interest in water quality given in Table 4.3 do not include those for common compounds that are highly water soluble and dissolve completely under most conditions. Solid compounds may occur in more than one form. Variation in crystalline structure, and degree of hydration in particular, may influence solubility. An amorphous mineral—one that has not formed the typical crystalline structure of the particular compound—has a different solubility than the crystalline form. It is not uncommon to find somewhat different solubility product constants and solubility data for different forms of the same compound.

The solubility of a compound is not always presented as a simple dissolution in water. Consider the dissolution of gibbsite [Al(OH)$_3$]:

$$Al(OH)_3 \rightleftharpoons Al^{3+} + 3OH^- \qquad K = 10^{-33}. \tag{4.16}$$

This dissolution also may be presented as

$$Al(OH)_3 + 3H^+ = Al^{3+} + 3H_2O \qquad K = 10^9. \tag{4.17}$$

Either way, the concentration of aluminum ion at equilibrium is the same for a given pH: $(Al^{3+}) = 10^{-6}$ M at pH 5 by both Eqs. 4.16 and 4.17.

Some examples showing how to estimate the concentrations of substances in water from chemical reactions and equilibrium constants are provided in Exs. 4.5, 4.6, and 4.7.

Table 4.3 Solubility product constants of selected compounds at 25 °C

Compound	Formula	K_{sp}	Compound	Formula	K_{sp}
Aluminum hydroxide	$Al(OH)_3$	10^{-33}	Lead carbonate	$PbCO_3$	$10^{-13.13}$
Aluminum phosphate	$AlPO_4$	10^{-20}	Lead hydroxide	$Pb(OH)_2$	10^{-20}
Barium carbonate	$BaCO_3$	$10^{-8.59}$	Lead sulfate	$PbSO_4$	$10^{-7.60}$
Barium hydroxide	$Ba(OH)_2$	$10^{-3.59}$	Lead sulfide	PbS	10^{-28}
Barium sulfate	$BaSO_4$	$10^{-9.97}$	Magnesium carbonate	$MgCO_3$	$10^{-5.17}$
Beryllium hydroxide	$Be(OH)_2$	$10^{-21.16}$	Magnesium fluoride	MgF_2	$10^{-10.29}$
Cadmium carbonate	$CdCO_3$	10^{-12}	Manganese carbonate	$MnCO_3$	$10^{-10.65}$
Cadmium sulfide	CdS	10^{-27}	Manganese hydroxide	$Mn(OH)_2$	$10^{-12.7}$
Calcium carbonate	$CaCO_3$	$10^{-8.3}$	Manganese sulfide	MnS	$10^{-10.52}$
Calcium fluoride	CaF_2	$10^{-10.46}$	Mercury carbonate	Hg_2CO_3	$10^{-16.44}$
Calcium hydroxide	$Ca(OH)_2$	$10^{-5.3}$	Mercury sulfate	Hg_2SO_4	$10^{-6.19}$
Calcium magnesium carbonate	$CaCO_3 \cdot MgCO_3$	$10^{-16.8}$	Mercury sulfide	HgS	$10^{-52.7}$
Calcium phosphate	$Ca_3(PO_4)_2$	$10^{-32.7}$	Nickel carbonate	$NiCO_3$	$10^{-6.85}$
Calcium sulfate	$CaSO_4 \cdot 2H_2O$	$10^{-4.5}$	Nickel hydroxide	$Ni(OH)_2$	$10^{-15.26}$
Cobalt carbonate	$CoCO_3$	10^{-10}	Nickel sulfide	NiS	$10^{-19.4}$
Cobalt sulfide	CoS	$10^{-21.3}$	Silver carbonate	Ag_2CO_3	$10^{-11.08}$
Copper hydroxide	Cu_2O	$10^{-14.7}$	Silver chloride	$AgCl$	$10^{-9.75}$
Copper sulfide	CuS	$10^{-36.1}$	Silver sulfate	Ag_2SO_4	$10^{-4.9}$
Iron carbonate	$FeCO_3$	$10^{-10.5}$	Strontium carbonate	S_rCO_3	$10^{-9.25}$
Iron fluoride	FeF_2	$10^{-5.63}$	Strontium sulfate	S_rSO_4	$10^{-6.46}$
Ferrous hydroxide	$Fe(OH)_2$	$10^{-16.31}$	Tin hydroxide	$Sn(OH)_2$	$10^{-26.25}$
Ferrous sulfide	FeS	$10^{-18.1}$	Zinc carbonate	$ZnCO_3$	$10^{-9.84}$
Ferric hydroxide	$Fe(OH)_3$	$10^{-38.5}$	Zinc hydroxide	$Zn(OH)_2$	$10^{-16.5}$
Ferric phosphate	$FePO_4 \cdot 2H_2O$	10^{-16}	Zinc sulfide	ZnS	10^{-25}

Ex. 4.5 *Concentrations of Ca^{2+} and SO_4^{2-} will be calculated for a saturated solution of gypsum ($CaSO_4 \cdot 2H_2O$) in distilled water.*

__Solution:__
The dissolution equation is

$$CaSO_4 \cdot 2H_2O \rightleftharpoons Ca^{2+} + SO_4^{2-} + 2H_2O$$

and from Table 4.3, *K is $10^{-4.5}$.*

Because gypsum and water both can be assigned a value of unity, the solubility product expression becomes

$$(Ca^{2+})(SO_4^{2-}) = 10^{-4.5}.$$

Each gypsum molecule dissolves into one calcium and one sulfate ion; hence, $(Ca^{2+}) = (SO_4^{2-})$ at equilibrium. This allows us to let $x = (Ca^{2+}) = (SO_4^{2-})$, and

$$(x)(x) = 10^{-4.5}$$

$$x = 10^{-2.25} \, M.$$

This molar concentration is equal to 225 mg/L Ca^{2+} and 540 mg/L sulfate.

Ex. 4.6 *The common ion effect will be illustrated by estimating the calcium concentration that could occur in water containing 1000 mg/L ($10^{-1.98}$ M) sulfate.*

__Solution:__

$$(Ca^{2+})(SO_4^{2-}) = K_{sp}$$

$$(Ca^{2+}) = \frac{K_{sp}}{(SO_4)}$$

$$(Ca^{2+}) = \frac{10^{-4.5}}{10^{-1.98}} = 10^{-2.52} \, M.$$

This is a calcium ion concentration of 121 mg/L as compared to 225 mg/L of calcium possible when gypsum dissolves in distilled water (Ex. 4.4).

Ex. 4.7 *The solubility of calcium fluoride (CaF_2) will be calculated.*

__Solution:__
From Table 4.3, *the K_{sp} for CaF_2 is $10^{-10.46}$, and the reaction is*

$$CaF_2 = Ca^{2+} + 2F^-.$$

The solubility product expression is

$$(Ca^{2+})(F^-)^2 = 10^{-10.46}.$$

Letting $(Ca^{2+}) = X$ and $(F^-) = 2X$,

$$(X)(2X)^2 = 10^{-10.46}$$

$$4X^3 = 10^{-10.46} = 3.47 \times 10^{-11}$$

$$X^3 = 8.68 \times 10^{-12} = 10^{-11.06}$$

$$X = 10^{-3.7} M = 0.0002\ M.$$

Thus, $(Ca^{2+}) = 0.0002\ M$ (8.0 mg/L) and $(F^-) = 0.0002\ M \times 2 = 0.0004\ M$ (7.6 mg/L).

Electrostatic Interactions

Electrostatic interactions among ions in solutions usually cause ions to react to a lesser degree than expected from measured molar concentrations, e.g., the reacting concentration is less than the measured concentration. The ratio of the reacting concentration:measured concentration is called the activity coefficient. For dilute solutions, such as most natural waters, the activity coefficient of a single ion may be calculated with the Debye-Hückel equation

$$\log \gamma_i = -\frac{(A)(Z_i)^2(I)^{1/2}}{1 + (B)(a_i)(I)^{1/2}} \tag{4.18}$$

where γ_i = the activity coefficient for the ion i, A and B are dimensionless constants for standard atmospheric pressure and different water temperatures (Table 4.4), Z_i = the valence of ion i, a_i = the effective size of ion i (Table 4.5), and I = the ionic strength. In Eq. 4.18, the 10^{-8} of a_i in the denominator is cancelled by the 10^{-8} of B in the numerator. Thus, the a_i and B values usually are substituted into the equation without their power of 10 factors.

The ionic strength of a solution may be calculated as

$$I = \sum_i^n \frac{(M_i)(Z_i)^2}{2} + \ldots + \frac{(M_n)(Z_n)^2}{2} \tag{4.19}$$

where M = the measured concentration of individual ions.

The activity of an ion is

Table 4.4 Values A and B for substitution into the Debye-Hückel equation at standard atmospheric pressure (Hem 1970)

Temperature (°C)	A ($\times 10^{-8}$)	B ($\times 10^{-8}$)
0	0.4883	0.3241
5	0.4921	0.3249
10	0.4960	0.3258
15	0.5000	0.3262
20	0.5042	0.3273
25	0.5085	0.3281
30	0.5130	0.3290
35	0.5175	0.3297
40	0.5221	0.3305

Table 4.5 Values for ion size (a_i) for use in the Debye-Hückel equation (Hem 1970)

$a_i \times 10^{-8}$	Ion
9	Al^{3+}, Fe^{3+}, H^+
8	Mg^{2+}
6	Ca^{2-}, Cu^{2+}, Zn^{2+}, Mn^{2+}, Fe^{2+}
5	CO_3^{2-}
4	PO_4^{3-}, SO_4^{2-}, HPO_4^{2-}, Na^+, HCO_3^-, $H_2PO_4^-$
3	OH^-, HS^-, K^+, Cl^-, NO_2^-, NO_3^-, NH_4^+

$$(M_i) = \gamma M_i \, [M] \tag{4.20}$$

where (M_i) = the activity and $[M]$ = measured molar concentration. Millimolar concentrations may be used instead of molar concentrations in Eqs. 4.19 and 4.20 if desired as illustrated in Ex. 4.8.

Ex. 4.8 *The ionic strength in a water sample will be estimated.*
   Solution:
The measured concentrations (mg/L) presented in the list below must be converted to millimolar concentration:

Ion	Measured concentration (mg/L)	Conversion factor (mg/mM)	Concentration (mM)
HCO_3^-	136	61	2.23
SO_4^{2-}	28	96	0.29
Cl^-	29	35.45	0.82
Ca^{2+}	41	40.08	1.02
Mg^{2+}	9.1	24.31	0.37
Na^+	2.2	23	0.10
K^+	1.2	3.91	0.03

   Substituting into Eq. 4.19:

$$I = \sum_i^n \frac{(2.23)(1)^2}{2} + \frac{(0.29)(2)^2}{2} + \frac{(0.82)(1)^2}{2} + \frac{(1.02)(2)^2}{2} + \frac{(0.37)(2)^2}{2}$$
$$+ \frac{(0.1)(1)^2}{2} + \frac{(0.03)(1)^2}{2}$$

$$I = 4.96 \; mM \; (0.00496 \; M).$$

The ionic strength of the solution can be used in the Debye-Hückel equation to estimate activity coefficients for individual ions as shown in Ex. 4.9.

Ex. 4.9 *The activities will be estimated for Mg^{2+} and Cl^- in the water from Ex. 4.8.*
Solution:
Using Eq. 4.18 and obtaining values for variables A, B, and a_i from Tables 4.4 and 4.5,

$$log\gamma_{Mg} = -\frac{(0.5085)(2)^2(0.00496)^{1/2}}{1 + (0.3281)(8)(0.00496)^{1/2}}$$

$$log \; \gamma_{Mg} = -0.12090$$

$$\gamma_{Mg} = 0.76.$$

$$log\gamma_{Cl} = -\frac{(0.5085)(1)^2(0.00496)^{1/2}}{1 + (0.3281)(3)(0.00496)^{1/2}}$$

$$log \; \gamma_{Cl} = -0.0335$$

$$\gamma_{Cl} = 0.93.$$

Activities will be calculated with Eq. 4.20,

$$\left(Mg^{2+}\right) = 0.76 \; (0.37 \; mM) = 0.28 \; mM$$

$$(Cl^-) = 0.93 \; (0.82 \; mM) = 0.76 \; mM.$$

Monovalent ion activities deviate less from measured concentrations than do divalent ion activities as seen in Ex. 4.9. The discrepancy between measured concentrations and activities also increases with increasing ionic strength.

The hydrogen ion concentration as calculated from pH measured with a glass electrode is an activity term and needs no correction. The activities of solids and water are taken as unity. Under conditions encountered in natural waters, the measured concentration of a gas in atmospheres may be used without correction to activity.

Ion Pairs

Although the Debye-Hückel equation is widely used in calculating activities of single ions, as illustrated in Exs. 4.8 and 4.9, a portion of cations and anions in solution are strongly attracted to each other and act as if they are un-ionized or of lesser or different charge than anticipated. Ions attracted in this manner are called ion-pairs, e.g., Ca^{2+} and SO_4^{2-} form the ion-pair $CaSO_4^0$, Ca^{2+} and HCO_3^- form $CaHCO_3^+$, and K^+ and SO_4^{2-} form KSO_4^-. The degree to which ions from ion-pairs in solution is a function of the equilibrium (formation) constant K_f for the particular ion pair. Formation of an ion pair is illustrated for the ion pair $CaHCO_3^+$:

$$CaHCO_3^+ = Ca^{2+} + HCO_3^- \tag{4.21}$$

$$\frac{(Ca^{2+})(HCO_3^-)}{CaHCO_3^+} = K_f. \tag{4.22}$$

The activity coefficients of ion-pairs are taken as unity. Formation constants for select ion-pairs of major ions in natural waters are given (Table 4.6). These constants can be used with the ion-pair equations to estimate ion pair concentration (Ex. 4.10).

Ex. 4.10 *The concentration of the magnesium sulfate ion pair ($MgSO_4^0$) will be calculated for water with 2.43 mg/L (10^{-4} M) magnesium and 9.6 mg/L (10^{-4} M) sulfate.*

Solution:
The ion pair equation from Table 4.6 *is*

$$Mg\,SO_4^0 \rightleftharpoons Mg^{2+} + SO_4^{2-} \qquad K_f = 10^{-2.23}$$

$$and \qquad \frac{(Mg^{2+})(SO_4^{2-})}{(MgSO_4^0)} = 10^{-2.23}$$

$$(MgSO_4^0) = \frac{(10^{-4})(10^{-4})}{10^{-2.23}} = 10^{-5.77}\ M\ (0.204\ mg/L).$$

Note: 0.204 mg/L $MgSO_4^0$ is equivalent to 0.041 mg/L Mg^{2+} and 0.163 mg/L SO_4^{2-}. Ions bound in ion pairs are indistinguishable from free ions by usual analytical methods. In this example, 1.69% of both magnesium and sulfate concentration was involved in ion pairing.

Table 4.6 Formation constants (K_f) at 25 °C and zero ionic strength for ion-pairs in natural waters (Adams 1971)

Reaction	K_f
$CaSO_4^{0} = Ca^{2+} + SO_4^{2-}$	$10^{-2.28}$
$CaCO_3^{0} = Ca^{2+} + CO_3^{2-}$	$10^{-3.20}$
$CaHCO_3^{+} = Ca^{2+} + HCO_3^{-}$	$10^{-1.26}$
$MgSO_4^{0} = Mg^{2+} + SO_4^{2-}$	$10^{-2.23}$
$MgCO_3^{0} = Mg^{2+} + CO_3^{2-}$	$10^{-3.40}$
$MgHCO_3^{+} = Mg^{2+} + HCO_3^{-}$	$10^{-1.16}$
$NaSO_4^{-} = Na^{+} + SO_4^{2-}$	$10^{-0.62}$
$NaCO_3^{-} = Na^{+} + CO_3^{2-}$	$10^{-1.27}$
$NaHCO_3^{0} = Na^{+} + HCO_3^{-}$	$10^{-0.25}$
$KSO_4^{-} = K^{+} + SO_4^{2-}$	$10^{-0.96}$

Analytical methods (specific ion electrodes excluded) do not distinguish between free ions and ion-pairs. Sulfate in solution might be distributed among SO_4^{2-}, $CaSO_4^{0}$, $MgSO_4^{0}$, KSO_4^{-}, and $NaSO_4^{-}$. While the total sulfate concentration may be measured, the free SO_4^{2-} concentration must be calculated. As actual ionic concentrations are always less than measured ionic concentrations in solutions containing ion-pairs, ionic activities calculated directly from analytical data, as done in Ex. 4.9, are not exact. Adams (1971) developed a method for correcting for ion-pairing in the calculation of ionic activities and demonstrated a considerable effect of ion-pairing on ionic strength, ionic concentrations, and ionic activities in solutions. The method of correcting for ion-pairing involves: (1) using measured ionic concentrations to calculate ionic strength assuming no ion-pairing, (2) calculating ionic activities assuming no ion-pairing, (3) calculating ion-pair concentrations with respective ion-pair equations, equilibrium constants, and initial estimates of ionic activities, (4) revising ionic concentrations and ionic strength by subtracting the calculated ion-pair concentrations, and (5) repeating steps 2, 3, and 4 until all ionic concentrations and activities remain unchanged with succeeding calculations. Adams (1971, 1974) discussed all aspects of the calculations and gave examples. The iterative procedure is tedious and slow unless it is programmed into a computer. The procedure was developed for soil solutions, but it is equally applicable to natural waters.

Boyd (1981) calculated activities of major ions in samples of natural water with analytical data uncorrected for ion-pairing and analytical data corrected for ion-pairing. Ion-pairing had little effect on ionic activity calculations for weakly mineralized water (I < 0.002 M). For more strongly mineralized waters, ionic activities corrected for ion-pairing were appreciably smaller than uncorrected ionic activities. Ion pairing by major ions is particularly significant in ocean and other saline waters. As a general rule, if waters contain less than 500 mg/L of total dissolved ions, ion-pairing may be ignored in calculating ionic activities unless highly accurate data are required. For simplicity, equilibrium considerations in this book will assume that activities and measured molar concentrations are the same.

Conclusions

Concentrations of water quality variables typically are measured rather than calculated. Nevertheless, knowledge of the principles of solubility and use of solubility products and equilibrium constants to calculate anticipated concentrations can be helpful in explaining observed concentrations and predicting changes in water quality. The concept of chemical equilibrium also can be especially useful in predicting changes in concentrations of water quality variables that may result from addition of substances to water by natural or anthropogenic processes.

References

Adams F (1971) Ionic concentrations and activities in soil solutions. Soil Sci Soc Am Proc 35:420–426

Adams F (1974) Soil solution. In: Carson WE (ed) The plant root and its environment. University Press of Virginia, Charlottesville, pp 441–481

Boyd CE (1981) Effects of ion-pairing on calculations of ionic activities of major ions in freshwater. Hydrobio 80:91–93

Garrels RM, Christ CL (1965) Solutions, minerals, and equilibria. Harper and Row, New York

Hem JD (1970) Study and interpretation of the chemical characteristics of natural water. Water-supply paper 1473, United States Geological Survey, United States Government Printing Office. Washington.

Dissolved Solids

5

Abstract

The total dissolved solids (TDS) concentration in freshwater is determined by passing water through a 2-μm filter, evaporating the filtrate to dryness, and reporting the weight of the solids remaining after evaporation in milligrams per liter. Salinity and electrical conductivity usually are used to assess the concentrations of ions in seawater, and electrical conductivity also may be used to determine the extent of mineralization of freshwater. The TDS concentrations in natural freshwaters typically range from about 20 to 1000 mg/L, and the solids consist mainly of bicarbonate (and carbonate at pH above 8.3), chloride, sulfate, calcium, magnesium, potassium, sodium, and silicate. The concentration of TDS in inland waters is controlled mainly by geological and climatic factors. The most weakly mineralized waters are found in areas with high-rainfall and heavily leached or poorly developed soils. The most strongly mineralized waters usually occur in arid regions. Examples of TDS concentrations in different regions are presented, and reasons for differences in TDS concentrations among regions and sources of water are discussed. Although the major dissolved inorganic substances in natural water are essential for life, minor dissolved constituents in water often have the greatest effect on aquatic organisms. The main effect of TDS concentration on animals and plants usually is related to osmotic pressure that increases with greater TDS concentration. The average TDS concentration in seawater is about 35,000 mg/L. Aquatic species are adapted to specific TDS ranges, and outside this range osmoregulatory difficulty occurs. Excessive TDS concentration also can make water unsuitable for domestic use, irrigation, and various other purposes. Desalination is increasingly used to increase water supply in arid regions. Bottled water has become very popular as a source of drinking water.

C. E. Boyd, *Water Quality*, https://doi.org/10.1007/978-3-030-23335-8_5

Introduction

Natural waters contain both gases and solids in solution. Solids differ from gases in that they remain as a residue when water evaporates. Solids originate mainly from dissolution of minerals and suspension of mineral and organic particles from soils and other geological formations, living aquatic microorganisms, and decaying remains of organisms. The portion of these solids in solution is called the dissolved solids. Chemists usually consider dissolved particles to be ≤ 0.5–1 µm in diameter, but the method for determining dissolved solids in waters typically relies on a 2.0-µm filter to separate dissolved from suspended solids (Eaton et al. 2005).

Most of the weight of dissolved solids results from inorganic particles, and waters with large dissolved solids concentrations often are said to be highly mineralized. Concentrations of dissolved solids in natural waters vary from 1–2 mg/L to over 100,000 mg/L. Waters in regions with acidic, weathered soils and geological formations of dissolution-resistant rocks may contain <50 mg/L of dissolved solids. Waters from areas with limestone deposits or calcareous soils may have 200–400 mg/L of dissolved solids, while in arid regions concentrations above 1000 mg/L are common. The ocean averages 34,500 mg/L dissolved solids, but closed-basin lakes (water flows in but not out) like the Dead Sea and the Great Salt Lake may contain over 250,000 mg/L.

Freshwater usually is considered to contain no more than 1000 mg/L of dissolved solids, and some inland waters are saline. Drinking water ideally should not exceed 500 mg/L of dissolved solids, but in some areas higher concentrations cannot be avoided.

The atmosphere contains particles of solids resulting from evaporation of spray from waves, dust, and combustion by-products. As a result, raindrops acquire small amounts of dissolved solids from the atmosphere. When rainwater reaches the earth, it contacts vegetation, soil and other geological formations and dissolves additional mineral and organic matter. Dissolved mineral matter consists largely of calcium, magnesium, sodium, potassium, chloride, sulfate, bicarbonate, and carbonate. Most natural waters also have a significant concentration of undissociated silicic acid. Hutchinson (1957) defined freshwaters as "dilute solutions of alkaline earth and alkali bicarbonate, carbonate, sulfate, and chloride with a quantity of silica." Seawater is essentially a sodium chloride solution containing sulfates and bicarbonates of potassium, calcium, and sodium. Both freshwater and seawater also contain small concentrations of other dissolved inorganic substances. The most important are hydrogen and hydroxyl ions, nitrate, ammonium, phosphate, borate, iron, manganese, zinc, copper, cobalt, and molybdenum. Hydrogen and hydroxyl ions determine pH and the others are nutrients. Natural waters also contain minor concentrations of other elements that are not nutrients. Minor elements, both nutrients and non-nutrients, may be toxic at elevated concentrations.

The degree of mineralization of inland water is highly variable from place to place and depends upon the solubility of minerals, length of time and conditions of contact of water with minerals, and concentration of substances through evaporation. While ocean water is rather similar in mineral content worldwide, water in estuaries can vary greatly in degree of mineralization. A direct way of assessing the mineralization of water is to measure the concentrations of the individual inorganic constituents. A large effort is required to make a total analysis of water, and it is more common to measure the total concentration of dissolved substances in water in a single analysis. In freshwater, this is possible by the total dissolved solids (TDS) analysis (Eaton et al. 2005). In seawater, the degree of mineralization is usually assessed by chloride concentration, salinity, or specific conductance. Mineralization of freshwater also is often assessed by specific conductance.

Soluble organic compounds reach water through the decay of dead plant and animal matter and by the excretion of organic matter by aquatic organisms. Thousands of organic compounds occur in water, and dissolved organic compounds usually are measured collectively for convenience. Dissolved organic matter concentrations are highest in waters that have had contact with organic residues or in which phytoplankton abound.

Dissolved Solids Measurement

Total Dissolved Solids

The method for determining the TDS concentration begins with filtration of a water sample through a filter with 2-μm or smaller openings. A measured volume of the filtrate is evaporated in a tared (weighed) evaporating dish, the dish and residue are cooled in a desiccator and weighed to constant weight (Fig. 5.1, Left; Ex. 5.1). The residue left upon evaporation is mainly from dissolved inorganic matter, but a variable portion is dissolved organic matter. This method does not give exact results, because carbon dioxide is lost from bicarbonate when calcium carbonate precipitates during evaporation:

$$Ca^{2+} + 2HCO_3^- \xrightarrow{\Delta} CaCO_3 \downarrow + H_2O \uparrow + CO_2 \uparrow . \tag{5.1}$$

The weight loss of solids during evaporation is about 38% of the portion of the TDS concentration resulting from calcium and bicarbonate. This is because molar weights of Ca^{2+} and $2HCO_3^-$ sum to 162.08 g and the molar weights of the volatilized CO_2 and H_2O sum to 62 g; thus, $(62 \div 162.08) \times 100 = 38\%$. A small amount of organic matter also may be lost by volatilization. Nevertheless, TDS concentration is indicative of the degree of mineralization of freshwaters. The TDS procedure does not work well for saline waters, because the large residue of hygroscopic salts absorbs moisture from the air making accurate weighing difficult.

Ex. 5.1 A 100-mL water sample that passed a 2-μm filter is dried at 105 °C in a 5.2000-g dish. The dish and residue weigh 5.2100 g. The TDS concentration will be estimated.

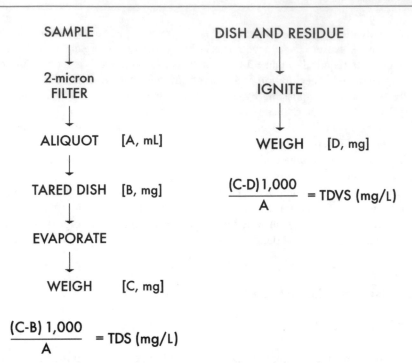

Fig. 5.1   Left. Schematic for total dissolved solids (TDS) analysis. Right. Schematic for conducting TDVS analysis following the total dissolved volatile solids (TDVS) analysis

Solution:

$$Weight\ of\ residue = (5.2100 - 5.2000)g = 0.0100\ g\ or\ 10\ mg$$

$$TDS = 10\ mg \times \frac{1,000\ mL/L}{100\ mL\ sample} = 100\ mg/L.$$

Salinity

The degree of mineralization of water is of considerable benefit in assessing the basic characteristics of a water body or source, and rapid, onsite methods for estimating TDS concentration are useful. A widely used surrogate for TDS concentration in seawater, estuarine water, and saline inland water is salinity.

Salinity is commonly defined as the concentration (g/L) of dissolved salts, but its scientific definition has undergone several modifications. Salinity of seawater was traditionally defined by an equation developed by Knudsen (1901) as follows:

$$\text{Salinity} = 0.030 + 1.805 \; Cl^- \tag{5.2}$$

where Cl^- is in g/L. After 1967, the Knudsen equation was modified by international agreement (Lyman 1969) as

$$\text{Salinity} = 1.80655 \; Cl^-. \tag{5.3}$$

Equations 5.2 and 5.3 are not applicable to inland saline water or freshwater, because the ratio of Cl^- to the total ion concentration often differs greatly from that of seawater (Livingstone 1963).

In 1978, the Joint Panel on Oceanographic Tables and Standards (UNESCO 1981a,b) recommended designating salinity in practical salinity units (psu). The practical salinity unit is dimensionless and based on the ratio of the electrical conductivity of seawater to electrical conductivity of a solution containing 32.4356 g KCl in 1 kg of distilled water (Lewis 1980). Common salinity (g/L) usually does not differ from practical salinity by more than 0.01% (Lewis and Perkin 1981). Absolute salinity was defined in 1985 by the Intergovernmental Oceanographic Commission (IOC) of the United Nations Educational, Scientific and Cultural Organization (UNESCO) (UNESCO 1985) as the ratio of dissolved minerals in seawater to the total mass of seawater (g/kg). Absolute salinity is slightly less than common salinity (g/L) because 1 L of seawater weighs more than 1 kg. At 20 °C, 1 L of seawater (35 g/L salinity) weighs 1.0088 kg. This difference is small, and practical salinity is almost identical in numerical value to absolute salinity. Hence, common salinity is also not greatly different from absolute salinity. A very rigorous definition of salinity, the reference-composition salinity scale (Millero et al. 2008) was recently adopted by IOC (Wright et al. 2011). The reference salinity (S_R) can be related to practical salinity (S) by the expression $S_R = (35.16504/35)$ g/kg $\times$ S. A sample with S = 35 g/L has a $S_R = 35.16504$ g/kg.

Differences in common salinity, practical salinity, absolute salinity, and reference salinity obviously are not great and important mainly to theoretical oceanographers. The common salinity (g/L) is sufficient for most water quality assessments.

Density of water increases with greater salinity (Table 1.4), and density can be measured with a hydrometer. A traditional hydrometer is a cylindrical air-filled bulb, cone-shaped at its bottom with a graduated stem (Fig. 5.2). Ballast in the bottom of the bulb causes the device to float upright. The distance that the stem protrudes above the water surface increases with greater density of the water. The stem of a salinity hydrometer is calibrated in salinity units which usually can be read to 0.1 salinity units (g/L). In practice, water is placed in a clear cylinder to facilitate reading the hydrometer.

The refractive index of water increases with greater salinity. The refractive index of pure water is 1.33300 at 20 °C (Baxter et al. 1911) while the corresponding value for seawater is 1.3394 (Austin and Halikas 1976). A refractometer can be calibrated against salinity concentration and used to estimate salinity. Laboratory refractometers can make precise and highly-accurate estimates of salinity, but

Fig. 5.2 A density
hydrometer for salinity
measurement

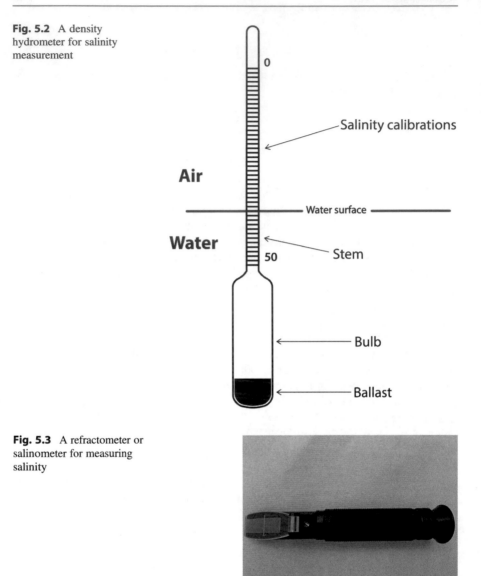

Fig. 5.3 A refractometer or
salinometer for measuring
salinity

simple, hand-held, salinity refractometers (Fig. 5.3) can measure salinities of 1–60 g/
L to one decimal place. Some salinity refractometers give salinity readings in parts
per thousand (ppt) which is essentially the same as grams per liter.

Hydrometers and salinometers are not highly accurate at salinities below 3 or
4 ppt and they are not useful as a surrogate for TDS concentration in freshwater.
They are widely used in marine waters and saline inland waters.

Specific Conductance

Electricity is defined as electric charges (electrons and protons), and these charges are a form of energy. The current (I) measured in amperes is the amount of charge flowing in an electrical circuit. The current is related to the voltage which is the difference in charge measured in volts (V) between two points in the circuit. Resistance (R) measured in ohms is the difficulty within the circuit to convey electricity. The relationship of these three variables is defined by Ohm's law (I = V/R). An unknown resistance can be measured with a Wheatstone bridge circuit that consists of four resistors (Fig. 5.4). Resistors R_1 and R_2 have known resistances, resistor R_A has an adjustable resistance, and the resistance R_X is the medium for which the resistance is determined by the Wheatstone bridge. By adjusting the resistance at R_A, the circuit can be balanced so that the same amount of current flows through the upper part (U) of the circuit containing R_X and R_2 as through the lower part (L) of the circuit containing R_1 and R_A. Balance will exist when there is no current flow through the galvanometer (G). When this state is achieved, the following relationships exist between current (I) and resistance (R):

$$I_U R_2 = I_L R_1$$

$$I_U R_X = I_L R_A.$$

By dividing the upper expression by the lower and solving for R_X, the resistance can be measured

$$\frac{I_U R_2}{I_U R_X} = \frac{I_L R_1}{I_L R_A}$$

Fig. 5.4 A Wheatstone bridge circuit for determining an unknown resistance (R_X). Resistances R_1 and R_2 are known resistances, and resistance R_A is an adjustable resistance

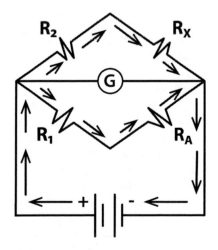

$$R_X = \frac{R_2 R_A}{R_1}. \tag{5.4}$$

Conductance is the reciprocal of resistance, and the principle of the Wheatstone bridge circuit can be used to measure the conductivity of water. Resistance and conductance depend upon the dimensions of the conductor, and the terms resistivity and conductivity are used to describe the resistance of unit length and cross-sectional area of the conductor

$$R = \rho \frac{L}{A} \quad \text{or} \quad \rho = R \frac{A}{L} \tag{5.5}$$

where L = length, A = cross-sectional area, and ρ = specific resistance of the conductor. With the dimensions of the conductor in centimeters, R will be in ohm/cm and ρ in ohm·cm. In SI units, R will be in siemens/cm (S/cm) and ρ in siemens·cm (S·cm). Because k is the reciprocal of resistance,

$$k = \frac{1}{\rho}. \tag{5.6}$$

The unit for k is reciprocal ohm·cm (1/ohm·cm) which usually is referred to as mho/cm—mho is ohm spelled backwards.

The resistivities (and conductivities) of different materials vary greatly. Metals such as copper used as wires in electrical circuits conduct electricity via free electrons that can move around, because they are not attached to cations of the metal. The resistance (R) is greater for an equal length of small diameter wire than for a larger diameter wire, but resistivity (ρ) is equal for both because ρ is a property of the metal and R is the product of ρ multiplied by (L/A). The same reasoning can be applied to conductance (K) and k (conductivity).

Pure water is not a good conductor of electricity, because electrical current in liquids is conducted by dissolved ions, but as the concentration of ions in water increases, its electrical conductivity (EC) increases. The measurement of electrical conductivity is usually an excellent indicator of the degree of mineralization of water.

The electrical conductivity of water is the conductance afforded by a cubic mass of water with edges of 1 cm. As a result of the specificity of the determination, electrical conductivity sometimes is referred to as specific conductance. The standard for electrical conductivity is 0.0100 N KCl solution which has a specific conductance of 0.001413 mho/cm at 25 °C. To avoid small decimal numbers, specific conductance is customarily reported in micromhos per centimeter, or in the case of 0.0100 N KCl, 1413 µmho/cm. The unit, µmho/cm, is replaced by its equivalent microsiemens per centimeter (µS/cm) in the International System (SI) of Units. The SI units can be easily confused, because conductivity meters may allow the display to be selected as siemens/m (S/m), millisiemens/m (mS/m), millisiemens/cm (mS/cm), microsiemens/cm (µS/cm) or other combinations of conductance and length. Also, some meters may have a single display choice, but different from $\mu S/cm$. The original use of micromhos per centimeters for EC is much superior, because there is less opportunity for mistakes in reporting and comparing data. Scientists love to make scientific units confusing.

The electrical conductivity decreases with decreasing temperature, and it can be adjusted to 25 °C (EC_{25}) with the equation

$$EC_{25} = \frac{EC_m}{1 + 0.0191(T - 25)} \tag{5.7}$$

where EC_m = EC at another temperature (T in °C). This equation assumes that specific conductance measurements have been corrected for the cell constant—the value necessary to adjust for variation in the dimensions of the electrode. Specific conductance increases with increasing temperature as illustrated in Ex. 5.2. Most modern conductivity meters have an automatic temperature compensator, and readings are displayed for 25 °C.

Ex. 5.2 *The specific conductance of 1203 μmho/cm for a sample at 22 °C will be adjusted to 25 °C.*

Solution:
Using Eq.5.7,

$$k_{25} = \frac{1,203\ \mu mho/cm}{1 + 0.0191\ (22°C - 25°C)} = 1,276\ \mu mho/cm.$$

All ions do not carry the same amount of current. The equivalent conductances (λ) of common ions at infinite dilution are provided in Table 5.1. The value of λ is equal to the conductance afforded by one equivalent weight of an ion. In the case of calcium, one equivalent weight of calcium (40.08 g/mol $\div$ 2) has a conductance of 59.5 mho, while one equivalent weight of potassium (39.1 g/mol $\div$ 1) has a conductance of 73.5 mho. The equivalent conductance of a particular ion is related to specific conductance as follows:

Table 5.1 Equivalent ionic conductance (λ) at infinite dilution and 25 °C for selected ions (Sawyer and McCarty 1967; Laxen 1977)

Cation	λ (mho·cm^2/eq)	Anion	λ (mho·cm^2/eq)
Hydrogen (H$^+$)	349.8	Hydroxyl (OH$^-$)	198.0
Sodium (Na$^+$)	50.1	Bicarbonate (HCO$_3^-$)	44.5
Potassium (K$^+$)	73.5	Chloride (Cl$^-$)	76.3
Ammonium (NH$_4^+$)	73.4	Nitrate (NO$_3^-$)	71.4
Calcium (Ca^{2+})	59.5	Sulfate (SO$_4^{2-}$)	79.8
Magnesium (Mg^{2+})	53.1		

$$k = \frac{N}{1,000} \tag{5.8}$$

where λ = equivalent conductance (mho·cm^2/eq) and N = the normality of the ion in solution. The specific conductance of a solution will equal the sum of the specific conductances of the individual ions. If specific conductance is calculated with Eq. 5.8, it usually will be larger than the measured specific conductance (Ex. 5.3).

Ex. 5.3 *The specific conductance of 0.01 N potassium chloride will be estimated and compared to the standard measured value of 1413 μmho/cm.*
Solution:

(i)

$$k_{K^+} = \frac{(73.5 \ mho \cdot cm^2/eq)(0.01 \ eq/L)}{1,000 \ cm^3/L} = 0.000735 \ mho/cm.$$

$$k_{Cl^-} = \frac{(76.3 \ mho \cdot cm^2/eq)(0.01 \ eq/L)}{1,000 \ cm^3/L} = 0.000763 \ mho/cm.$$

$$k_{KCl} = k_{K^+} + = k_{Cl^-} = 0.001498 \ mho/cm = 1,498 \ \mu mho/cm.$$

(ii) *This is greater than the standard measured value of 1413 μmho/cm.*

The reason that measured specific conductance is less than the estimated value lies in electrostatic interactions among anions and cations in solution which neutralize a portion of the charge on the ions and reduces their ability to conduct electricity (see Chap. 4). The effect becomes greater as ionic concentration increases as illustrated in Table 5.2 for standard KCl solutions. The electrostatic effect is even greater in solutions containing divalent and trivalent ions.

Because all ions do not have the same equivalent conductance, water samples containing the same concentration of total ions may have different specific conductance values. For example, a water containing mainly potassium, chloride, and sulfate would have a higher specific conductance than a water containing the same amount of total ions but with calcium, magnesium, and bicarbonate predominating (compare equivalent conductances of the ions in Table 5.1). Nevertheless, for a particular water body, there usually is a positive correlation between specific conductance and TDS concentration as illustrated in Fig. 5.5. The relationship also is fairly constant in a surface water of a particular physiographic region. This results because there usually is a fairly constant proportionality among the ions in surface waters from a particular region irrespective of TDS concentration and TDS = K × EC, and K = TDS/EC. The value of K varies from about 0.5 to 0.9 among water sources (Walton 1989), and van Niekerk et al. (2014) reported K values of 0.50–0.80 (average = 0.68) for 38 sites representing 14 rivers in South Africa. The K for seawater is about 0.7, and it usually varies little from place to place.

Table 5.2 Relationship between potassium chloride concentration and measured specific conductance at 25 °C

Concentration	Specific conductance (μmho/cm)	
(N)	Measured	Calculated
0	0	
0.0001	14.94	14.98
0.0005	73.90	
0.001	147.0	149.8
0.005	717.8	
0.01	1413	1488
0.02	2767	
0.05	6668	
0.1	12,900	14,980
0.2	24,820	
0.5	58,640	
1.0	111,900	150,800

Calculated specific conductance is provided for selected concentrations (Eaton et al. 2005)

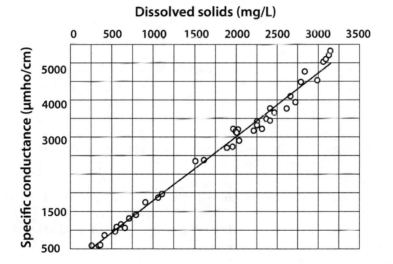

Fig. 5.5 Dissolved solids determined by water analysis compared with specific conductance of daily samples from Gila River at Bylas, Arizona, in a 12-month period. (Hem 1970)

Modifications of conductivity meters have been made to allow results to be displayed in TDS or salinity concentration. The accuracy of such devices depends on how close the ratio of TDS/EC or salinity/EC in a particular water agrees with the ratio used in the meter for translating from EC to TDS concentration or salinity. Some meters have the option for selecting the ratio in case it is known.

The most weakly mineralized water is distilled water with specific conductance usually <5 μmho/cm. Rainwater normally has specific conductance <50 μmho/cm. Inland surface waters seldom have conductances >500 μmho/cm in humid areas; but, in waters of arid regions, specific conductance often is >5000 µmho/cm. Drinking waters normally have specific conductances of 50–1500 µmho/cm. The upper limit for specific conductance of freshwater is usually assumed to be around 1500 µmho/cm, while the specific conductance of seawater is around 50,000 µmho/cm.

Alkalinity and Hardness

In humid regions, the dominant cations in freshwaters will be calcium and magnesium, while bicarbonate and carbonate will be the predominant anions—as was the case for data from the sample used in Ex. 5.4. The calcium and magnesium often are measured together by a titration procedure and reported as total hardness concentration. The alkaline substances in water (mainly HCO_3^- and CO_3^{2-}) also can be determined together by acid titration and the results reported as total alkalinity. The relative degree of mineralization of such waters can be ascertained from the concentrations of these two variables, because they often represent a large portion of the TDS concentration.

Cation-Anion Balance

The principle of electrical neutrality requires the sum of the equivalent weights of positively-charged ions (cations) to equal the sum of the equivalent weights of negatively-charged ions (anions). In most natural waters, major ions make up most of the total weight of ions, and charge is essentially neutral among major ions. This often allows a check on the validity of water analyses. In an accurate analysis, the sum of the measured milliequivalents of major cations and major anions should be nearly equal as illustrated in Ex. 5.4.

Ex. 5.4 *An analysis of a water sample reveals 121 mg/L bicarbonate (HCO_3^-), 28 mg/L sulfate (SO_4^{2-}), 17 mg/L chloride (Cl^-), 39 mg/L calcium (Ca^{2+}), 8.7 mg/L magnesium (Mg^{2+}), 8.2 mg/L sodium (Na^+), and 1.4 mg/L potassium (K^+). The accuracy of the results will be checked.*
   Solution:
   Anions

121 mg HCO_3/L ÷ 61 mg HCO_3/meq = 1.98 meq HCO_3/L
28 mg SO_4/L ÷ 48 mg SO_4/meq = 0.58 meq SO_4/L
17 mg Cl/L ÷ 35.45 mg Cl/meq = 0.48 meq Cl/L
Total = 3.04 meq anions/L

   Cations

39 mg Ca/L ÷ 20.04 mg Ca/meq = 1.95 meq Ca/L
8.7 mg Mg/L ÷ 12.16 mg Mg/meq = 0.72 meq Mg/L

8.2 mg Na/L ÷ 23 mg Na/meq = 0.36 meq Na/L
1.4 mg K/L ÷ 39.1 mg K/meq = 0.04 meq K/L
Total = 3.07 meq cations/L

Charge balance (anions/cations)

$$\frac{3.04}{3.07} \times 100 = 99.02\%$$

The anions and cations balance closely, and even though concentrations of all ions were not measured, the analysis probably was quite accurate.

The anion-cation balance concept also allows construction of pie diagrams to illustrate equivalent proportions of ions in a water (Fig. 5.6). One-half of each pie is for anions and the other half for cations. The proportion of the pie given to an ion is proportional to that ion's concentration in milliequivalents per liter. For comparing the degree of mineralization of different waters, diameters of the pies can be made proportional to the total milliequivalents of ions in a sample.

Ex. 5.5 *A pie diagram will be made with the data from Ex. 5.4.*
    Solution:
    There are 360° in a circle, so 180° each will be given to cations and anions. The number of degrees of the pie (circle) allowed to an anion or cation depends on its ratio to the total anions or total cations multiplied by 180° (one-half of the pie):

$HCO_3^- = (1.98 \text{ meq}/3.04 \text{ meq})180° = 117.2°$
$SO_4^{2-} = (0.58 \text{ meq}/3.04 \text{ meq})180° = 34.3°$
$Cl^- = (0.48 \text{ meq}/3.04)180° = 28.5°$

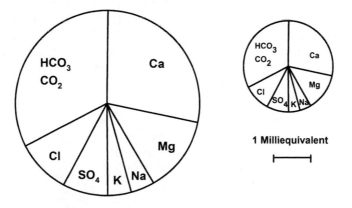

Fig. 5.6 Pie diagrams for the proportions of major ions in two waters. The diameters of the circles are proportional to the milliequivalents per liter of total ions

$Ca^{2+} = (1.95 \ meq/3.07)180° = 114.3°$
$Mg^{2+} = (0.72 \ meq/3.07)180° = 42.2°$
$Na^+ = (0.36 \ meq/3.07)180° = 21.1°$
$K^+ = (0.04 \ meq/3.07)180° = 2.4°.$

The data are plotted as a pie diagram in Fig. 5.6.

Anion-cation balance may not always provide irrefutable proof of the accuracy of a water analysis. The author once sent a sample of water from the Pacific Ocean along the coast of New Caledonia, a rather remote island a considerable distance from the northwest coast of Australia, to a laboratory for analyses of major ions. The specific conductance measured at sampling with a portable meter was 49,500 µmho/cm—a reasonable specific conductance for normal seawater. The laboratory reported a similar specific conductance, and the anion-cation percentage of 98% suggested an accurate analysis of major ions. However, upon adding, the total measured concentration of major ions provided by the laboratory was 48,000 mg/L. This would correspond to an electrical conductivity of approximately 68,600 µmhos/cm (48,000 mg/L ÷ 0.7). Further investigation revealed that the instrument used for measuring concentrations of anions and cations in the analysis was incorrectly calibrated and had given erroneously high values for all major ions.

Major Ions in Natural Waters

Rainwater

The atmosphere contains salts of marine origin, dust, products of combustion, and nitrate produced by electrical activity in clouds. Particles in the atmosphere are swept out by rain, and rainwater contains dissolved mineral matter. Average concentrations of seven mineral constituents in rainwater collected from 18 locations across the conterminous US are provided (Table 5.3). Concentrations of constituents varied greatly among locations with the highest concentrations for variables being 6–220 times greater than the lowest. The highest total concentrations of several variables—especially sodium and chloride—were from locations near the ocean. For example,

Table 5.3 Concentrations of seven constituents in rainwater from 18 locations across the conterminous US (Carroll 1962)

Constituent	Concentration (mg/L)	
	Mean ± standard deviation	Range
Sodium	2.70 ± 5.95	0.10–22.30
Potassium	0.29 ± 0.31	0.07–1.11
Calcium	1.48 ± 1.71	0.23–6.50
Chloride	3.24 ± 7.10	0.13–22.58
Sulfate	1.48 ± 1.32	0.03–5.34
Nitrate	2.31 ± 1.17	0.81–4.68
Ammonium	0.43 ± 0.52	0.05–2.21

sodium concentration at coastal locations of Brownsville, Texas and Tacoma, Washington was 22.3 and 14.5 mg/L, respectively, while at inland sites of Columbia, Missouri and Ely, Nevada, the respective values were 0.33 and 0.69 mg/L. The average specific conductance of US rainfall calculated from ion concentrations reported in Table 5.3 and assuming a pH of 6 was 25.7 μmhos/cm.

Average concentrations of the six major ions in rainwater over the entire United States were lower than concentrations of these ions typically found in inland surface waters. Ammonium and especially nitrate, however, were of higher average concentration than often found in surface waters—including the ocean. Rainfall in highly populated and industrialized regions may have particularly high sulfate concentrations because of contamination of the atmosphere with sulfur dioxide from combustion of fossil fuels.

Rainfall at a location varies in composition from storm to storm because of differences in origins of air masses and the duration of rainfall events. Initial rainfall tends to be more concentrated in impurities than rainfall occurring later in a storm event, because the initial rainfall tends to sweep the majority of particles from the atmosphere.

Inland Waters

Inland surface waters consist of runoff from watersheds (overland runoff), streams, lakes, reservoirs, ponds, and swamps. Overland runoff has brief contact with soil, but this provides the first opportunity for increasing its mineral content above that of rainwater. Streams carry overland flow and groundwater seepage (base flow). Groundwater usually is more concentrated in mineral matter than overland flow, because it has resided for months to years within geological formations. Groundwater often is depleted of dissolved oxygen and charged with carbon dioxide as a result of infiltration through the soil. Carbon dioxide lowers the pH of water, and many minerals are more soluble at low pH and low oxygen content (low redox potential). The presence of carbon dioxide will increase the capacity of water to dissolve limestone, calcium, silicate, feldspars, and some other minerals. Moreover, stream water has contact with soil and geological material in the streambed. These factors contribute to higher concentrations of mineral matter in stream water than in overland flow, but stream water may be less concentrated in minerals than groundwater. Streams often flow into standing water bodies where the residence time is greater than in streams. This provides longer contact of water with bottom soils and greater opportunity for dissolution of minerals.

Each mineral has a given solubility depending on temperature, pH, concentrations of carbon dioxide and dissolved oxygen, and redox potential of the water with which it has contact. If contact is long enough, minerals will dissolve until equilibrium is reached between the concentrations of dissolved ions and the solid phase minerals (Li et al. 2012). In humid climates where there has been extensive weathering, soils may be highly leached and contain only small quantities of soluble minerals. In areas where there are abundant outcrops of rocks and weathering has not produced deep soils, rainwater falling on the surface has little opportunity to dissolve minerals. Of course, limestone is quite soluble in comparison to many types of rocks, and rainwater contacting limestone has greater opportunity for mineralization. In more arid areas, there is not enough rainfall to leach minerals from the soil profile. Salts may accumulate in surface layers of soil in arid regions or at intermediate depths in soil profiles of semi-arid regions. This provides a greater opportunity for mineralization of overland flow than is afforded in many humid regions.

Large rivers have extensive drainage basins, and water comes from a variety of geological and climatic zones resulting in an averaging effect upon ionic composition. Livingstone (1963) gave the average composition of the waters for the larger rivers by continent and computed world average concentrations of major constituents of river water (Table 5.4). The world average TDS concentration was 120 mg/L, but average TDS concentrations for continents ranged from 59 mg/L for Australian rivers to 182 mg/L for European rivers. Variation in individual ionic constituents was even greater than for TDS. On a global basis, bicarbonate was the dominant anion and calcium was the dominant cation.

Silica concentration was greater than concentrations of several major ions in average river water of the continents (Table 5.4). Minerals containing silicon are common in the earth's crust and include quartz, micas, feldspars, amphiboles, and other rocks. The common form of silicon is silicon dioxide (called silica) of which common sand is comprised. Nearly 28% of the earth's crust is made of silicon, and 40% of minerals contain this element. These minerals dissolve to release silicic acid (H_4SiO_4) as shown below for the feldspar albite ($NaAlSi_3O_3$) and sand (SiO_2):

$$2NaAlSi_3O_3 + 2CO_2 + 11H_2O = 2Na^+ + 2HCO_3^- + 4H_4SiO_4$$
$$+ Al_2SiO_5(OH)_4. \tag{5.9}$$

Sand also dissolves slightly in water to form silicic acid as follows:

Table 5.4 Mean concentrations of major constituents (mg/L) in waters of major rivers in different continents (Livingstone 1963)

Continent	$HCO_3^- CO_3^{2-}$	SO_4^{2-}	Cl^-	Ca^{2+}	Mg^{2+}	Na^+	K^+	SiO_2	Sum[a]
North America	68	20	8	21	5	9	1.4	9	142
South America	31	4.8	4.9	7.2	1.5	4	2	11.9	69
Europe	95	24	6.9	31.1	5.6	5.4	1.7	7.5	182
Asia	79	8.4	8.7	18.4	5.6	(9.3)[b]		11.7	142
Africa	43	13.5	12.1	12.5	3.8	11	–	23.2	121
Australia	32	2.6	10	3.9	2.7	2.9	1.4	3.9	59
World average	58	11.2	7.8	15.0	4.1	6.3	2.3	13.1	120

[a]Sum is approximately equal to total dissolved solids (TDS)
[b]Sum of sodium and potassium

$$SiO_2 + 2H_2O = H_4SiO_4. \qquad (5.10)$$

Silicic acid dissociates as shown below:

$$H_4SiO_4 = H^+ + H_3SiO_4^- \qquad K = 10^{-9.46} \qquad (5.11)$$

$$H_3SiO_4^- = H^+ + H_2SiO_4^{2-} \qquad K = 10^{-12.56}. \qquad (5.12)$$

The ratio of the parent silicic acid to ionized $H_3SiO_4^-$ is dependent upon pH or more specifically the hydrogen ion concentration:

$$\frac{(H_3SiO_4^-)}{(H_4SiO_4)} = \frac{10^{-9.46}}{(H^+)}. \qquad (5.13)$$

Dissociation of silicic acid is favored by elevated pH. The pH must be 9.46 before the ratio has a value of 1.0. The pH of surface waters usually is below 9, and undissociated silicic acid predominates.

Small streams have small drainage basins and concentrations of ionic constituents in their waters reflect geological and climatic conditions within a limited area. For example, streams near Auburn, Alabama (USA) usually have <100 mg/L TDS, but there is one stream draining from an area where there is limestone that has a TDS concentration around 200 mg/L. The ionic composition of water from small streams in a region is much more variable than is the composition of waters of larger rivers in the region which have more diverse conditions within their larger drainage basins. Concentrations of major mineral constituents in some relatively small rivers (Table 5.5) illustrate this variation. The Moser River of Nova Scotia has very dilute water because it flows through an area where the geological formations consist of hard, highly insoluble rocks and soils are not highly developed or fertile. The Etowah River in Georgia has fairly dilute water, because it flows through areas where weathering and leaching of soils has been extensive because of high rainfall and temperature. But, the Etowah River is more highly mineralized than the Moser River. The Withlacoochee River in Florida also flows through an area with weathered and highly leached soils, but there are limestone formations in its drainage basin. Dissolution of limestone results in a higher degree of mineralization of the water than in either the Moser or Etowah Rivers. The Republican River in Kansas is in an area of relatively low rainfall and soils are deep and fertile. The water is more highly mineralized than stream waters in regions of greater rainfall where soils are either poorly developed or highly leached. The Pecos River in New Mexico flows through an arid region, and its waters are highly mineralized as compared to the other four rivers. Sulfate, chloride, and sodium are greatly enriched in the Pecos River. This results from the high concentrations of soluble salts that accumulate in soils as the result of the rate of evaporation exceeding that of rainfall in arid regions. Limestone (calcium and magnesium carbonates) is not as soluble as the sulfate and chloride salts of alkali and alkaline earth elements found in soils of arid regions.

Table 5.5 Concentrations of major mineral constituents in some relatively small rivers in the United States and Canada

Continent	$HCO_3^- CO_3^{2-}$	SO_4^{2-}	Cl^-	Ca^{2+}	Mg^{2+}	Na^+	K^+	SiO_2	TDS
Moser River, Nova Scotia	0.7	4.3	6.1	3.6	2.5	$(5.5)^a$		3.0	27
Etowah river, Georgia	20	3.1	1.4	3.8	1.2	2.5	1.0	10	43
Withlacoochee River, Florida	118	23	10	44	3.8	5.0	0.3	7.6	213
Republican River, Kansas	244	64	7.6	53	15	34	10	49	483
Pecos river, New Mexico	139	1620	755	497	139	488	10	20	3673

Values are in milligrams per liter (Livingstone 1963)
[a]Sum of sodium and potassium

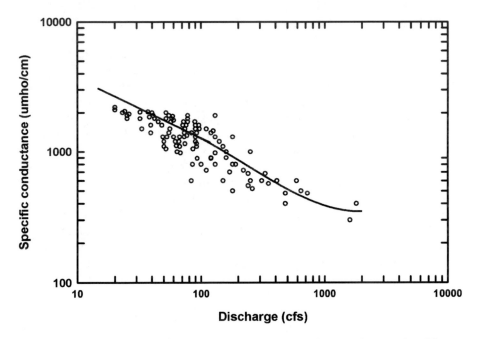

Fig. 5.7 Specific conductance of daily samples and daily mean discharge, San Francisco River at Clifton, Arizona, October 1, 1943 to September 30, 1944. (Hem 1970)

The composition of stream flow also tends to vary with discharge. Ionic concentrations increase as discharge decreases and *vice versa*. This phenomenon is particularly noticeable for small streams in arid regions (Fig. 5.7) and results from runoff from rainstorms diluting the more highly mineralized base flow.

Open basin lakes have inflow and outflow, and their chemical compositions are generally similar to river waters in the same region (Table 5.6). Lakes in Nova

Table 5.6 Concentrations of major mineral constituents in some selected lakes (Livingstone 1963)

Lake	$HCO_3^-\ CO_3^{2-}$	SO_4^{2-}	Cl^-	Ca^{2+}	Mg^{2+}	Na^+	K^+	TDS
Average of nine lakes in Nova Scotia	2.7	5.4	5.2	2.3	0.5	3.2	0.6	20
Lake Ogletree, Auburn, Alabama[a]	25.6	3.2	1.2	3.1	1.7	2.6	1.4	46
Lake Okeechobee, Florida	136	28	29	41	9.1	2.2	1.2	277
Lake Balmorheahak, Texas	159	555	560	38	0.1	(642)[b]		1970
Little Borax Lake, California	8166	10	905	8	24	3390	731	13,600
Dead Sea	240	540	208,020	15,800	41,960	34,940	7560	315,040

Values are in milligrams per liter
[a]From C. E. Boyd, unpublished data
[b]Sum of sodium and potassium

Scotia, Alabama, and Florida are quite similar in TDS concentration and major ion concentrations to those found in river waters of the same regions (see Table 5.5). The lake in west Texas (Table 5.6) is in an arid region and similar in composition to the Pecos River (see Table 5.5) which is in the same arid region. Closed basin lakes have stream inflow but no stream outflow and ions concentrate through evaporation. Little Borax Lake in California and the Dead Sea that lies between Israel and Jordan (Table 5.6) are closed basins.

Minerals dissolve in water until equilibrium between certain solid phase minerals and dissolved ions is reached, and mineral precipitation occurs. The first substance to precipitate is calcium carbonate and further concentration of ions leads to calcium sulfate precipitation. In humid regions, calcium and bicarbonate tend to be dominate ions, but as climate becomes drier, magnesium, sodium, potassium, sulfate, and chloride increase relative to calcium and bicarbonate. In extremely concentrated natural waters, sodium and chloride will be the dominant ions. Of course, many possible intermediate mixtures of ions exist among the three extremes of waters in which bicarbonate-carbonate, sulfate-chloride, or chloride dominate.

The composition of pond waters reflects the composition of soils in a specific area. Of course, if the climate is dry, concentration of ions through evaporation may overshadow the influence of soil chemical composition in the area. Some data on the average composition of water from ponds in several physiographic regions in Mississippi and Alabama (USA) are provided (Table 5.7). Annual rainfall in all regions is between 130 and 160 cm/year, and annual mean air temperature differs by no more than 2 or 3 °C. However, because of differences in soil chemical composition, concentrations of major constituents in the pond waters vary. The Yazoo Basin in Mississippi has fertile, heavy clay soils containing free calcium carbonate and having a high cation exchange capacity. Soils of the Black Belt Prairie and

Table 5.7 Concentrations of major constituents in waters of manmade ponds in different physiographic regions of Alabama and Mississippi

Physiographic region	$HCO_3^-CO_3^{2-}$	SO_4^{2-}	Cl^-	Ca^{2+}	Mg^{2+}	Na^+	K^+	TDS
Mississippi Yazoo Basin	307	–	14.2	61.7	20.8	–	5.1	–
Alabama								
Black Belt Prairie	51.1	4.3	6.8	19.7	1.5	4.3	1.5	94.4
Limestone Valleys and Uplands	42.2	4.2	6.6	11.9	4.7	4.2	3.2	112.0
Appalachian Plateau	18.9	6.6	3.2	5.0	2.8	2.9	1.7	60.2
Coastal Plain	13.2	1.8	5.5	3.4	1.1	2.9	2.8	44.3
Piedmont Plateau	11.6	1.4	2.6	2.7	1.4	2.6	1.4	34.5

Values are in milligrams per liter (Arce and Boyd 1980)

Limestone Valley and Uplands of Alabama are not as fertile as those in the Yazoo Basin soils, but they often contain limestone. The other three physiographic regions have less fertile, highly leached soils that do not contain limestone. The waters in Table 5.7 are primarily dilute calcium and magnesium bicarbonate solutions with variable concentrations of other major ions.

Groundwater can be highly variable in concentrations of major constituents. Several aquifers may occur beneath the ground surface in a particular area and each formation may have different characteristics. Analyses were made of waters from over 100 wells within an 80 km radius in west-central Alabama (Boyd and Brown 1990). There were five distribution patterns among the major ions in the well waters (Fig. 5.8). The most highly mineralized waters (A) (average TDS = 1134 mg/L) had modest proportions of bicarbonate and calcium, large proportions of chloride and sodium, and relatively small proportions of other major ions. Three patterns were observed in the moderately mineralized water (average TDS of 247–372 mg/L). In one type of water (B), bicarbonate was the chief anion and sodium was the primary cation; other ions were relatively scarce. In most moderately mineralized waters (C), bicarbonate was the dominant anion and calcium was the main cation. In other waters of moderate mineralization (D), chloride was proportionally greater than bicarbonate, and calcium and magnesium were roughly proportional to sodium and potassium. In weakly mineralized waters (E) (average TDS = 78 mg/L), bicarbonate was the dominant anion, but sulfate was proportionally greater than in the other four types. Calcium and magnesium were roughly proportional to sodium and potassium.

The water containing primarily sodium and chloride and 1134 mg/L of total dissolved solids (A in Fig. 5.8) came from an aquifer that was influenced by subsurface deposits of sodium chloride that are found in some areas of Alabama.

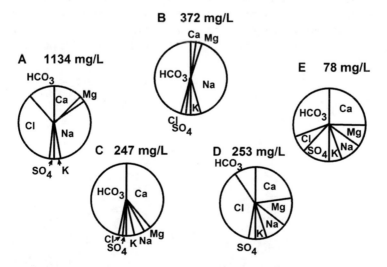

Fig. 5.8 Distribution of major ions in samples of groundwater of different total dissolved solids concentrations from wells in west-central Alabama (Boyd 1990)

The water containing primarily bicarbonate and sodium (B) was the result of an interesting phenomenon sometimes found on coastal plains and known as natural softening of groundwater (Hem 1970). Groundwater high in bicarbonate and sodium concentrations occurs in areas where surface formations contain limestone and underlying aquifers have solids which have adsorbed an abundance of sodium. The aquifers contain sodium because they were filled with seawater during an earlier geological time. Seawater was replaced by freshwater as uplifting of the coastal plain occurred. Rainwater infiltrating downward is charged with carbon dioxide and dissolves limestone to increase concentrations of bicarbonate, calcium, and magnesium. Upon reaching the aquifers, calcium in the infiltrating water is exchanged for sodium in aquifer solids resulting in a dilute sodium bicarbonate solution.

Coastal and Ocean Water

The ocean is essentially saturated with dissolved inorganic substances and is more or less at equilibrium with respect to these substances. The ocean is quite similar in chemical composition worldwide (Table 5.8). The variation in composition is likely quite similar to that observed for salinity. For example, the northern Pacific Ocean has areas where salinity is 31–33 ppt, while some areas in the Atlantic Ocean have salinities of 36–37 ppt. The Mediterranean and Red Seas have areas with salinity above 38 ppt (World Ocean Atlas 2018) (https://www.nodc.noaa.gov/OC5/woa18/woa18data.html).

Ocean waters and river waters mix together in estuaries. Because most river waters are much more dilute in ions than ocean waters, ionic proportions in estuarine waters generally reflect those of ocean waters. The salinity of estuarine water at a given location in an estuary may vary greatly with water depth, time of day, and freshwater inflow. In coastal reaches of rivers, a density wedge of salt water may occur in the bottoms of rivers causing depth stratification of salinity. Tidal action causes changes in salinity by causing salt water to flow into and out of estuaries.

Table 5.8 The average composition of major ions in seawater (Goldberg 1963)

Constituent	mg/L
Cl^-	19,000
Na^+	10,500
SO_4^{2-}	2700 (900 mg S/L)
Mg^{2+}	1350
Ca^{2+}	400
K^+	380
HCO_3^-	142 (28 mg C/L)
Br^-	65
Sr^{2+}	8
SiO_2	3 (as Si)
$B(OH)_3$, $B(OH)_4^-$	4.6 (as B)
F^-	1.3

Fig. 5.9 Relationship between rainfall and salinity in coastal water near Guayaquil, Ecuador

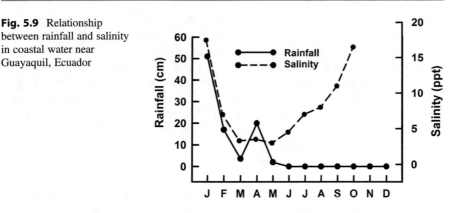

Large inputs of freshwater following rainy weather can dilute the salinity in estuaries as shown in Fig. 5.9. The author once visited a small estuary in Honduras soon after a hurricane with much rain had passed two days earlier. Salinity profiles were taken at several places revealing that only freshwater was present at the time. In estuaries with limited water exchange with the sea, salinity may increase in the dry season in response to less freshwater inflow and concentration of ions by evaporation. Such estuaries may have salinities of 40–60 g/L or even more, but in the rainy season, salinity will decline.

Dissolved Organic Matter in Natural Waters

Natural waters also contain dissolved organic matter from decomposition of plant and animal remains. Water bodies receive organic matter from both external and internal sources which gradually breaks down as a result of physical, chemical, and biological processes, and a fraction of this material becomes small enough to be soluble. Aquatic organisms also can excrete soluble organic compounds directly into the water. Organic matter in soil and water is a mixture of plant and animal remains in various stages of decomposition, compounds synthesized chemically and biologically during decomposition, and microorganisms of decay and their remains.

Organic matter in water has not been investigated extensively, but much that is known about terrestrial soil organic matter is generally applicable to organic matter in aquatic systems. In fact, most organic matter in aquatic systems is in the sediment and behaves similarly to soil organic matter. Organic matter is either nonhumic or humic in nature. The nonhumic substances consist of carbohydrates, proteins, peptides, amino acids, fats, waxes, resins, pigments, and other compounds of relatively low molecular weight. These compounds are readily decomposable by microorganisms and do not occur in large concentrations in either soil or water.

Humus is a product of synthesis and decomposition by microorganisms. It exists as a series of acidic, yellow to black macromolecules of unknown but high molecular weight. Soil organic matter usually is 60–80% humus. Humus chemistry is not well understood, but it consists of a heterogeneous mixture of molecules that form polymers. Molecular weights of polymers in humus range from several hundred to more than 300,000 g/mole. One hypothesis holds that in humus formation, polyphenols derived from lignin and other compounds synthesized by microorganisms polymerize along or with amino compounds to form polymers of variable molecular weights with functional acidic groups. Humus often is considered to be composed of three classes of compounds: humic acids, fulvic acids, and humin. Fulvic acids have higher oxygen contents, lower carbon contents, lower molecular weights, and higher degrees of acidity than humic acids. Fulvic acids are yellowish, whereas humic acids are dark brown or black. The properties of humin are poorly defined. Humin differs from fulvic and humic acids by being insoluble in alkali. Humic substances decompose very slowly and they accumulate in soil. Humin is not water soluble, but humic acids are soluble in water at pH > 1.0 and fulvic acids are soluble at all pHs (Steelink 2002). Because of their solubility, humic and fulvic acids are found in natural waters and especially in those that receive inflow from forested areas or contain large populations of higher aquatic plants. Some of the most highly humic water bodies in areas with peat bogs (Druvietis et al. 2010). These authors reported that such lakes in Latvia had Secchi disk visibilities as low as 0.4 m as a result of humic substances.

Most relatively clear waters will have less than 10 mg/L dissolved organic matter, but nutrient enriched waters with abundant phytoplankton often have considerably more dissolved organic matter. Highest concentrations are found in waters stained with humic substances such as those of bog lakes and swamps. Dissolved organic matter causes color (usually the color of coffee or tea) and restricts light penetration. Organic compounds chelate metals and increase the concentrations of trace metals that can occur in water. Some polluted waters receive exceptionally high inputs of organic matter in effluents.

One way of determining dissolved organic matter concentration [often called the total dissolved volatile solids (TDVS) concentration] is to ignite the residue from the TDS analysis at 450–500 °C to burn off the organic matter. The weight loss is the dissolved organic matter or TDVS. The determinations of TDS and TDVS often are combined into the single methodology outlined in Fig. 5.1, Right. The inorganic fraction of the TSS often is called the total dissolved fixed solids (TDFS), because it is not lost during ignition, and TDFS = TDS – TDVS.

Ex. 5.6 *The dish and dry residue from the TDS analysis that weighed 5.2100 g (Ex. 5.1) are ignited, and after cooling, the dish and residue weigh 5.2080 g. The TDVS concentration will be estimated.*

   Solution:

$$TDVS = (5.2100 - 5.2080) \ g \times 1,000 \ mg/g \times \frac{1,000 \ mL/L}{100 \ mL \ sample}$$

$TDVS = 20 \ mg/L.$

The dissolved organic matter concentration may be estimated using a carbon analyzer that combusts the organic matter and measures the amount of carbon dioxide released. Such an instrument is expensive; a cheaper way is by sulfuric acid-potassium dichromate digestion. In this procedure, a sample of water filtered to remove the suspended particles is treated with concentrated sulfuric acid (H_2SO_4) and excess, standard potassium dichromate ($K_2Cr_2O_7$). It is then held at boiling temperature for 2 hours in a reflux apparatus. Certain other reagents are sometimes added as catalysts or to act as inhibitors of chloride of other interfering substances (Eaton et al. 2005). The organic matter is oxidized by the $Cr_2O_7^{2-}$ ion

$$2Cr_2O_7^{2-} + 3 \text{ organic } C^0 + 16H^+ = 4Cr^{3+} + 3CO_2 + 8H_2O. \tag{5.14}$$

The amount of dichromate remaining at the end of the digestion is determined by back-titration with standard ferrous ammonium sulfate ($FeNH_4SO_4$). The milliequivalents of $Cr_2O_7^{2-}$ consumed in the reaction are equal to the milliequivalents of dissolved organic carbon in the sample as illustrated in Ex. 5.7.

Ex. 5.7 *Estimation of dissolved organic carbon concentration by dichromate digestion.*
  Situation: *A 20-mL water sample that passed a 2-μm filter is treated with concentrated H_2SO_4, 10.00 mL of 0.025 N $K_2Cr_2O_7$, and refluxed at boiling temperature for 2 hours. The digestate requires 4.15 mL 0.025 N $FeNH_4SO_4$ (FAS) for reducing the remaining $Cr_2O_7^{2-}$. The blank titration requires 9.85 mL of the FAS solution.*
  Solution:
The unconsumed $K_2Cr_2O_7$ is estimated:

   Blank $(0.025 \text{ meq}/mL)(9.85 \text{ mL FAS}) = 0.246 \text{ meq}$

   Sample $(0.025 \text{ meq}/mL)(4.15 \text{ mL FAS}) = 0.104 \text{ meq}.$

The amounts of $K_2Cr_2O_7$ consumed are:

  Blank $[(0.025 \text{ meq}/L)(10.00 \text{ mL } K_2Cr_2O_7)] - 0.246 \text{ meq} = 0.004 \text{ meq}$

  Sample $[(0.025 \text{ meq}/L)(10.00 \text{ mL } K_2Cr_2O_7)] - 0.104 \text{ meq} = 0.146 \text{ meq}.$

Organic matter in the sample equaled the amount of $K_2Cr_2O_7$ consumed or (0.146–0.004) meq = 0.142 meq. The equivalent weight of organic carbon in the reaction depicted in Eq. 5.14 can be determined from the observation that three carbon atoms with valence of 0 lost four electrons each to obtain a valence of 4+ in carbon dioxide. The equivalent weight of carbon in the reaction is obtained by dividing the atomic weight of carbon (12) by the number of electrons transferred per carbon atom (4).
Thus, the amount of dissolved organic carbon in 20 mL of the water sample was

$$0.142 \ meq \ K_2Cr_2O_7 \times 3 \ mg \ C/meq = 0.426 \ mg,C \ in \ the \ 20$$
$$- \ mL \ sample \ or \ 21.3 \ mg \ C/L.$$

Dissolved Solids and Colligative Properties

Vapor Pressure

The colligative properties of solutions are physical changes caused by the solute that depends on the amount of solute but not on the type of solute. These changes include decreased vapor pressure, decreased freezing point, increased boiling point, and increased osmotic pressure with greater solute concentration. The TDS concentration effects the colligative properties of water.

The French chemist, François Raoult, discovered in the late eighteenth century that at the same temperature the vapor pressure of a solution is less than the vapor pressure of its pure solvent. This phenomenon, known as Raoult's law, is expressed as

$$P_{sol} = XP^o \tag{5.15}$$

where P_{sol} = vapor pressure of the solution, X = mole fraction of pure solvent [moles solvent/(moles of solute + moles of solvent)], and P^o = the vapor pressure of pure solvent.

The phenomenon results because solute molecules occupy space among the solvent molecules. The solution surface does not consist totally of solvent molecules, and the rate of diffusion of the solvent molecules into the air above the solution will be less than it would be for the pure solvent exposed to air. As the solute concentration increases, vapor pressure decreases.

Freezing and Boiling Points

Water boils when its vapor pressure equals atmospheric pressure. A solute reduces the vapor pressure of water, and to effect boiling, the solution must be raised to a higher temperature than would pure water (see Ex. 5.8).

At the freezing point, the movements of water molecules slow enough to allow hydrogen bonding to arrange the solid crystalline lattice of ice. Of course, water molecules are continuously exchanged between the solid and liquid phases of water, but at the freezing point, there is no net exchange of molecules between the two phases. Just as solute molecules occupy space at the water surface and lessen the rate at which water molecules enter the air (lower the vapor pressure), solute molecules occupy space in liquid water. But, solute molecules do not enter the crystalline structure of ice. The overall effect is that the presence of solutes in water slows the rate at which water molecules enter the solid phase. This results in a decrease in the temperature at which molecular movement becomes slow enough to allow equilibrium in exchange of water molecules between solid and liquid phases, i.e., it results in freezing point depression (see Ex. 5.8).

The effect of the solute results from the solute particles, and some substances dissolve into ions rather than molecules. Thus, 1 mole of salt (NaCl) gives 2 moles of particles (1 mol of Na^+ and 1 mol of Cl^-). Sugar (sucrose) dissolves as molecules, and 1 mole of sugar gives 1 mole of particles. The phenomenon is the basis for the van't Hoff factor which is used to adjust ionic concentrations to particle concentrations.

The equations for estimating freezing point depression and boiling point elevation are

$$\Delta T_{fp} = (i)(K_{fp})(m) \tag{5.16}$$

$$\Delta T_{bp} = (i)(K_{bp})(m) \tag{5.17}$$

where ΔT_{fp} and ΔT_{bp} = the changes in freezing and boiling points (°C), respectively; I = van't Hoff factor; K_{fp} and K_{bp} = proportionality constants for freezing and boiling point changes, respectively; and m = molality of the solution (m = g solute/g solution). Typical values for K_{fp} and K_{bp} for water are 1.86 °C/m and 0.512 °C/m, respectively.

Ex. 5.8 *The vapor pressure, depression of freezing point, and elevation of boiling point in a 1-m aqueous salt solution will be estimated for 20 °C and standard pressure.*

Solution:

The 1 m KCl solution consists of 74.55 g (1 mole) KCl and 1000 g water (55.56 moles). The van't Hoff factor is 2 (KCl = K^+ + Cl^-). The mole fraction, X = (55.56 moles) ÷ (2 moles + 55.56 moles) = 0.965. Using Eq. 5.15 and P^o from Table 1.2, vapor pressure of the solution is:

$$P_{sol} = (0.965)(17.535 \ mm) = 16.92 \ mm.$$

The freezing point reduction is calculated with Eq. 5.16

$$\Delta T_{fp} = (2)(1.86°C/m)(1 \ m) = 3.72°C.$$

The elevation of the boiling point is obtained with Eq. 5.17

$$\Delta T_{bp} = (2)(0.512°C/m)(1 \ m) = 1.02°C.$$

While Ex. 5.8 shows substantial changes in freezing and boiling points of a 1 m KCl solution, this solution is quite concentrated in TDS concentration as compared to freshwater. Freshwater freezes and boils at about the same temperature as pure water, but the TDS concentration of seawater and some inland saline waters is great enough to have a significant effect on boiling and freezing points. Seawater typically freezes at −1.9 °C and boils at around 100.6 °C.

Osmotic Pressure

The osmotic pressure of a solution is the amount of pressure or force needed to prevent the flow of water molecules across a semipermeable membrane from a concentrated to a less concentrated solution as illustrated in Fig. 5.10. The semiper-meable membrane is permeable to solvent molecules, but not to solute particles. Assuming water is the solvent, it will pass through a semipermeable membrane from a dilute solution to a more concentrated solution. A simple way to view this phenomenon is that the membrane is bombarded continuously on both sides by molecules of both water and solute. On the side of the membrane facing the dilute solution, more water molecules will strike the membrane surface than on the side of the membrane facing the more concentrated solution, because there are more water molecules per unit volume in the dilute solution than in the concentrated solution. There will be net movement of water molecules from the dilute solution to the concentrated solution until equilibrium (equal concentration) is attained between the two sides. The idea can be expressed in terms of pressure, because pressure can be applied to the concentrated side to stop the net movement of water molecules across the membrane (Fig. 5.10). The amount of pressure necessary to accomplish this feat is the osmotic pressure of a solution. The equation for osmotic pressure is

$$\pi = \frac{nRT}{V} \tag{5.18}$$

where π = osmotic pressure (atm), n = number of moles of solute in the solution, R = ideal gas constant constant (0.082 L·atm/mole·°K), T = absolute temperature (°C + 273.15), and V = volume of solution (L). Because n/V is equal to the molar concentration of the solute, Eq. 5.18 becomes

Fig. 5.10 Illustration of the concept of osmotic pressure

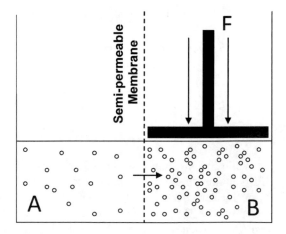

$$\pi = CRT \tag{5.19}$$

where C = the molar concentration of the solute. Some solutes dissociate into two or more particles per molecule, and van't Hoff's factor is used to account for this phenomenon,

$$\pi = iCRT. \tag{5.20}$$

The osmotic pressure of 0.1 M NaCl and sugar solutions will be calculated in the following example.

Ex. 5.9 *The osmotic pressure of 0.1 M solutions of sugar and sodium chloride will be calculated for 25 °C.*
Solution:
Sugar does not ionize when it dissolves; a 0.1 M sugar solution is 0.1 M in particles. Sodium chloride dissolves into ions (NaCl = Na⁺ + Cl⁻), and van't Hoff's factor is 2.
Using Eq. 5.20,

$$Sugar = (0.1\ mole/L)(0.082\ L \cdot atm/mole \cdot °K)(273.15°K + 25) = 2.44\ atm$$

$$NaCl = 2(0.1\ mole/L)(0.082\ L \cdot atm/mole \cdot °K)(273.15°K + 25) = 4.89\ atm.$$

In applying the osmotic pressure concept (Fig. 5.10) to aquatic animals, body fluids of aquatic animals represent one solution, the surrounding water is the other solution, and the part of the animal that separates the two solutions can be thought of as the membrane. Freshwater animals have body fluids more concentrated in ions than the surrounding water; they are hypersaline or hypertonic to their environment. Saltwater species have body fluids more dilute in ions than the surrounding water; they are hyposaline or hypotonic to their environment. The freshwater fish tends to accumulate water because it is hypertonic to the environment, and it must excrete water and retain ions to maintain its osmotic balance (Fig. 5.11). Because the saltwater fish is hypotonic to the environment, it loses water. To replace this water, the fish takes in saltwater; but to prevent the accumulation of excess salt, it must excrete salt (Fig. 5.11). Each species has an optimum salinity range. Outside of this range the animal must expend considerably more energy than normal for osmoregulation at the expense of other processes such as growth. The effect of salinity in utilization of food energy by fish is shown (Table 5.9). Of course, if salinity deviates too much from optimum, the animal will die because it cannot maintain homeostasis.

Plants also can have problems with osmoregulation if the water available to them has excessive concentration of ions. The influence of dissolved solids on the quality of irrigation water is predicted from the sodium adsorption ratio (SAR)

Fig. 5.11 Osmoregulation
by freshwater versus
marine fish

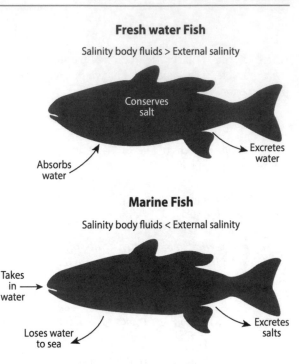

Fresh water Fish

Salinity body fluids > External salinity

Conserves
salt

Excretes
water

Absorbs
water

Marine Fish

Salinity body fluids < External salinity

Takes
in
water

Excretes
salts

Loses water
to sea

Table 5.9 Effect of
salinity on recovery of food
energy as growth in the
common carp (Wang et al.
1997)

Salinity (ppt)	Food energy recovered as fish growth (%)
0.5	33.4
2.5	31.8
4.5	22.2
6.5	20.1
8.5	10.4
10.5	−1.0%

$$SAR = \frac{(Na)}{0.5(Ca + Mg)^{0.5}} \tag{5.21}$$

where Na, Ca, and Mg = concentrations in meq/L. Some general standards for use of irrigation water based on TDS and SAR are provided (Table 5.10).

Removal of Dissolved Solids

It may be necessary to remove dissolved solids from water to produce potable water or for water of low conductivity for use in analytical laboratories or in industrial processes. The natural way of producing relatively pure water is evaporation

Table 5.10 General standards for total dissolved solids (TDS) and sodium adsorption ratio (SAR) in irrigation water

Salt tolerance of plants	TDS (mg/L)	SAR
All species, no detrimental effects	500	2–7
Sensitive species	500–1000	8–17
Adverse effects on many common species	1000–2000	18–45
Use on tolerant species on permeable soils only	2000–5000	46–100

followed by condensation of water vapor in the atmosphere and rainfall (see Chap. 3). This natural process is mimicked by distillation in which water is converted to vapor by heating it in a chamber and then condensing the vapor and capturing the condensate as illustrated in Fig. 5.12. Dissolved solids also may be removed from water by ion exchange and reverse osmosis.

Ion Exchange

Relatively small volumes of water with very low concentrations of dissolved substances for use in chemical, biological, and medical laboratories or for certain industrial purposes were traditionally produced with small-scale distillation units. Such water is presently produced mainly by ion exchange. There are standards for distilled water such as those of the International Standards Organization (ISO), the American Society of Testing Materials (ASTM), and the United States Pharmacopeia (USP). The conductivity standards of these organizations are provided (Table 5.11). Distilled or deionized water that meets the standards is quite pure. However, because of the ionization of water ($H_2O = H^+ + OH^-$, $K_w = 10^{-14}$), a conductivity below about 0.055 μmhos/cm is not possible.

A deionizer consists of a column packed with an anion exchange resin and a cation exchange resin. These resins typically are high molecular weight polymers; the anion resin exchanges hydroxyl ion with anions in the water, while the cation resins exchanges hydrogen ion with cations in the water. An example of an anion exchange resin is a polymer with many quaternary ammonium sites. Nitrogen in the quaternary ammonium group is linked to the polymer and the three methyl groups of the quaternary ammonium impart a positive charge to the nitrogen. The resin is treated with hydroxide resulting in the charge on the quaternary ammonium sites being satisfied with hydroxyl ions. When water is passed through the resin, hydroxyl ions are exchanged for anions in the water as illustrated below:

$$Polymer - N(CH_3)_3^+ \, OH^- + Cl^- \rightarrow Polymer - N(CH_3)_3^+ Cl^- + OH^-. \quad (5.22)$$

A common cation resin consists of a polymer with many sulfonic acid groups on it. The sulfur in sulfonic acid groups has two double-bonded oxygens and one single-bonded oxygen with a negative charge. Hydrogen ion neutralizes the charge on the

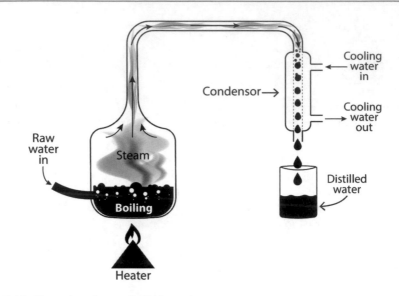

Fig. 5.12 Illustration of water distillation unit

Table 5.11 American Society of Testing Materials (ASTM), International Standards Organization (ISO) and US Pharmacopeia (USP) standards for specific conductivity of distilled or deionized water

Standards	Specific conductivity (μmhos/cm)
ASTM	
Type I	<0.06
Type II	<1.0
Type III	<2.5
Type IV	<5.0
ISO	
Grade 1	<0.1
Grade 2	<1.0
Grade 3	<5.0
USP	
USP (purified)	<1.3
USP (for injection)	<1.3

single-bonded oxygen, and when water is passed through the resin, the hydrogen ions are exchanged for cations in the water as follows:

$$\text{Polymer} - SO_3^- H^+ + K^+ \rightarrow \text{Polymer} - SO_3^- K^+ + H^+. \qquad (5.23)$$

In the case of divalent ions, polymers have many quaternary ammonium or sulfonic acid groups, and the hydroxyl or hydrogen ions on two nearby sites are exchanged for a divalent anion or divalent cation. Hydrogen and hydroxyl ions released into the water during the process combine to form water. When ion exchange resins become saturated with anions or cations, they may be backwashed with an alkaline or an acidic solution, respectively, to recharge them for further use.

Dissolved organic matter that is charged—usually it will bear a negative charge—can be removed by ion exchange. Activated carbon filtration can be used to remove dissolved organic carbon that is not charged. Activated carbon is highly porous charcoal with a huge surface area—up to 2000 m^2/g. Water can be passed through a column of activated carbon and the charcoal will adsorb many dissolved organic substances from water.

Desalination

In many respects, the entire human race suffers from the problem of the Ancient Mariner who shot the albatross resulting in a curse on him and his sailing ship: *"water, water everywhere, nor any drop to drink"* (*The Rime of the Ancient Mariner* by Samuel Taylor Coleridge). The ocean holds an immense quantity of water, but it is not suitable for most human purposes because of its high salinity. Desalination of ocean water is promoted as a means of increasing the supply of freshwater for human use. There are several processes for desalination, but most desalination plants rely on either distillation or reverse osmosis. Multistage flash distillation allows heat recovered from condensing steam in the first stage vaporization of water to be used to vaporize water in the second stage, and the process is repeated for several stages making the overall process more energy efficient. In reverse osmosis, pressure is used to force water through membranes that exclude the dissolved solids.

According to the International Desalination Association (idadesal.org), as of June 2015, there were 18,246 desalination plants in over 150 countries that produced about 86.8 million m^3/day (32 $km^3/year$) and supplied water for about 300 million people.

Desalination occurs naturally when water freezes in the ocean, because dissolved ions are not included in the structure of ice and displaced into the liquid phase. Icebergs and glaciers represent relatively pure water, and it has been proposed that these masses of frozen freshwater could be towed to coastal cities as a freshwater source. This idea has yet to be brought to fruition.

Bottled Water

Bottled water production has become a huge global industry. Municipal water systems produce water similar in microbiological and chemical quality to bottled water, but the public has taken a liking to bottled water. An estimated 400 billion liters (0.4 km^3) worth ≈$US 240 billion were consumed globally in 2017 (https://www.statista.com/statistics/387255/global-bottle-water-consumption/).

Water for bottling can be taken from springs, wells, or municipal water supplies, but most is purified by distillation, deionization, or reverse osmosis. Fluorine may be added to bottled water, and most brands have been subjected to microbial purity tests. The chemical composition of bottled water varies among types and brands, and

Table 5.12 Types of bottled water

Type of water	Special feature
Bicarbonated	Bicarbonate >600 mg/L
Sulfated	Sulfate >200 mg/L
Chlorinated	Chloride >200 mg/L
Calcic	Calcium >150 mg/L
Magnesic	Magnesium >50 mg/L
Fluorinated	Fluorine >1 mg/L
Sodic	Sodium >200 mg/L
Ferruginous	Iron >1 mg/L
Lightly mineralized	TDS <50 mg/L
Oligomineral water	TDS >50 to <500 mg/L
Mediomineral water	TDS >500 to <1500 mg/L
Mineral water	>1500 mg/L

Source: https://www.lenntech.com/mineral-water/bottled-water.htm.

it often is supplied on the bottle label. The general types of bottled water are listed in Table 5.12.

In locations where the water supply is not sanitary or the water has objectionable appearance, taste, or odor, bottled water is an excellent choice. In many parts of the world there are no health or other benefits to choosing bottled water over municipal tap water. Bottled water industry requires a tremendous number of plastic bottles, many of which will likely not be recycled. The bottled-water industry uses more energy and emits more carbon emissions per unit volume of water delivered to the consumer than do municipal water treatment plants.

Effects of Dissolved Solids

Excessive TDS concentrations interfere with various beneficial uses of water. As the TDS concentration rises in inland waters, osmotic pressure increases, and there will be fewer and fewer aquatic species that can tolerate the osmotic pressure. There is much variation in TDS concentrations in aquatic ecosystems in different edaphic and climatic regions, and the species composition of aquatic communities is influenced by tolerance to dissolved solids (or tolerance to salinity, specific conductance, or osmotic pressure). Ecosystems with a good, mixed, freshwater fish fauna usually have TDS concentrations below 1000 mg/L, but many species of freshwater fish and other organisms may tolerate $\geq$5000 mg/L TDS. Nevertheless, to assure protection of aquatic life, TDS concentrations should not be allowed to increase above 1000 mg/L in freshwater ecosystems. Several states in the United States limit "end of pipe" discharges into natural water bodies to 500 mg/L of TDS. In water bodies of naturally greater TDS, higher TDS concentration limits usually are allowed in discharge. Concentrations of TDS naturally fluctuates in estuaries, and species occurring there are adapted to these changes. The TDS concentration usually is not an issue in effluents discharged directly into the ocean—an exception is reverse osmosis seawater desalination plants that discharge water much more concentrated than ocean water in salts.

The main physiological influence to humans of high TDS concentration is a laxative effect caused by sodium and magnesium sulfates. Sodium also has adverse effects on individuals with high blood pressure and cardiac disease and pregnant women with toxemia. Normally, an upper limit for chlorides and sulfates of 250 mg/ L each is suggested. Excessive solids also impart a bad taste to water, and this taste is caused primarily by chloride. Problems with encrustations and corrosion of plumbing fixtures also occur when waters have excessive TDS concentration. If possible, public water supplies should not have TDS concentration above 500 mg/L, but some may have up to 1000 mg/L or more. Livestock water can contain up to 3000 mg/L total dissolved solids without normally causing adverse effects on animals.

Conclusions

The total dissolved solids concentration is one of the most important variables for assessing the suitability of inland water for domestic, agricultural and industrial water sources. Specific conductance is probably the most versatile method for assessing concentrations of dissolved solids in freshwater and salinity is the most widely used indicator of dissolved solids in estuarine and ocean water. Dissolved solids concentration may negatively affect organisms, because if the concentration is outside the acceptable range, osmoregulatory problems occur. Desalination has become a significant source of freshwater in many arid countries located along the coast. Bottled drinking water has become a huge industry in many countries including those with a safe supply of municipal water.

References

Arce RG, Boyd CE (1980) Water chemistry of Alabama ponds. Bulletin 522, Alabama Agricultural Experiment Station, Auburn University

Austin RW, Halikas G (1976) The index of refraction of seawater. US Department of Commerce Technical Information Series AD-A024800, Washington

Baxter GP, Burgess LL, Daudt HW (1911) The refractive index of water. J Am Chem Soc 33:893–901

Boyd CE (1990) Water quality in ponds for aquaculture. Alabama Agricultural Experiment Station, Auburn University

Boyd CE, Brown SW (1990) Quality of water from wells in the major catfish farming area of Alabama. In: Tave D (ed) Proceedings 50th Anniversary Symposium Department of Fisheries and Allied Aquacultures, Auburn University

Carroll D (1962) Rainwater as a chemical agent of geologic processes—a review. Water-supply paper 1535-G, United States Geological Survey, United States Government Printing Office, Washington.

Druvietis I, Springe G, Briende A, Kokorite I, Parele E (2010) A comparative assessment of the bog aquatic environment of the Ramsar site of Teici Nature Reserve and North Vidzeme Biosphere Reserve, Latvia. In: Klavins M (ed) Mires and Peats. University of Latvia Press, Salaspils, Latvia

Eaton AD, Clesceri LS, Rice EW, Greenburg AE (eds) (2005) Standard methods for the examination of water and wastewater, 21st edn. American Public Health Association, Washington

Goldberg ED (1963) Chemistry—the oceans as a chemical system. In: Hill MN (ed) Composition of sea water, comparative and descriptive oceanography, Vol. II. The sea. Interscience Publishers, New York, pp 3–25

Hem JD (1970) Study and interpretation of the chemical characteristics of natural water. Water-supply paper 1473, United States Geological Survey, United States. Government Printing Office, Washington

Hutchinson GE (1957) A treatise on limnology, vol 1, geography, physics, and chemistry. Wiley, New York

Knudsen M (1901) Hydrographical tables. G. E. C. Gad, Copenhagen

Laxen DPH (1977) A specific conductance method for quality control in water analysis. Water Res 11:91–94

Lewis EL (1980) The practical salinity scale 1978 and its antecedents. J Ocean Eng 5:3–8

Lewis EL, Perkin RG (1981) The practical salinity scale 1978: conversion of existing data. Deep-Sea Res 28:307–328

Li L, Dong S, Tian X, Boyd CE (2012) Equilibrium concentrations of major cations and total alkalinity in laboratory soil-water systems. J App Aqua 25:50–65

Livingstone DA (1963) Chemical composition of rivers and lakes. Professional Paper 440-G, United States Geological Survey, United States Government Printing Office, Washington

Lyman J (1969) Redefinition of salinity and chlorinity. Lim and Ocean 14:28–29

Millero FJ, Rainer F, Wright DG, McDougall TJ (2008) The composition of standard seawater and the definition of the reference-composition salinity scale. Deep-Sea Res 55:50–72

Sawyer CN, McCarty PL (1967) Chemistry for sanitary engineers. McGraw-Hill, New York

Steelink C (2002) Investigating humic acids in soil. Anal Chem 74:326–333

UNESCO (1981a) The practical salinity scale 1978 and the international equation of state of seawater 1980. UNESCO Technical Papers in Mar Sci 36, Paris

UNESCO (1981b) Background papers and supporting data on the practical salinity scale 1978. UNESCO Technical Papers in Mar Sci 37, Paris

UNESCO (1985) The international system of units (SI) in oceanography. UNESCO Technical Papers in Mar Sci 45, Paris

van Niekerk H, Silberbauer MJ, Maluleke M (2014) Geographical differences in the relationship between total dissolved solids and electrical conductivity in South African rivers. Water South Africa 40:133–137

Walton NRG (1989) Electrical conductivity and total dissolved solids—what is their precise relationship? Desal 72:275–292

Wang J-O, Liu H, Po H, Fan L (1997) Influence of salinity on food consumption, growth, and energy conversion efficiency of common carp (Cyprinus Carpio) fingerling. Aqua 148:115–124

Wright DG, Pawlowicz R, McDougall TJ, Feistel R, Marion GM (2011) Absolute salinity, "density salinity," and the reference-composition salinity scale: present and future use in the seawater standard TEOS-10. Ocean Sci 7:1–26

Suspended Solids, Color, Turbidity, and Light

6

Abstract

Natural waters contain suspended particulates that increase turbidity, impart apparent color, and interfere with light penetration. These particles originate from erosion, vegetative debris from watersheds, and microorganisms produced in water bodies. Suspended particles vary from colloids that remain suspended indefinitely to larger silt and sand particles held in suspension by turbulence. Settling velocities of particles in still water are estimated by the Stoke's law equation and depend mainly on particle diameter and density—large, dense particles settle the fastest. Organic particles settle slowly because of their low density, but planktonic organisms also have adaptions that lessen settling velocity. Solar radiation of all wavelengths is quenched quickly—about 50% is reflected or converted to heat within the first meter. Within the visible spectrum, clear water absorbs red and orange light most strongly followed by violet, and by yellow, green, and blue, but suspended particles in water tend to absorb all wavelengths. Sedimentation of suspended solids can result in gradual "filling in" of water bodies and destruction of benthic organisms. Turbidity and color in water bodies lower photosynthesis rates and reduce productivity. Clear water is more pleasing to the eye, preferred for recreational water uses, and greatly favored for drinking purposes. Erosion control, sedimentation, chemical coagulation, and filtration techniques are used in color and turbidity control.

Introduction

Natural waters contain particles not in true solution but held in temporary or permanent suspension against the force of gravity. These particles are called suspended solids, but many speak of them as particulate matter. Strictly speaking the latter term should not be used, because all particles, dissolved or suspended, are matter. Larger suspended particles continuously settle in relation to their density and size, but turbulence may prevent or slow their settling rate. Suspended particles

Table 6.1 Typical sizes and particle densities of particles in water

Particle	Maximum dimension or width (μm)	Particle density (g/cm^3)
Water molecules	0.000282	1.0
Dissolved inorganic ions	0.0004–0.0006	Varies
Dissolved organic compounds	0.005–0.05	Varies
Bacteria	0.2–10	1.02–1.1
Clay	0.5–2	2.7–2.8
Phytoplankton	2–2000	1.02–1.20
Silt	2–50	2.65–2.75
Particles visible to naked eye	>40	–
Sand	50–2000	2.6–2.7
Zooplankton	100–2500	1.02–1.20
Organic detritus	0.2–2500	0.8–1.0

along with certain dissolved substances impart color to water and interfere with the passage of light making water less transparent and increasing its turbidity. Suspended matter may be of both organic and inorganic origin and includes living organisms, detritus, and soil particles. Turbidity restricts light penetration and has a powerful limiting effect upon growth of aquatic plants. Suspended particles also settle to bottoms of water bodies resulting in severe bottom habitat degradation. Waters for drinking purposes and for many industrial uses are treated to remove excessive turbidity and color.

Suspended Particles

Sources

Suspended soil particles in water bodies usually originate from rainfall erosion on watersheds, stream bed erosion by flowing water, shoreline erosion by waves, and resuspension of sediment in lakes and ponds. Organic particles originate from leaf fall, suspension of organic particles on watersheds by overland flow, growth of microscopic organisms in water bodies (plankton and bacteria), and remains of dead aquatic organisms (detritus).

Suspended particles exhibit a wide range in size and density (Table 6.1). Larger and denser particles such as sand and coarse silt tend to settle to the bottom quickly. Turbidity results from fine silt and clay particles that do not settle rapidly, and water containing an abundance of soil particles appears "muddy." Phytoplankton, zooplankton, and detritus also make waters turbid. In unpolluted, natural waters, the total amount of suspended organic matter is usually <5 mg/L, but in nutrient rich waters with abundant plankton, concentrations of organic particles may be >50 mg/L.

Settling Characteristics

There are two size classification schemes for mineral soil particles based on particle diameter (Table 6.2). Particles seldom are spherical, and separation of particles for size classification by sieve analysis depends on the maximum dimension of a particle. Organic particles also are classified according to particle size. The settling velocity of a spherical particle is related to its size (diameter) and density. As particles settle in response to gravity, their downward motion is opposed by buoyant and drag forces. The buoyant force is equal to the weight of water displaced by the settling particle, and this force increases in direct proportion to particle volume. The net gravitational force is the difference in gravitational and buoyant forces, and it causes the particle to settle. As the velocity of a settling particle increases, the viscous friction force opposing downward motion of the particle relative to the water increases. This force is the drag force. When the drag force becomes as great as the net gravitational force (Fig. 6.1), the particle settles at a constant velocity known as the terminal settling velocity. Variables influencing settling of particles in water are combined in the Stoke's law equation for estimating terminal settling velocity:

Table 6.2 United States Department of Agriculture (USDA) and International Society of Soil Science (ISSS) classifications of soil particles based on sieve analysis

Particle fraction name	USDA (mm)	ISSS (mm)
Gravel	>2	>2
Very coarse sand	1–2	–
Course sand	0.5–1	0.2–2
Medium sand	0.25–0.5	–
Fine sand	0.1–0.25	0.02–0.2
Very fine sand	0.05–0.1	–
Silt	0.002–0.05	0.002–0.02
Clay	<0.002	<0.002

Fig. 6.1 Forces acting on a settling particle

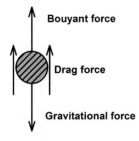

Bouyant force

Drag force

Gravitational force

Net gravitational force = Gravitational force - Bouyant force

When Net gravitational force = Drag force, particle settles at constant velocity

$$v_s = \frac{g\left(\rho_p - \rho_w\right)D^2}{18\,\mu} \tag{6.1}$$

where v_s = terminal settling velocity (m/sec), g = gravitational acceleration (9.81 m/sec^2), ρ_p = particle density (kg/m^3), ρ_w = density of water (kg/m^3), D = particle diameter (m), and μ = viscosity of water (Newton·sec/m^2).

Sedimentation theory was developed for an ideal, spherical particle, but it is applied to suspended particles in general. The greater the diameter and density of a particle the faster it settles (Ex. 6.1). The viscosity and density of water decrease with increasing temperature (Table 6.3), and particles settle faster in warm water than in cool water. Particle density of organic matter is much less that of sand, silt, and clay (Table 6.1). Mineral particles settle faster than organic particles of similar size.

Ex. 6.1 *Terminal settling velocities of 0.0001-mm diameter clay particles and 0.02-mm diameter silt particles will be compared at 30 °C.*

   Solution:
   Using Eq. 6.1 and obtaining the density and viscosity of water at 30 °C from Table 6.3,

$$v_s\ silt = \frac{(9.81\ m/sec)(2,700 - 995.7)\ kg/m^3\ \left(2 \times 10^{-5}\ m\right)^2}{18\left(0.798 \times 10^{-3}\ N \cdot sec/m^2\right)}$$

$$v_s\ silt = 5.2 \times 10^{-4}\ m/sec$$

$$v_s\ clay = \frac{(9.81\ m/sec)(2,700 - 995.7)\ kg/m^3\ \left(1 \times 10^{-7}\ m\right)^2}{18\left(0.798 \times 10^{-3}\ N \cdot sec/m^2\right)}$$

Table 6.3 Density (ρ) and dynamic viscosity (μ) of water

Temp. (°C)	Density (kg/m^3)	Dynamic viscosity ($\times 10^{-3}$ N·sec/m^2)
0	999.8	1.787
5	999.9	1.519
10	999.7	1.307
15	999.1	1.139
20	998.2	1.022
25	997.0	0.890
30	995.7	0.798
35	994.0	0.719
40	992.2	0.653

$$v_s \, clay = 1.16 \times 10^{-8} \, m/sec.$$

Silt particles, because of their greater diameter, sink much faster than clay particles.

Bacterial cells seldom sink, because most have densities less than water. Phytoplankton cells are quite small, but many of them still tend to sink. Many species have adaptations to reduce or prevent settling such as gas vacuoles to increase buoyancy or projections to increase drag force. Some species have flagella, cilia, or other methods of motility to avoid sinking. These adaptations are known as form resistance, a concept expressed as the rate at which a particle sinks as compared to the rate that a sphere of equal density would sink (Padisák et al. 2003). For planktonic algae, form resistance factors range from 0.476 to 2.008. The larger the value of form resistance, the slower plankton settle. Stokes law equation usually is not applicable to the settling rate of microorganisms. In natural waters, turbulence also plays a critical role in particle settling. Turbulent conditions may keep a particle in suspension much longer than would be expected from calculations made with the general form of the Stokes law equation.

Colloids

Colloidal particles range in size from 0.001 to 1 μm. Although larger than most molecules, colloidal particles are so small that they remain in water even though they are not in true solution. Colloidal particles have several properties that are different from those of dissolved molecules and suspended particles. A narrow beam of light passes through a true solution, and its path cannot be observed. In a colloidal suspension, the path of the light is visible because colloids are large enough to scatter light. This effect of light scattering by colloids is known as the Tyndall effect in honor of the British physicist, John Tyndall, who first described it. In a colloidal suspension viewed under a dark field microscope, colloidal particles appear as bright points that move irregularly against a dark background. Random movement of colloidal particles results from their bombardment by molecules of water. Random motion of colloidal particles is called Brownian movement. Colloidal particles cannot be removed from water by filtration unless special filters with very small apertures are used. Colloidal particles have a tremendous surface area relative to volume, and this imparts to them a large surface adsorptive capacity. Most colloids carry either net positive or net negative charges, but all colloids of the same type have the same charge and repel each other. Colloidal clay particles (<200 μm) found in natural waters are negatively charged. Neutralization of the charge on colloids allows dispersed particles to come together, and they aggregate and precipitate from solution. At municipal water supply plants, suspended, negatively-charged clay particles often are precipitated from water by adding aluminum ion to neutralize the charge on their surfaces.

Color

Large bodies of water often appear blue because their surfaces reflect the color of the sky. The true color of natural water results from the unabsorbed light rays remaining from sunlight after it has passed into the water. Because water absorbs light more strongly at the red end of the visible spectrum than at the blue end, blue and blue-green light is scattered back and visible. This gives clear waters a blue hue, but dissolved and suspended particles also absorb light and influence color. True color in water includes color resulting from the water itself and substances dissolved in it.

Phytoplankton blooms color water various shades of green, blue-green, yellow, brown, red, and even black. Suspended mineral particles also can differ greatly in color (various shades of black, red, yellow, gray, and white). Tannins and lignins usually impart a yellow-brown, tea-like appearance to water, but when their concentrations are high the water may appear black. Well water may contain traces of suspended iron and manganese oxides that are yellowish or black in color, respectively. Color disappears from waters following the removal of suspended particles is called apparent color, while the true color remains.

Turbidity, Light, and Photosynthesis

The upper, illuminated layer of a water body in which plants grow is called the photic zone—the bottom of this layer is usually considered to receive 1% of incident light. The photic zone in the open ocean is about 200 m thick. Freshwater lakes range in nutrient status from oligotrophic (nutrient poor with scant plant growth) to eutrophic (nutrient rich with much plant growth). The thickness of the photic zone based on lake trophic status is: ultraoligotrophic, 50–100 m; oligotrophic, 5–50 m; mesotrophic, 2–5 m; eutrophic, <2 m (Wetzel 2001). Lakes with turbidity from suspended soil particles or tannins and lignins may be of low productivity yet have a photic zone as shallow as that of a eutrophic water body. The photic zone in small water bodies such as fish ponds may extend to the bottom if waters do not have a plankton bloom.

Light penetration in water was discussed in Chap. 2 in relation to energy, heat, and temperature, but the degree of brightness of visible light also is of importance. The original unit for visible light was candlepower, and 1 candlepower was equal to the light produced by a particular type and size of candle (standard candle). The basic unit used today is the candela (cd) defined in SI units is a luminous intensity equal to 1/60 of the luminous intensity per square centimeter of a blackbody radiating at $2046°K$. In addition, 1 cd also is equivalent to 1/683 W/steradian (sr). A steradian is the solid geometry equivalent of the radian. One radian isolates a segment of the circumference of a circle equal to its radius, while a steradian isolates an area on the surface of sphere equal to πr^2. A circle has 2π radians so the area associated with 1 radian is $1/2\pi$ or $\approx 16\%$ of the area of a circle. The surface area of a sphere is $4\pi r^2$, a sphere is 4π steradian, and 1 radian is $1/4\pi$ or $\approx 8\%$ of the surface area of a sphere.

The lumen (lm) is a unit of luminous flux equal to the amount of light given out through a solid angle by a source with an intensity of 1 cd in all directions. The lux (lx) is the brightness of light at a surface, and equals 1 lm/m^2 or to 1 $cd \cdot sr/m^2$. Direct sunlight has a brightness of 30,000–100,000 lux as compared to 320–500 lx for an office. A 200-watt light bulb provides about 1600 lx.

The visible spectrum consists of wavelengths between 380 and 750 nm (0.38–0.75 μm), and the different colors of the spectrum are depicted in Table 6.4. In clear water, visible light undergoes extinction (quenching) quickly—only about 25% passing the ocean surface reaches 10 m depth. The rate of light absorption by pure water is in the order: red and orange > violet > yellow, green, and blue. As a result, the spectrum is rapidly altered as light penetrates to greater depth. Natural waters contain dissolved and suspended substances that further interfere with light penetration. Phytoplankton absorb light within the visible range with the greatest absorption between 600 and 700 nm in the red and orange range, and their lowest adsorption is in the green and yellow range of 500–600 nm. Humic substances have a particularly strong preference for blue and violet light. Inorganic particles tend to absorb light across the entire spectrum. Dissolved salts in water do not interfere with light penetration.

Light for photosynthesis has been measured in various units ranging from watts (a power unit) to lux (an illuminance unit), but photosynthetically active radiation (PAR) is the most meaningful light with respect to photosynthesis. Light absorption peaks for plant pigments are: chlorophyll a, 430 nm and 665 nm; chlorophyll b, 453 nm and 642 nm; carotenoids, 449 nm and 475 nm, but these pigments absorb over a wider range and even absorb some green light. Thus, the PAR is considered to be the amount of light in the wavelength band 400–700 nm.

Photosynthetically available light can be placed on the equivalent of a molar basis by referring to Avogadro's number (6.022×10^{23}) of photons as 1 molecular weight of light. Underwater light meters measure photosynthetically active radiation (PAR) as the photosynthetic photon flux (PPF) in micromoles per meter square per second ($\mu M/m^2/sec$). On a bright day in summer in mid-latitudes, the peak input of PAR to a water body often is around 1400–1500 $\mu M/m^2/sec$.

Temperatures of turbid waters tend to be higher than those of clear waters under similar conditions. This results because suspended solids in turbid water adsorb heat. Afternoon surface water temperature was 31 °C in a small pond at the inception of a

Table 6.4 The spectrum of visible light

Color	Wavelength[a] (nm)
Violet	380–450
Blue	450–495
Green	495–570
Yellow	570–590
Orange	590–620
Red	620–750

[a]nm = 0.001 μm (or microns)

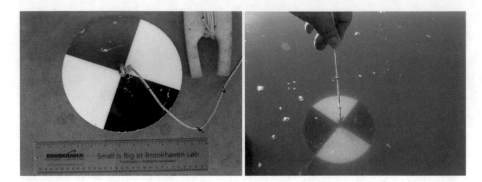

Fig. 6.2 Left: A Secchi disk; Right: Lowering Secchi disk into the water

phytoplankton bloom and 35 °C at the peak of this algal bloom. Water temperatures at a depth of 60 cm in the two water bodies were almost the same on both days (Idso and Foster 1974).

The Secchi disk is a useful tool for assessing transparency of water bodies. It is a 20-cm diameter disk painted with alternate black and white quadrants, weighted under the bottom, and attached at its upper face to a calibrated line (Fig. 6.2). The depth to which this disk is visible in the water is the Secchi disk visibility. The Secchi disk visibility provides a fairly reliable estimate of the extinction coefficient

$$K = \frac{1.7}{Z_{SD}} \tag{6.2}$$

where Z_{SD} = Secchi disk visibility (m).

People often express their desire for clear water bodies, but this preference is not held by all, or at least not for all water bodies. There is an ancient proverb stated in slightly different forms that contains the following truism: *"water that is too clear has no fish."* Many cultures apparently came up with this proverb independently.

Assessment of Turbidity and Color

Direct Turbidity Measurement

The standard way of measuring turbidity for many years was the Jackson turbidimeter. This instrument consisted of a calibrated glass tube, a tube holder, and a candle. The glass tube inside the holder was mounted directly over the candle, and the water sample for turbidity measurement was slowly added to the tube until the candle flame was no longer discernable. The glass tube was calibrated against a silica standard, and 1 mg SiO_2/L equaled 1 Jackson turbidimeter unit (JTU). The turbidity in JTU was indicated by the depth of water in the calibrated tube necessary to obscure the candle flame. The Jackson turbidimeter would not read turbidities less than 25 JTU, and it has been replaced by turbidity meters that employ the principles of nephelometry.

A nephelometer is an instrument in which a beam of light is passed through a water sample and the amount of light scattered at 90° to the light beam is measured. The amount of scattered light increases with greater turbidity. The standard for calibrating a nephelometer for turbidity analysis is a suspension of formazin made by combining 5.0 mL volumes of a hydrazine sulfate solution (1 g/100 mL) and a solution of hexamethylenetetramine (10 g/100 mL) and diluting to 100 mL with distilled water. This solution has a turbidity of 400 nephelometer turbidity units (NTU), and it can be further diluted to calibrate the nephelometer (Eaton et al. 2005). Most natural waters have turbidities <50 NTU, but values can range from <1 NTU to >1000 NTU. There is no direct relationship in natural waters between turbidity measurements made by nephelometry or by the Jackson turbidimeter.

A standard spectrophotometer also can be used to measure turbidity (Kitchner et al. 2017) by comparing absorbance at 750 nm with a standard curve prepared from formazin. The results are reported as formazin attenuation units (FAU), and there is good agreement between FAU units and FTU units measured by nephelometry.

Suspended Solids

The total suspended solids (TSS) concentration is determined by filtering a sample through a tared glass fiber filter, the filter and residue are dried at 102 °C, and the weight gain of the filter is caused by the suspended solids retained on it (Fig. 6.3). The suspended organic matter concentration can be determined by igniting the residue from the TSS analysis. The loss of weight upon ignition is equal to the organic component of the residue. This fraction of the TSS concentration is called the total volatile suspended solids (TVSS). Example 6.2 illustrates the calculations of TSS and TVSS analyses. Of course, filtered and unfiltered portions of a water sample can be evaporated to dryness in tared dishes, and the difference in weights of residues in the two dishes is the TSS concentration (see Fig. 5.1).

Ex. 6.2 *Calculation of TSS and TVSS concentrations.*
  Solution:
  A 100-mL water sample is passed through a glass fiber filter weighing 1.25000 g. Upon drying, the filter weighs 1.25305 g. The TSS concentration will be estimated.

$$TSS = \frac{(1.25305 - 1.25000)g \left(10^3 \ mg/g\right)(1,000 \ mL/L)}{100 \ mL} = 30.5 \ mg/L.$$

The glass fiber filter and residue from TSS analysis weighed 1.25211 g after it was ignited at 550 °C. The TVSS concentration will be calculated.

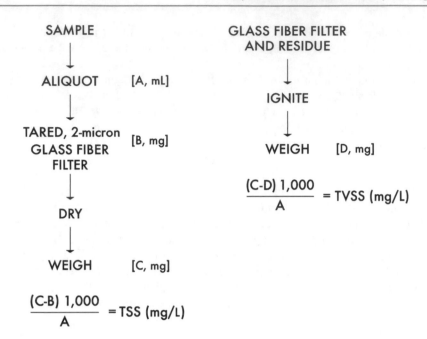

$$\frac{(C\text{-}B)\ 1{,}000}{A} = TSS\ (mg/L)$$

Fig. 6.3 Schematic of the protocol for total suspended solids (TSS) and total volatile suspended solids (TVSS)

$$TVSS = \frac{(1.25305 - 1.25211)g\ \left(10^3\ mg/g\right)(1{,}000\ mL/L)}{100\ mL} = 9.4\ mg/L.$$

The difference between the TSS concentration and the TVSS concentration represents the suspended inorganic solids. This fraction usually is known as the total fixed suspended solids (TFSS); the TFSS concentration in Ex. 6.2 is 21.1 mg/L.

Settleable Solids

The volume of solids that will rapidly settle from water is sometimes estimated. This can be done with a 1-L, inverted cone called an Imhoff cone (Fig. 6.4) that has graduations to allow the volume of sediment to be measured visually. Water is poured into the cone and allowed to set undisturbed for 1 hour. After 1 hour, the volume of settleable solids in milliliters per liter is read from calibration marks in the bottom of the cone. The volume of settleable solids must be 1 mL or greater for accurate measurement. The settleable solids analysis usually is made only for effluent with large amounts of coarse, suspended particles.

Fig. 6.4 Imhoff cones with settleable solids in bottoms

Secchi Disk

The Secchi disk is used to estimate water clarity as mentioned above. However, there is usually a close relationship between Secchi disk visibility and the concentration of suspended particles in water. The source of turbidity associated with a reduction in Secchi disk visibility usually can be determined from the apparent color of the water body and the appearance of suspended solids particles.

Suspended Solids Removal

The most common methods of controlling suspended solids is watershed protection to prevent erosion and resulting TSS contamination of overland flow, and erosion control in water bodies to avoid shoreline erosion and sediment resuspension. Suspended solids may be removed from effluents by passing them through settling basins. Chemical coagulation can be used to remove suspended solids from small water bodies.

Settling Basins

The typical settling basin design is given in Fig. 6.5, and it can be seen that inflow must equal outflow once the basin is filled by inflow. The terminal settling velocity of a particle (v_s) is determined by Eq. 6.1. The minimum particle size for which a

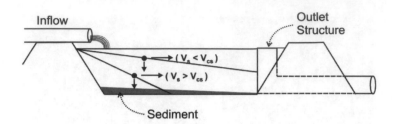

Fig. 6.5 This basic diagram of a settling basin illustrates the effect of settling velocity on the removal of suspended solids; v_s and v_{cs} = terminal and critical settling velocities, respectively

particular settling basin is designed to remove varies with the application, but a range of 0.01–0.1 mm is probably typical.

The critical settling velocity (v_{cs}) is the minimum velocity at which a particle must settle to be removed (Fig. 6.5). The value for v_{cs} can be determined as

$$v_{cs} = \frac{D}{T_{HR}} \tag{6.3}$$

where v_{cs} is in m/sec, D = depth (m), and T_{HR} = hydraulic retention time (sec). Because $T_{HR} = V/Q$,

$$v_{cs} = \frac{D}{V/Q} \tag{6.4}$$

where V = settling basin volume (m³) and Q = inflow (m³/sec). By assigning a depth of 1 m to the basin, Eq. 6.4 becomes

$$v_{cs} = \frac{1}{A/Q} = \frac{Q}{A}. \tag{6.5}$$

In Ex. 6.1, it was found that 0.02-mm soil particles have $v_s = 5.2 \times 10^{-4}$ m/sec. This settling velocity will be used in the settling basin area calculation below (Ex. 6.3).

Ex. 6.3 *The areas of a 1-m and a 1.2-m deep basin to remove soil particles ≥ 0.02 mm from a maximum inflow of 0.5 m³/sec will be calculated.*
   Solution:
Using Eq. 6.5 for 1-m deep basin,

$$5.2 \times 10^{-4} \, m/\sec = \frac{0.5 \, m^3/\sec}{A}$$

$$A = 962 \ m^2.$$

The 1.2-m deep basin could be smaller by a fraction of 1/1.2 (0.83). Its area would be 798 m^2.

Settling basins fill with sediment over time, and they normally are built 1.5 or more times larger than the minimum area. Of course, basins also can be cleaned of sediment periodically in order to maintain the design T_{HR}.

Chemical Coagulants

The most common chemical coagulant for removing suspended solids from water is aluminum sulfate ($Al_2SO_4 \cdot 14H_2O$) commonly called alum. Suspended clay particles have negative charges and mutually repel. By increasing the cation concentration in water, the charges on clay particles will be neutralized allowing them to flocculate into clumps heavy enough to settle. The effectiveness of cations in coagulating clay particles increases with greater valence, and the trivalent aluminum ion often is used. Treatment rates are determined by evaluating a series of alum concentrations in small containers of the turbid water. Normally, 15–30 mg/L of alum are necessary to remove turbidity.

Alum treatment must be done with caution, because aluminum has an acidic reaction in water.

$$Al_2(SO_4)_3 14H_2O = 2Al^{3+} + 3SO_4^{2} + 14H_2O \qquad (6.6)$$

$$2Al^{3+} + 6H_2O = 2Al(OH)_3 + 6H^+. \qquad (6.7)$$

Each milligram of alum will produce enough acidity to neutralize about 0.5 mg/L alkalinity (expressed as equivalent $CaCO_3$)

$$3CaCO_3 + 6H^+ \rightarrow 3Ca^{2+} + 3H_2O + 3CO_2. \qquad (6.8)$$

Alkalinity can be increased by treating with lime before the alum treatment is applied. Alum is widely used for clearing water of turbidity, but ferric sulfate and other iron and aluminum compounds also have been applied for this purpose.

Water quality criteria usually have some reference to TSS concentration, turbidity, and color. In order to protect aquatic ecosystems, TSS, turbidity, and color should not change by more than 10% of the seasonal mean concentration, or the compensation point for photosynthetic activity should not be reduced by more than 10%. Water quality criteria for effluents may contain limits for TSS, and an upper limit of 25 mg/L usually will provide a good level of protection against excessive turbidity and sedimentation in aquatic ecosystems, and fair protection may be achieved by a limit of 50 mg/L. Some effluent discharge permits may prohibit a turbidity plume.

The maximum allowable turbidity concentration for municipal water supply usually is about 5 NTU with a requirement for 95% of samples to have less than 1 NTU. Turbidity usually is removed by chemical coagulation and sedimentation or by filtration. Such treatment will remove coarse and colloidal particles, but it will not remove true color of dissolved matter. High-dose chlorination sometimes will remove tannic substances, but it may be necessary to resort to activated carbon filters, anion exchange, or potassium permanganate treatment.

Color

Although many associate clear water with high purity and turbid water with contamination, clear water could contain harmful microorganisms, and the possibility for contamination with pathogens should always be considered in potable water. Turbidity and color in water is an important aesthetic issue. Water users want clear drinking water, and do not desire that their fabrics be stained during washing or that sinks, tubs, showers, and glassware be stained by exposure to water with a high degree of color. While there are no mandated standards for color, the recommended allowable limit is 15 color units.

The usual method of measuring color in water is relatively primitive in comparison with methods for determining concentrations of other variables. It relies upon comparison of natural water color to a series of standard color solutions in clear tubes (Nessler tubes) of 17.5 cm or 20 cm in height and with volumes of 50 or 100 mL, respectively. A solution containing 500 color units is prepared by dissolving 1.246 g potassium chloroplatinate and 1.00 g cobaltous chloride with 100 mL nitric acid and diluting to 1000 mL with distilled water (Eaton et al. 2005). Standards containing from 0 (distilled water) to 70 color units are prepared and placed in Nessler tubes. The water sample is filtered and placed in a Nessler tube and a color unit is assigned to the sample after comparison with the standards. The water sample may be diluted with distilled water if necessary to assign a color reading to a highly colored sample. A spectrophotometric color method is available; it reveals the hue of the color but not the intensity (Eaton et al. 2005).

Conclusions

Suspended matter in water can result in high turbidity and excessive sedimentation. Sedimentation in water bodies makes affected areas shallower, and it may destroy benthic organisms and fish eggs. Turbidity seldom has direct toxic or mechanical effects on fish and other aquatic organisms, but it restricts light penetration and lessens productivity of water bodies. Turbidity and color in public water supplies is undesirable because consumers want clear water. Turbidity in water also detracts from its value for recreational purposes and makes it less aesthetically pleasing.

References

Eaton AD, Clesceri LS, Rice EW, Greenburg AE (eds) (2005) Standard methods for the examination of water and wastewater. American Public Health Association, Washington

Idso SB, Foster JM (1974) Light and temperature relations in a small desert pond as influenced by phytoplanktonic density variations. Water Resources Res 10:129–132

Kitchner GB, Wainwright J, Parsons AJ (2017) A review of the principles of turbidity measurement. Prog Phys Geo 41:620–642

Padisák J, Soróczki-Printér E, Rezner Z (2003) Sinking properties of some phytoplankton shapes and the relation of form resistance to morphological diversity of plankton—an experimental study. Hydrobio 500:243–257

Wetzel RG (2001) Limnology, 3rd edn. Academic, San Diego

Dissolved Oxygen and Other Gases

7

Abstract

Concentrations of dissolved oxygen and other dissolved atmospheric gases at saturation in water vary with the partial pressures and solubilities of the gases, and the temperature and salinity of the water. Concentrations of dissolved oxygen and other dissolved gases at saturation decrease with greater elevation (lower barometric pressure), higher salinity, and rising temperature. Their concentrations in water may be expressed in milligrams per liter, but it also can be reported in milliliters per liter, percentage saturation, oxygen tension, or other units. The rate of diffusion of dissolved oxygen and other gases into water bodies is related to various factors, but the most important is the concentration of each gas already present in the water, the area of contact between the air and water, and the amount of turbulence in the water. Gases enter water from the air until saturation concentrations are reached, but if saturation concentrations in the water are in excess of concentration, gases diffuse into the air. Dissolved oxygen is of utmost importance in water quality, because it is essential for aerobic respiration. Absorption of oxygen by fish and other aquatic animals is controlled by the pressure of oxygen in the water rather than the dissolved oxygen concentration in milligrams per liter. Low oxygen pressure can stress or even kill aquatic organisms. Excessive dissolved gas (including oxygen) concentrations in water can lead to gas bubble trauma in fish and other aquatic animals.

Introduction

Aerobic organisms, whether they live on land or in water, must have oxygen or they will suffocate. Almost everyone also knows that while water (H_2O) is mainly made of oxygen (88.89%), aquatic animals cannot utilize the oxygen from water molecules; they require molecular oxygen (O_2) that is dissolved in water. A fish in a lake with water depleted of dissolved oxygen is literally suffocating in a lake of oxygen.

C. E. Boyd, *Water Quality*, https://doi.org/10.1007/978-3-030-23335-8_7

Dissolved oxygen diffuses into natural waters from the atmosphere or it is released into the water by the photosynthesis of aquatic plants. Molecular oxygen is only one component of air; the atmosphere is a mixture of gases to include nitrogen, oxygen, argon, carbon dioxide, water vapor, and traces of a few other gases. Gases are soluble in water, and an equilibrium state between atmospheric gases and dissolved gases exists. Surface waters usually are near equilibrium (saturation) with nitrogen, argon, and trace gases, but oxygen and carbon dioxide concentrations vary because of biological processes. Groundwater may vary in gas concentrations because of both biologic and geologic factors.

Dissolved oxygen is important in water quality because its presence is necessary for maintaining oxidized conditions, and it is the terminal electron acceptor in aerobic respiration. Carbon dioxide is produced as a byproduct of respiration, but it is also a source of carbon for photosynthesis, a major determinant of pH in natural waters, and potentially harmful to aquatic animals. Dissolved nitrogen usually is at a higher concentration in water than other gases, but its influence on biological and chemical processes is much less than that of oxygen or carbon dioxide. Other gases usually have little influence on water quality.

This chapter focuses primarily on gas solubility in water with emphasis on dissolved oxygen.

Atmospheric Gases and Pressure

The atmosphere is the layer of gases that surrounds the earth to a total thickness of about 300 km. The concentration of gases in the atmosphere decreases considerably beyond the limit of the troposphere which has a thickness of about 12 km. The atmosphere near the earth's surface contains 78.08% nitrogen (N_2) and 20.95% oxygen (O_2) and small concentrations of other gases (Table 7.1). The atmosphere contains particles of matter and therefore has weight, and the density of air decreases with rising temperature (Table 7.2), Molecules of air are not attached together, and when heated, they exhibit greater movement and require more space.

The weight of the atmosphere above the earth's surface is the atmospheric pressure (often called the barometric pressure). The reference point for atmospheric pressure is mean sea level (MSL) and 0 °C. The pressure under such conditions is referred to as standard temperature and pressure (STP), and this amount of pressure

Table 7.1 Average percentages by volume of atmospheric gases in the troposphere

Gas	Volume %	Gas	Volume %
Nitrogen	78.08	Helium	0.0005
Oxygen	20.95	Methane	0.00017
Water vapor	Depends on humidity	Hydrogen	0.00005
Argon	0.93	Nitrous oxide	0.00003
Carbon dioxide	0.04	Ozone	0.000004
Neon	0.0018	Various others	–

Table 7.2 Weight of air at standard atmospheric pressure and different temperatures

Temperature (°C)	Density (kg/m^3)	Temperature (°C)	Density (kg/m^3)
0	1.2922	25	1.1839
5	1.2690	30	1.1644
10	1.2466	35	1.1455
15	1.2250	40	1.1270
20	1.2041		

Table 7.3 Several ways of expressing standard atmospheric pressure

Dimension	Abbreviation	Amount
Atmosphere	Atm	1.000
Feet of water	ft H$_2$O	33.8958
Inches of water	in H$_2$O	405.512
Inches of mercury	in Hg	29.9213
Meters of water	m H$_2$O	10.331
Millimeters of mercury	mm Hg	760
Pascals (also Newton's per meter square)	Pa (also N/m^2)	101,325
Kilopascals	kPa	101,325
Hectopascals	hPa	1,013.25
Pound per square inch	psi	14,696
Kilograms per meter square	kg/m^2	1.033
Bars	bar	1.01325
Millibars	mbar	1,013.25

is considered to be 1 atmosphere (atm). There are many other units of measure for atmospheric pressure (Table 7.3), but in this chapter, atmospheres and millimeters of mercury (mm Hg) will be used as pressure units.

The traditional method for measuring atmospheric pressure with a barometer is shown in Fig. 7.1. This device consists of a tube closed at its upper end and evacuated of air mounted vertically with its open end extending into a dish of liquid. Water is not used as the liquid, because the force of the atmosphere acting down on its surface at sea level would cause it to rise in the column to a height of 10.331 m. Such a long water column for the measurement of atmospheric pressure is avoided by using mercury (Hg) in the barometer. Mercury is 13.594 times denser than water, and standard atmospheric pressure measured with a mercury barometer is 760 mm Hg. Today, there are alternatives to the mercury barometer for measuring atmospheric pressure. The most common—the aneroid barometer—is basically a box with an elastic top that is partially exhausted of air. A pointer is attached to register the degree of compression of the top caused by the pressure of the atmosphere.

If a barometer is unavailable, atmospheric pressure can be estimated from elevation as illustrated in Ex. 7.1 using the following equation from Colt (2012):

Fig. 7.1 A schematic view of
a traditional mercury
barometer

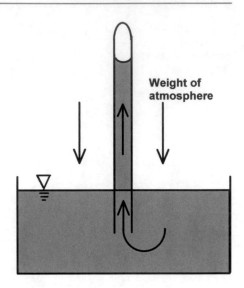

$$\log_{10} BP = 2.880814 - \frac{h}{19,748.2} \qquad (7.1)$$

where BP = barometric pressure (mm Hg) and h = elevation (m).

Ex. 7.1 *The atmospheric pressure at 500 m will be estimated using Eq. 7.1.*
  Solution:

$$log_{10} BP = 2.880814 - \frac{500 \ m}{19,748.2}$$

$$= 2.880814 - 0.025319 = 2.855495$$

$$BP = antilog \ 2.855495 = 717 \ mm \ Hg.$$

The effect of increasing altitude on barometric pressure as calculated with Eq. 7.1
is illustrated in Fig. 7.2. The atmospheric pressure is 50% less at 5,945 m than it is at
sea level.

Partial Pressure

Mankind has known for centuries that life required the breathing of air, but people
did not realize that it was the oxygen that they needed from air. The 18th century
English clergyman and chemist, Joseph Priestly, remarked following his famous
experiment, *"I have procured air between five and six times as good as the best
common air that I ever met with"*. He had separated oxygen from the air.

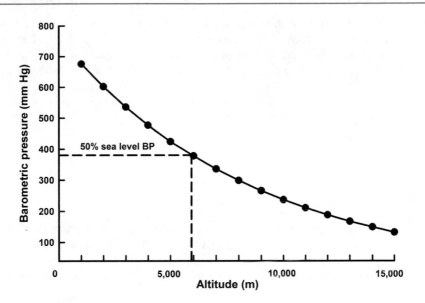

Fig. 7.2 Standard barometric pressures at different elevations

According to Dalton's Law of Partial Pressures, the total pressure caused by a mixture of gases that do not react with each other is the sum of the partial pressures of the individual gases in the mixture. The atmospheric pressure or barometric pressure (BP) consists of the sum of the partial pressures of the individual atmospheric gases:

$$BP = P_{N_2} + P_{O_2} + P_{Ar} + P_{CO_2} + P_{H_2O} + P_{\text{Other gases}} \qquad (7.2)$$

where P = partial pressures of the gases indicated by subscripts. Dalton's Law also states that the partial pressure of each gas is directly proportional to its volume percentage in the mixture.

The partial pressure of a gas in atmospheres is equal to the decimal fraction of its volume percentage multiplied by the total gas pressure. At STP, the partial pressures of the most abundant atmospheric gases on a dry-air basis are

$$P_{N_2} = (760)(0.7808) = 593.4 \text{ mm}$$

$$P_{O_2} = (760)(0.2095) = 159.2 \text{ mm}$$

$$P_{Ar} = (760)(0.0093) = 7.07 \text{ mm}$$

$$P_{CO_2} = (760)(0.00040) = 0.30 \text{ mm.}$$

Completely dry air seldom is found in the atmosphere. The water vapor pressure of air varies tremendously at different places and at different times, and saturation vapor pressure increases with increasing air temperature (Table 1.2). Water vapor dilutes other gases slightly. The saturation vapor pressure at 0 °C is 4.38 mm (0.0058 atm), and the concentration of other atmospheric gases at 100% relative humidity will be diluted by a factor of 0.994 [(760 mm − 4.38 mm) ÷ 760 mm], e.g., if dry air contains 20.95% oxygen at STP, moist air at STP would contain 20.82% oxygen (20.95% oxygen × 0.994). The dilution increases at higher temperature for which the factors are: 5 °C, 0.991; 10 °C, 0.989; 15 °C, 0.983; 20 °C, 0.977; 25 °C, 0.969; 30 °C, 0.958; 35 °C, 0.945.

Gas Solubility

The solubility of a gas in water, according to Henry's Law of gas solubility, is directly proportional to the partial pressure of that gas above the water. Gases do not have equal solubilities at the same partial pressure, and the Henry's Law constant is the ratio of the gas concentration in water at equilibrium to its concentration in air above the water.

$$K_H = \frac{P}{C} \tag{7.3}$$

where K_H = Henry law constant for a particular gas (L·atm/mol), P = partial pressure of the gas (atm, C = concentration of the gas in water (mol/L). Values of K_H for oxygen are provided in Table 7.4.

Ex. 7.2 *The solubility of oxygen from moist air at 760 mm in water at 0 °C and at 25 °C will be calculated using the Henry law constant:*
   Solution:
   Using Eq. 7.3 and K_H from Table 7.4,
   At 0 °C:

$$C = \frac{P}{K_H} \quad C = \frac{0.2095 \ atm}{455.0 \ L \cdot atm/mol} = 0.00046 \ mol/L$$

Table 7.4 Henry law constants (K_H) for solubility of oxygen in freshwater at 760 mm Hg and different temperatures

Temperature (°C)	K_H (L·atm/mole)	Temperature (°C)	K_H (L·atm/mole)
0	455.0	25	769.2
5	510.3	30	845.1
10	568.5	35	925.7
15	631.0	40	1,011.0
20	697.5		

$$0.00046 \ mol/L \times 32 \ g \ O_2/mol = 0.01472 \ g/L = 14.72 \ mg/L$$

Multiplying 14.72 mg/L by the dilution factor of 0.994 for moist air gives 14.63 mg/L.

At 25 °C:

$$C = \frac{0.2095}{769.2} = 0.000272 \ mol/L$$

$$0.000272 \times 32 = 0.00872 \ g/L = 8.72 \ mg/L$$

Multiplying by the dilution factor of 0.969 for moist air gives 8.45 mg/L.

The K_H values at 25 °C for nitrogen and carbon dioxide are 1,600.0 L·atm/mol and 29.76 L·atm/mol, respectively.

Gas solubility often is presented in terms of Henry law, but most tables presenting equilibrium concentrations of dissolved gases at different temperatures are made using Bunsen's absorption coefficients. Bunsen coefficients (β) may be expressed in volume (mL or L) real gas/volume water or in milligrams gas per liter of water per millimeter of pressure (mg/L·mm Hg). Colt (2012) presents values of β for the atmospheric gases at different temperatures and salinities. Some selected β values are provided in Table 7.5.

The use of Bunsen's coefficients for calculating solubility of oxygen in water is illustrated in Ex. 7.3.

Ex. 7.3 The dissolved oxygen concentration at equilibrium will be calculated for 0 °C and 760 mm Hg using Bunsen's coefficients.
__Solution:__

Table 7.5 Bunsen's absorption coefficients for nitrogen, oxygen, and carbon dioxide in freshwater at different temperatures. Coefficients are given in terms of volume (mL real gas/mL·atm) or in terms of weight (mg/L gas·mm Hg)

Temperature	Bunsen absorption coefficients					
	Nitrogen		Oxygen		Carbon dioxide	
(°C)	(volume)	(weight)	(volume)	(weight)	(volume)	(weight)
0	0.02370	0.03934	0.04910	0.09240	1.7272	4.4924
5	0.02118	0.03485	0.04303	0.08091	1.4265	3.7105
10	0.01897	0.03120	0.03817	0.07177	1.1947	3.1073
15	0.01719	0.02828	0.03426	0.06443	1.0135	2.6363
20	0.01576	0.02592	0.03109	0.05846	0.8705	2.2641
25	0.01459	0.02401	0.02850	0.05358	0.7562	1.9669
30	0.01365	0.02246	0.02635	0.04955	0.6641	1.7273
35	0.01289	0.02121	0.02458	0.04622	0.5891	1.5324
40	0.01228	0.02021	0.04344	0.04344	0.5277	1.3726

Using volume/volume coefficients for a real gas (Table 7.5)

1. *According to Avogadro's law relating volume to mass of a gas, 1 mole oxygen weighs 31.999 g and occupies 22.4 L at STP. Thus, the density of oxygen at STP is*

$$\frac{32 \ g \ O_2/mol}{22.4 \ L/mol} = 1.4285 \ g/L = 1.4285 \ mg/mL.$$

2. *The Bunsen's coefficient at 0 °C is 0.04910 mL/mL. Multiplying by density gives*

$$0.04910 \ mL/mL \times 1.4285 \ mg/mL = 0.07014 \ mg \ O_2/mL \ or \ 70.14 \ mg \ O_2/L.$$

3. *The calculation in step 2 is for pure oxygen, but the atmosphere is 20.95% oxygen. Thus, the concentration of oxygen in water would be*

$$70.14 \ mg/L \times 0.2095 = 14.694 \ mg/L.$$

4. *The concentration from step 3 is for dry air. Previously the dilution factor of 0.994 was calculated for 0°C, and the saturation concentration from moist air is*

$$14.694 \times 0.994 = 14.61 \ mg/L.$$

The calculation is much easier using the weight over volume version of the Bunsen's coefficient (Table 7.5). The value of β simply has to be multiplied by the partial pressure of oxygen in the air. From Table 7.5, $\beta = 0.09240$ mg/L·mm Hg, and as shown earlier at STP, the partial pressure of oxygen is 159.2 mm.

$$(0.09240 \ mg/L \cdot mm \ Hg)(159.2 \ mm \ Hg) = 14.71 \ mg/L.$$

Adjusting with the dilution factor of 0.994 for moist air, we get 14.62 mg/L.

The same approach illustrated in Ex. 7.3 can be used to calculate solubilities of nitrogen, argon, and carbon dioxide. The resulting solubilities of these three gases at STP are 23.05 mg/L for nitrogen, 1.32 mg/L for carbon dioxide, and 0.89 mg/L for argon. Nitrogen has the greatest concentration in water despite being the least soluble, because it comprises a much greater volume percentage of the atmosphere than do other gases.

Colt (1984) reported the Bunsen coefficients for atmospheric gases in water of different temperatures and salinities. However, he also calculated the atmospheric gases from moist air in water for the normal temperature and salinity ranges encountered in water quality efforts. The solubilities of oxygen, nitrogen, and carbon dioxide at selected temperatures and salinities are provided in Tables 7.6, 7.7, and 7.8,

Table 7.6 The solubility of oxygen (mg/L) as a function of temperature and salinity (moist air, barometric pressure = 760 mm Hg) (Benson and Krause 1984)

Temp. (°C)	Salinity (g/L) 0	5	10	15	20	25	30	35	40
0	14.602	14.112	13.638	13.180	12.737	12.309	11.896	11.497	11.111
1	14.198	13.725	13.268	12.825	12.398	11.984	11.585	11.198	10.825
2	13.813	13.356	12.914	12.487	12.073	11.674	11.287	10.913	10.552
3	13.445	13.004	12.576	12.163	11.763	11.376	11.003	10.641	10.291
4	13.094	12.667	12.253	11.853	11.467	11.092	10.730	10.380	10.042
5	12.757	12.344	11.944	11.557	11.183	10.820	10.470	10.131	9.802
6	12.436	12.036	11.648	11.274	10.911	10.560	10.220	9.892	9.573
7	12.127	11.740	11.365	11.002	10.651	10.311	9.981	9.662	9.354
8	11.832	11.457	11.093	10.742	10.401	10.071	9.752	9.443	9.143
9	11.549	11.185	10.833	10.492	10.162	9.842	9.532	9.232	8.941
10	11.277	10.925	10.583	10.252	9.932	9.621	9.321	9.029	8.747
11	11.016	10.674	10.343	10.022	9.711	9.410	9.118	8.835	8.561
12	10.766	10.434	10.113	9.801	9.499	9.207	8.923	8.648	8.381
13	10.525	10.203	9.891	9.589	9.295	9.011	8.735	8.468	8.209
14	10.294	9.981	9.678	9.384	9.099	8.823	8.555	8.295	8.043
15	10.072	9.768	9.473	9.188	8.911	8.642	8.381	8.129	7.883
16	9.858	9.562	9.276	8.998	8.729	8.468	8.214	7.968	7.730
17	9.651	9.364	9.086	8.816	8.554	8.300	8.053	7.814	7.581
18	9.453	9.174	8.903	8.640	8.385	8.138	7.898	7.664	7.438
19	9.261	8.990	8.726	8.471	8.222	7.982	7.748	7.521	7.300
20	9.077	8.812	8.556	8.307	8.065	7.831	7.603	7.382	7.167
21	8.898	8.641	8.392	8.149	7.914	7.685	7.463	7.248	7.038
22	8.726	8.476	8.233	7.997	7.767	7.545	7.328	7.118	6.914
23	8.560	8.316	8.080	7.849	7.626	7.409	7.198	6.993	6.794
24	8.400	8.162	7.931	7.707	7.489	7.277	7.072	6.872	6.677

(continued)

Table 7.6 (continued)

Temp. (°C)	Salinity (g/L)								
	0	5	10	15	20	25	30	35	40
25	8.244	8.013	7.788	7.569	7.357	7.150	6.950	6.754	6.565
26	8.094	7.868	7.649	7.436	7.229	7.027	6.831	6.641	6.456
27	7.949	7.729	7.515	7.307	7.105	6.908	6.717	6.531	6.350
28	7.808	7.593	7.385	7.182	6.984	6.792	6.606	6.424	6.248
29	7.671	7.462	7.259	7.060	6.868	6.680	6.498	6.321	6.148
30	7.539	7.335	7.136	6.943	6.755	6.572	6.394	6.221	6.052
31	7.411	7.212	7.018	6.829	6.645	6.466	6.293	6.123	5.959
32	7.287	7.092	6.903	6.718	6.539	6.364	6.194	6.029	5.868
33	7.166	6.976	6.791	6.611	6.435	6.265	6.099	5.937	5.779
34	7.049	6.863	6.682	6.506	6.335	6.168	6.006	5.848	5.694
35	6.935	6.753	6.577	6.405	6.237	6.074	5.915	5.761	5.610
36	6.824	6.647	6.474	6.306	6.142	5.983	5.828	5.676	5.529
37	6.716	6.543	6.374	6.210	6.050	5.894	5.742	5.594	5.450
38	6.612	6.442	6.277	6.117	5.960	5.807	5.659	5.514	5.373
39	6.509	6.344	6.183	6.025	5.872	5.723	5.577	5.436	5.297
40	6.410	6.248	6.091	5.937	5.787	5.641	5.498	5.360	5.224

Table 7.7 The solubility of nitrogen (mg/L) as a function of temperature and salinity (moist air, barometric pressure = 70 mm Hg)

Temp. (°C)	Salinity, parts per thousand (ppt)				
	0	10	20	30	40
0	23.04	21.38	19.85	18.42	17.10
5	20.33	18.92	17.61	16.40	15.26
10	18.14	16.93	15.81	14.75	13.77
15	16.36	15.31	14.32	13.40	12.54
20	14.88	13.96	13.09	12.28	11.52
25	13.64	12.82	12.05	11.33	10.65
30	12.58	11.85	11.17	10.52	9.91
35	11.68	11.02	10.40	9.82	9.26
40	10.89	10.29	9.73	9.20	8.70

Table 7.8 Solubility of carbon dioxide (mg/L) in water at different temperatures and salinities exposed to moist air containing 0.04% carbon dioxide at a total air pressure of 760 mm Hg

Temperature (°C)	Salinity ppt								
	0	5	10	15	20	25	30	35	40
0	1.34	1.31	1.28	1.24	1.21	1.18	1.15	1.12	1.09
5	1.10	1.08	1.06	1.03	1.01	0.98	0.96	0.93	0.89
10	0.93	0.91	0.87	0.85	0.83	0.81	0.79	0.77	0.75
15	0.78	0.77	0.75	0.73	0.70	0.68	0.66	0.65	0.64
20	0.67	0.65	0.63	0.62	0.61	0.60	0.58	0.57	0.56
25	0.57	0.56	0.54	0.53	0.52	0.51	0.50	0.49	0.48
30	0.50	0.49	0.48	0.47	0.46	0.45	0.44	0.43	0.42
35	0.44	0.43	0.42	0.41	0.40	0.39	0.39	0.38	0.37
40	0.39	0.38	0.37	0.36	0.36	0.35	0.35	0.34	0.33

respectively. The solubility of a gas decreases as temperature increases. This results because the movement of molecules of both water and gas increase as a function of temperature, and as temperature increases the more rapidly moving molecules collide more frequently to hinder the entrance of gas molecules into the water. Dissolved salts hydrate in water and a portion of the water molecules in a given volume of water are bound so tightly to salt ions through hydration that they are not free to dissolve gases. In reality, salinity has no effect on the solubility of a gas, but increasing the salinity decreases the amount of free water available to dissolve gases in a given volume of water.

Gas Solubility Tables

Oxygen solubilities in Table 7.6 are for moist air at 760 mm pressure, and they may be adjusted to a different barometric pressure with the following equation:

$$DO_s = DO_t \frac{BP}{760} \qquad (7.4)$$

where DO_s = concentration of dissolved oxygen at saturation corrected for pressure, and DO_t = dissolved oxygen concentration at saturation at 760 mm (Table 7.6).

Ex. 7.4 *Estimate the solubility of dissolved oxygen in freshwater at 26 °C when BP is 710 mm.*

Solution:
The solubility of dissolved oxygen in freshwater at 26 °C from Table 7.6 is 8.09 mg/L. Thus,

$$DO_s = 8.09 \times \frac{710}{760} = 7.56 \; mg/L.$$

Because barometric pressure declines with increasing elevation, the solubility of gases is less for the same temperature at a higher elevation than at sea level.

Ex. 7.5 *How much less is the solubility of oxygen in freshwater (20 °C) at 3,000 m than at sea level?*

Solution:
From Fig. 7.2 or Eq. 7.1, BP is about 536 mm at 3,000 m, and the solubility of oxygen in freshwater at 20 °C and 760 mm is 9.08 mg/L (Table 7.6). Thus,

$$DO_s = 9.08 \times \frac{536}{760} = 6.40 \; mg/L.$$

The solubility of oxygen will be 2.68 mg/L less at 3,000 m than at sea level.

Oxygen solubility data in Table 7.6 are for the water surface. The pressure holding a gas in solution at some depth beneath the surface is greater than the barometric pressure by an amount equal to the hydrostatic pressure (see Chap. 1). The hydrostatic pressure expressed in millimeters of mercury per meter of depth for different water temperatures and pressures are provided (Table 7.9). To estimate the gas solubility at some depth below the water surface, the total pressure must be used instead of barometric pressure in the calculation as shown below:

$$DO_s = DO_t \frac{BP + HP}{760} \qquad (7.5)$$

where HP = hydrostatic pressure in mm Hg/m depth (Table 7.9).

Ex. 7.6 *Estimate the solubility of dissolved oxygen in water at 25 °C with a salinity of 20 ppt at a depth of 4.5 m when BP is 752 mm.*

Solution:
From Table 7.6, the solubility of oxygen at 760 mm, 20 ppt salinity, and 25 °C is 7.36 mg/L. The specific weight of water at 25 °C and 20 ppt salinity from Table 7.9 is 74.44 mm/m. The computation is as follows:

$$DO_s = 7.36 \; mg/L \times \frac{752 \; mm + (4.5 \; m)(74.44 \; mm/m)}{760 \; mm} = 10.53 \; mg/L.$$

Table 7.9 The specific weight (mm Hg/m depth) as a function of temperature and salinity (Colt 1984)

Temp.	Salinity (g/L)								
(°C)	0	5	10	15	20	25	30	35	40
0	73.54	73.84	74.14	74.44	74.73	75.03	75.33	75.62	75.92
5	73.55	73.85	74.14	74.43	74.72	75.01	75.30	75.59	75.88
10	73.53	73.82	74.11	74.39	74.68	74.97	75.25	75.54	75.83
15	73.49	73.77	74.05	74.34	74.62	74.90	75.18	75.47	75.75
20	73.42	73.70	73.98	74.26	74.54	74.82	75.10	75.38	75.66
25	73.34	73.62	73.89	74.17	74.44	74.72	75.00	75.27	75.55
30	73.24	73.51	73.78	74.06	74.33	74.60	74.88	75.15	75.43
35	73.12	73.39	73.66	73.93	74.20	74.48	74.75	75.02	75.30
40	72.98	73.25	73.52	73.79	74.06	74.34	74.61	74.88	75.15

Dissolved oxygen concentrations may sometimes be expressed in milliliters per liter. To convert milligrams of oxygen to milliliters of oxygen, the density of dissolved oxygen must be determined for the existing temperature and pressure. The density of oxygen can then be computed if the volume of 1 mole of oxygen is known for existing conditions.

The law that describes the relationships about pressure, temperature, volume, and number of moles of an ideal gas is called the ideal gas law or the universal gas law. The common expression of the law is the universal gas law equation

$$PV = nRT \qquad (7.6)$$

where P = pressure (atm), V = volume (L), n = number of moles of gas, R = universal gas law equation (0.082 L·atm/mol·°K), and T = absolute temperature (°A = 273.15 + °C).

The universal gas law equation is a useful tool for solving both theoretical and practical problems related to gases. This equation will be derived in Ex. 7.7 in order to reinforce the readers' understanding of the gas laws.

Ex. 7.7 *The universal or ideal gas law equation will be derived.*
   Solution:
   The number of molecules (n) in a mass of a gas depends upon its volume (V), temperature (T), and pressure (P). If P is doubled without a change in V and T, there will be twice as many molecules in the same volume, i.e., n = kP. The number of molecules of a gas also varies directly with V if T and P are constant, i.e., V = kP. If V and P are constant, an increase in T will increase P and gas molecules would have to be removed to maintain constant V. This reveals that the number of molecules varies inversely with temperature, i.e., n = 1/T. We can summarize the relationships in the following expression:

$$n = \frac{kPV}{T}$$

where units for n, P, V, and T are the same as already defined in Eq. 7.6.

Because it is known that 1 mol gas occupies 22.4 L at STP, the previous expression becomes

$$n = k \; \frac{(1 \; atm)(22.4 \; L/mol)}{273^\circ K} = k \big(0.082 \; L \cdot atm/mol^\circ K\big).$$

Assign 1 mol to n, k becomes 1/0.082 L·atm/mole·°K, and we may rearrange the equation as follows:

$$\frac{1}{0.082} PV = nT$$

$$and \; PV = n \; 0.082 \; T.$$

The quantity 0.082 L·atm/mol·°K is known as the universal or ideal gas law constant. There are other values for this constant depending upon the unit of pressure used. The final form of the universal gas law equation usually is given as

$$PV = nRT.$$

The gas law equation can be used for many purposes; an example is the conversion of the weight of a dissolved gas in water to the volume of the gas in water (Ex. 7.8).

Ex. 7.8 A water contains 7.54 mg/L dissolved oxygen. The pressure is 735 mm and the temperature is 30°C. The dissolved oxygen concentration will be converted to milliliters per liter.
   Solution:
The volume of 1 mol oxygen is

$$\left(\frac{735 \; mm}{760 \; mm}\right)(V) = (1 \; mol)\big(0.082 \; L \cdot atm/mol^\circ K\big)\big(303.15^\circ A\big)$$

$$V = 25.7 \; L/mol.$$

One mole is 32 g of oxygen, so the density is

$$\frac{32 \; g/mol}{25.7 \; L/mol} = 1.245 \; g/L \; or \; 1.245 \; mg/mL.$$

The concentration in milliliters per liter is

$$\frac{7.54 \ mg/L}{1.245 \ mg/mL} = 6.06 \ mL/L.$$

Percentage Saturation and Oxygen Tension

A water may contain more or less dissolved oxygen than the saturation concentration for existing conditions because of biological activity. The percentage saturation of water with dissolved oxygen is calculated as follows:

$$PS = \frac{DO_m}{DO_s} \times 100 \qquad (7.7)$$

where PS = percentage saturation and DO_m = the measured concentration of dissolved oxygen (mg/L). A calculation of PS is made in Ex. 7.9.

Ex. 7.9 Calculate the percentage saturation at the surface of a freshwater body containing 10.08 mg/L dissolved oxygen when the water temperature is 28°C and BP is 732 mm.

Solution:

The concentration of dissolved oxygen at saturation is

$$DO_s = 7.81 \ mg/L \times \frac{732}{760} = 7.52 \ mg/L.$$

The percentage saturation is

$$PS = \frac{10.08}{7.52} \times 100 = 134\%.$$

Suppose the following morning, the body of water in Ex. 7.9 contained only 4.5 mg/L dissolved oxygen and the temperature and barometric pressure were 24 °C and 740 mm, respectively. The percentage saturation would now be only 55%.

Fish and other aquatic organisms that breathe with gills respond to the pressure of dissolved oxygen in the water rather than to its concentration. Physiologists often refer to the pressure of dissolved oxygen as the oxygen tension. The tension of dissolved oxygen refers to the pressure of oxygen in the atmosphere required to hold an observed concentration of dissolved oxygen in the water. The partial pressure of oxygen in air is 159.2 mm at STP. To estimate oxygen tension, the partial pressure of oxygen in the atmosphere is multiplied by the factor DO_m/DO_s—oxygen tension is closely linked to percentage saturation.

Ex. 7.10 *Estimate the oxygen tension in water with 10 ppt salinity and 24 °C where dissolved oxygen is 9.24 mg/L and BP is 760 mm.*

Solution:

The concentration at saturation is 7.93 mg/L. Thus,

$$Oxygen\ tension = \frac{9.24\ mg/L}{7.93\ mg/L} \times 159.2\ mm = 185.5\ mm.$$

Waters saturated with dissolved oxygen and at different temperatures will have the same oxygen tension. A freshwater at 20 °C with 9.08 mg/L dissolved oxygen has the same oxygen tension as a freshwater at 10 °C with 11.28 mg/L dissolved oxygen. This results because the factor DO_m/DO_s is unity for both. Aquatic organisms would be exposed to the same oxygen pressure (tension) even though the oxygen concentration would be greater for the cooler water. It also is significant to note that when water holding a certain concentration of a gas warms without loss of gas to the air, the gas tension and the percentage saturation increase. If a water holding 6.35 mg/L dissolved oxygen at 760 mm and 18 °C warms to 23 °C at the same atmospheric pressure and without loss of oxygen, the value for oxygen tension increases from 107 mm to 118 mm and saturation increases from 67% to 74%.

The ΔP

There are various reasons why waters become supersaturated with gases. Cold rainwater and snow melt become supersaturated with gases when they percolate downward to the water table and warm. In warm weather, water from wells or springs often will be cooler than ambient temperature at ground level, and supersaturated with gas. When water falls over a high dam, air bubbles may be entrained. The water will plunge to a considerable depth beneath the surface in the pool behind the dam into which it falls. The resulting increase in hydrostatic pressure raises the dissolved oxygen concentration at saturation, and when the water ascends and flows into shallower areas, it will be supersaturated with gases (Boyd and Tucker 2014).

Air leaks in or improper submergence of the intakes of pumps can result in gas supersaturation. Air bubbles can be sucked into the water, and as a result of the pressure increase caused by the pump, discharge will be supersaturated with gases (air). Aquatic plants often cause gas supersaturation by releasing copious amounts of dissolved oxygen during photosynthesis.

The difference between the total gas pressure (TGP) and the total pressure at a given depth is called ΔP. At the water surface, ΔP can be expressed as

$$\Delta P = TGP - BP. \tag{7.8}$$

The total gas pressure can be determined from the partial pressures (tensions) of the individual gases in water

$$\Delta P = (P_{O_2} + P_{N_2} + P_{Ar} + P_{CO_2}) - BP. \qquad (7.9)$$

For practical purposes, P_{Ar} and P_{CO_2} can be omitted from Eq. 7.9, but the analysis for dissolved nitrogen concentration is difficult. A relatively inexpensive instrument known as a saturometer may be used to measure ΔP directly.

Some of the ΔP is offset by hydrostatic pressure at depths below the water surface. The actual ΔP to which aquatic animals are exposed is called the uncompensated ΔP:

$$\Delta P_{uncomp.} = \Delta P - HP. \qquad (7.10)$$

Values for the hydrostatic pressure (HP) for different depths and salinities can be found in Table 7.9.

Gases dissolve in the blood of fish, shrimp, and other aquatic animals and reach equilibrium with external conditions. If water is supersaturated with gases, the blood of animals living in the water also becomes saturated. The blood of an aquatic animal at equilibrium with dissolved gases at a particular temperature will become supersaturated with gases if the animal moves to warmer water. If supersaturating gases are not rapidly lost to the surrounding water through the gills, gas bubbles can form in the blood.

The occurrence of gas bubbles in the blood causes gas bubble trauma—often called gas bubble disease. Gas bubble trauma leads to stress and mortality. Eggs may float to the surface, larvae and fry may exhibit hyperinflation of the swim bladder, cranial swelling, swollen gill lamellae, and other abnormalities. A common symptom of gas bubble trauma in juvenile and adult fish is gas bubbles in the blood that are visible in surface tissues on the head, in the mouth, and in fin rays. The eyes of affected fish also may protrude.

Aquatic animals exposed to ΔP values of 25 to 75 mm Hg on a continuous basis may present symptoms of gas bubble trauma, and some of the affected animals may die if exposure is prolonged for several days. Acute gas bubble trauma may lead to 50–100% mortality at greater ΔP.

Gas bubble trauma is most common in shallow water or in organisms that cannot escape the surface water by sounding. At greater depths, the hydrostatic pressure increases the equilibrium concentration of gases in water and thereby lowers the ΔP (Ex. 7.11)

Ex. 7.11 *Calculation of reduction in ΔP with depth.*
   Situation:
   A freshwater water body has a temperature of 28 °C, and a ΔP of 115 mm Hg at the surface. The water is thoroughly mixed and gas concentrations and temperature are the same at all depths. The uncompensated ΔP will be estimated for 1.5 m depth.
   Solution:
   Using Eq. 7.10 and obtaining the hydrostatic pressure per meter of water depth (by extrapolation between 25 ° and 30 °C) from Table 7.9, the uncompensated ΔP will be calculated.

$$\Delta P_{uncomp.} = 115 \ mm - (73.28 \ mm/m \times 1.5 \ m) = 5.1 \ mm.$$

Supersaturation of surface waters with dissolved oxygen during afternoons is a common occurrence. This condition usually does not cause harm to aquatic animals because the period of supersaturation is limited to a few hours in the afternoon and early evening, the dissolved oxygen concentration decreases with depth, and most animals can go to greater depth (sound) where the combination of lower dissolved concentration and greater hydrostatic pressure lessens ΔP.

Gas Transfer

Most of the information on gas transfer between air and water has been developed for oxygen, but principles presented for oxygen are applicable to other gases. In natural waters, dissolved oxygen concentrations are constantly changing because of biological, physical, and chemical processes. The air above water has a constant percentage of oxygen, even though the partial pressure of oxygen in air may vary because of differences in atmospheric pressure. When water is at equilibrium with atmospheric oxygen, there is no net transfer of oxygen between air and water. Transfer of oxygen from air to water will occur when water is undersaturated with dissolved oxygen, and oxygen will diffuse from the water to the air when water is supersaturated with oxygen. The driving force causing net transfer of oxygen between air and water is the difference in oxygen tension. Once equilibrium is reached, oxygen tension in the air and in the water are the same and net transfer ceases. The oxygen deficit or oxygen surplus may be expressed as

$$D = DO_s - DO_m \tag{7.11}$$

$$S = DO_m - DO_s \tag{7.12}$$

where D = oxygen deficit (mg/L), S = oxygen surplus (mg/L), DO_s = solubility of oxygen in water at saturation under existing conditions (mg/L), and DO_m = measured concentration of dissolved oxygen (mg/L).

Oxygen enters or leaves a body of water at the air-water interface. For the thin film of water in contact with air, the greater D or S, the faster oxygen will enter or leave the film. For undisturbed water, the net transfer of oxygen will depend upon the value of D or S, the area of the air-water interface, the temperature, and time of contact. Even when D or S is great, the rate of net transfer is slow because the surface film quickly reaches equilibrium and further net transfer requires oxygen to diffuse from the film to the greater body of water, or from the greater body of water to the film. Natural waters are never completely quiescent, and the rate of oxygen transfer increases with greater turbulence.

The rate of change of dissolved oxygen concentration over time can be expressed as

$$\frac{dc}{dt} = \frac{k}{F} \frac{A}{V} (C_s - C_m) \qquad (7.13)$$

where dc/dt = rate of change in concentration, k = diffusion coefficient, F = liquid film thickness, A = area through which the gas is diffusing, V = volume of water into which the gas is diffusing, C_m = saturation concentration of gas in solution, and C_s = concentration of gas in solution. Gases will be removed from solution ($-dc/dt$) in those applications in which C_m is greater than C_s.

Gas transfer can be accelerated by reducing liquid film thickness (F) and by increasing the surface area (A) through which gas diffuses. Because it is difficult to measure A and F, the ratios A/V and k/F often are combined to establish an overall transfer coefficient (K_La)

$$\frac{dc}{dt} = K_La(C_s - C_m). \qquad (7.14)$$

The overall transfer coefficient reflects conditions present in a specific gas-liquid contact system. Important variables include basin geometry, turbulence, characteristics of the liquid, extent of the gas-liquid interface and temperature. Temperature affects viscosity, which, in turn, influences k, F, and A. Values of K_La can be corrected for the effects of temperature, using the following expression:

$$(K_La)_T = (K_La)_{20}(1.024)^{T-20} \qquad (7.15)$$

where $(K_La)_T$ = overall gas transfer coefficient at temperature T, $(K_La)_{20}$ = overall gas transfer coefficient at 20 °C, and T = liquid temperature (°C). Although each gas species in a contact system has a unique value of K_La, relative values for a specific gas pair are inversely proportional to their molecular diameters

$$(K_La)_1/(K_La)_2 = (d)_2/(d)_1 \qquad (7.16)$$

where d = diameter of the gas molecule. The K_La determined experimentally for one gas can be used to predict K_La values for other gas species. However, the major atmospheric gases have similar molecular diameters—nitrogen, 0.314 nannometer (nm), oxygen, 0.29 nm, and carbon dioxide, 0.28 nm.

The air-water interface varies with turbulence and neither variable can be estimated accurately. Nevertheless, it is possible to calculate the gas transfer coefficient empirically between time 1 and time 2 because integration of Eq. 7.14 gives

$$K_La = \frac{\ln D_1 - \ln D_2}{t_2 - t_1}. \qquad (7.17)$$

where K_La = gas transfer coefficient (hr^{-1}), t = time, and subscripts 1 and 2 = at time 1 and at time 2. Various units may be used in Eq. 7.17, but for convenience K_La often is expressed as hr^{-1} (1/hr). A plot of the natural logarithm of D against time for a thoroughly mixed body of water, or the natural logarithm of D as calculated from

Fig. 7.3 Graphical illustration of the method for determining the transfer coefficient (K_La)

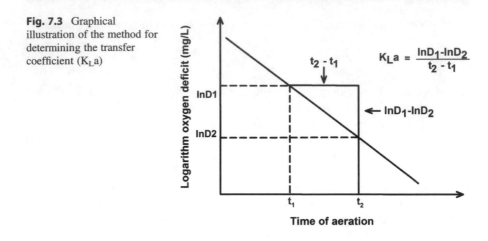

dissolved oxygen profiles for a less than thoroughly mixed body of water, gives a straight line for which K_La is the slope (Fig. 7.3).

Students often are confused about the exact meaning of the K_La_{20} term which is given as times per hour (hr^{-1}). A simple way to view K_La_{20} is that it indicates how many times in 1 hour that the method of aeration could raise the dissolved oxygen concentration in the test tank volume from 0 mg/L to saturation.

The standard oxygen transfer rate for 20 °C may be estimated as

$$SOTR = (K_La_{20})(C_s)(V) \tag{7.18}$$

where SOTR = standard oxygen transfer rate (g O_2/hr), C_s = dissolved oxygen concentration (g/m^3) at 20 °C and saturation, and V = water volume (m^3). Note: Remember g/cm^3 = mg/L. A sample calculation is provided in Ex. 7.12.

Ex. 7.12 *The K_La_T for 23 °C is 0.0275/hour for re-aeration by diffusion on a windy day of 30 m^3 of freshwater in a 20-m^2 tank. The SOTR will be estimated. Assume BP = 760 mm.*

Solution:

From Eq. 7.15,

$$K_La_{20} = 0.0275 \div 1.024^{23-20} = 0.0256/hr$$

And, using Eq. 7.18

$$SOTR = (0.0256/hr)(8.56\ g/m^3)(30\ m^3) = 6.57\ g\ O_2/hr.$$

The oxygen diffused through a 20 m^2 surface, and in terms of surface area,

$$SOTR = \frac{6.57\ g\ O_2/h}{20\ m^2} = 0.328\ g\ O_2/m^2/hr.$$

The methodology for determining oxygen transfer rate described above often is used to determine the SOTR of mechanical aerators used in wastewater treatment (American Society of Civil Engineers 1992). Dividing the SOTR by the power applied to the aeration device provides an estimate of the standard aeration efficiency of the aerator.

Measurements of SOTR in laboratory tanks range from 0.01 to 0.10 g oxygen/m^2 per hour in still water to as high as 1.0 g oxygen/m^2 per hour in turbulent water. Accurate measurement of SOTR is difficult in natural waters because biological processes add or remove oxygen during the period of measurement, and circulation patterns are difficult to assess. Schroeder (1975) reported SOTR values of 0.01 to 0.05 g oxygen/m^2 per hour for each 0.2 atm (152 mm) saturation deficit. Welch (1968) measured SOTR values of 0.1 to 0.5 g oxygen/m^2 per hour at 100% departure from saturation for a pond. The highest SOTR values were for windy days.

Gas exchange with the air is particularly rapid when waves break forming white caps. In small lakes and ponds, water surfaces are usually rather calm and oxygen transfer is impeded by lack of disturbance of the water surface. The following equation (Boyd and Teichert-Coddington 1992) can be used to relate wind speed to rate of wind aeration in small ponds:

$$\text{WRR} = (0.153\text{X} - 0.127)\left(\frac{\text{C}_\text{s} - \text{C}_\text{m}}{9.08}\right)(1.024^{\text{T}-20}) \qquad (7.19)$$

where WRR = wind re-aeration rate (g O_2/m^2 per hour) and X = wind speed 3 m above the water surface (m/sec).

Ex. 7.13 *The wind re-aeration rate for a wind speed of 3 m/sec and the oxygen added to a pond by diffusion during the night (12 hours) will be estimated for a 1.5 m deep pond at 25 °C when the barometric pressure is 760 mm Hg and dissolved oxygen concentration averages 5 mg/L during the night.*
   Solution:

(i) *By substitution of data on oxygen concentrations and wind speed into Eq. 7.19, we get*

(ii) *WRR* $= [(0.153)(3\ m/sec) - 0.127]\left[\dfrac{8.24 - 5.00}{9.08}\right](1.024)^5$

$$WRR = 0.133\ g\ O_2/m^2/hr.$$

(iii) *There is 1.5 m^3 water beneath 1 m^2 of surface, so*

$$\frac{0.133\ g\ O_2/m^2/hr}{1.5\ m^3/m^2} = 0.089\ g\ O_2/m^3/hr$$

or 0.089 mg $O_2/L/hr$.

In 12 hours, the amount of dissolved oxygen entering the pond by diffusion from the air would be

$$0.089 \text{ g } O_2/m^3/hr \times 12 \text{ hr} = 1.07 \text{ mg}/L.$$

Stream Re-aeration

Several equations for estimating stream re-aeration rate have been proposed. One such equation (O'Connor and Dobbins 1958) that was developed for streams with water velocities between 0.15 and 0.5 m/sec and average depths between 0.3 and 9 m follows:

$$k_s = 3.93 \nu^{0.5} H^{-1.5} \tag{7.20}$$

where k_s = stream re-aeration coefficient (day^{-1}), ν = velocity (m/sec), and H = average depth (m). The k_s can be adjusted for temperature and oxygen deficit in the stream by the equation

$$k'_s = k_s \left(\frac{C_s - C_m}{C_s} \right) 1.024^{T-20} \tag{7.21}$$

where k'_s = adjusted k_s. The daily oxygen input from diffusion from the atmosphere can be estimated using Eq. 7.18 in which the volume term (V) is taken as stream flow past a given point in cubic meters per day.

Stream re-aeration coefficients are lowest for sluggish rivers—usually being below 0.35 day^{-1} (0.35/day). Large rivers with moderate to normal velocities have stream re-aeration coefficients between 0.35 and 0.7 day^{-1}, while swift flowing rivers may have re-aeration constants of 1.0 or slightly greater. The higher rates of re-aeration are for streams with rapids and waterfalls.

Concentrations of Dissolved Oxygen

Diffusion of oxygen from air to water and *vice versa* causes dissolved oxygen concentrations in natural waters to tend towards equilibrium—but, they seldom are at equilibrium. This results because biological activity changes dissolved oxygen concentrations faster than diffusion can produce equilibrium for existing conditions. The biological processes that influence dissolved oxygen concentrations are photosynthesis by green plants and respiration by all aquatic organisms. During daylight, photosynthesis usually occurs more rapidly than does respiration, and dissolved oxygen concentration increases. Afternoon dissolved oxygen concentration will typically be above saturation in waters with healthy plant communities. At night, photosynthesis stops, but respiration continues to use oxygen and causes dissolved oxygen concentration to decline to less than saturation. Dissolved oxygen concentration also tends to decline with increasing water depth because there is less illumination in deeper waters.

Oxygen and Carbon Dioxide in Respiration

Absorption of oxygen is a greater challenge to aquatic animals than to terrestrial animals. At 25 °C and standard atmospheric pressure, the volume of molecular oxygen in freshwater at saturation is 6.29 mL/L, while in the atmosphere (20.95% oxygen) the volume of oxygen at the same temperature and pressure is about 209.5 mL/L (Ex. 7.14). Air in this instance contains 33 times greater volume of oxygen than occurs in oxygen-saturated surface water. The absorption of equal amounts of oxygen by an aquatic animal and a terrestrial animal requires the aquatic animal to pump a much larger volume of water across the gills than the land animal needs to breathe. Assuming the fish in Ex. 7.14 could extract all of the dissolved oxygen from water, it would have to pump 24.3 L of water per hour to satisfy its respiratory need for molecular oxygen. The same amount of oxygen is contained in 0.73 L of air. Neither the fish nor the land animal can remove all of the molecular oxygen from water or air, respectively, but Ex. 7.14 illustrates why oxygen availability is of greater concern in aquatic environments than in terrestrial ones.

Ex. 7.14 The volume of water that must be pumped across the gills of a fish as compared to the amount of air breathed by a land animal to supply the same amount of oxygen for respiration will be calculated.
   Solution:
   The volumes of oxygen in freshwater and in air at 25 °C and 1 atm pressure are: Volume of 1 mol of oxygen is

$$PV = nRT$$

$$V = (1 \ mol)(0.082 \ L \cdot atm/mol\degree A)(298.15\degree A)$$

$$V = 24.45 \ L/mol.$$

Density of oxygen at 25 °C and 1 atm is

$$\frac{32 \ g/mol}{24.45 \ L/mol} = 1.309 \ g/L \ (or \ mg/mL).$$

Freshwater at 25 °C and 1 atm holds 8.24 mg/L dissolved oxygen at saturation (Table 7.6), which on a volume of oxygen basis is

$$\frac{8.24 \ mg/L}{1.309 \ mg/mL} = 6.29 \ mL/L.$$

Suppose a 1-kg fish uses 200 mg O₂/hour, for which the required volume of water to contain this much oxygen is

$$200 \ mg \ O_2/hr \div 1.309 \ mg \ O_2/mL = 152.8 \ mL \ O_2/hr$$

and

$$\frac{152.8 \ mL \ O_2/hr}{6.29 \ mL \ O_2/L} = 24.3 \ L \ water/hr.$$

Air contains 20.95% oxygen on volume basis, and like oxygen, 1 mol of air at 25 °C and 1 atm occupies 24.45 L.

Volume of oxygen in 1 mol air (25 °C, 1 atm) is

$$24.45 \ L/mol \times 0.2095 = 5.122 \ L.$$

Weight of 5.122 L of oxygen is

$$5.122 \ L \times 1.309 \ g \ O_2/L = 6.70 \ g \ O_2.$$

Weight/volume concentration of oxygen in air is

$$\frac{6.70 \ g \ O_2}{24.45 \ L} = 0.274 \ g \ O_2/L \ air = 274 \ mg \ O_2/L \ air.$$

Volume air to provide 200 mg O₂ is

$$\frac{200 \ mg \ O_2}{274 \ mg \ O_2} = 0.73 \ L.$$

The fish must pump 24.3 L of water to expose the gills to the same amount of oxygen as the land animal would breathe into its lungs in 0.73 L of air.

When oxygen is removed from water, it must be replaced by diffusion from the air or by release of oxygen from photosynthesis, both of which are relatively slow processes. Air movement usually replaces very quickly the oxygen removed by land animals during respiration. Humans can congregate in large numbers to view outdoor sights or events with no danger of suffocation. The usual recommended standing capacity for humans at outdoor events is 3-5 individuals/m². This is a human biomass of 200-400 kg/m², and much greater than the standing crops of aquatic animals found in natural water bodies.

The concentration of atmospheric oxygen seldom changes in the air surrounding terrestrial ecosystems. In water bodies, there is typically a daily oscillation of dissolved oxygen concentration with lowest levels at night and highest levels in daytime. Many other events, e.g., cloudy weather, plankton density, wind velocity, sudden algal die-offs, etc., can cause dissolved oxygen concentrations to decline. It should be clear that the availability of dissolved oxygen in water bodies is a much more critical factor than is the availability of oxygen from the air in terrestrial habitats.

Aquatic organisms absorb molecular oxygen into their blood from the water and use it in respiration, but they must expel carbon dioxide produced in respiration into the water. The gills of fish and crustaceans provide a surface across which gases in water may enter the blood and *vice versa*. Blood cells of fish contain hemoglobin, and those of crustaceans carry hemocyanin; these pigments combine with oxygen to allow the blood to carry more oxygen than can be carried in solution. Both pigments are metalloproteins: hemoglobin has a porphyrin ring with iron in its center; hemocyanin has a porphyrin ring with copper in its center. The hemolymph (blood fluids) and hemoglobin are loaded with oxygen at the gills where oxygen tension is high and oxygen is unloaded at the tissues where oxygen use in respiration results in low oxygen tension as illustrated for hemoglobin (Hb) below:

$$Hb + O_2 \rightarrow HbO_2 \ (in\ gills) \tag{7.22}$$

$$HbO_2 \rightarrow Hb + O_2 \ (in\ tissues). \tag{7.23}$$

The effect of oxygen tension in the water on the loading and unloading of oxygen by hemoglobin—the oxyhemoglobin dissociation curve—is illustrated in Fig. 7.4. The curve usually is sigmoid for warmwater species and hyperbolic for coldwater

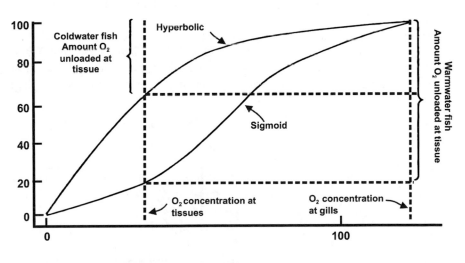

Fig. 7.4 Oxyhemoglobin saturation curves for warmwater and coldwater fish

species. As a result, warmwater species have a greater capacity than coldwater species to unload oxygen at the tissue level. This is a major reason why coldwater species require a higher dissolved oxygen concentration than do warmwater species.

Carbon dioxide from respiration dissolves in the hemolymph at the tissues and is transported in venous blood to the gills where it diffuses into the water. A high carbon dioxide concentration in the water inhibits the diffusion of carbon dioxide from the blood. Accumulation of carbon dioxide in the blood depresses blood pH leading to several negative physiological consequences.

As carbon dioxide concentration increases in the water, it also interferes with the loading of the hemoglobin with oxygen in the gills (Fig. 7.5). A high carbon dioxide concentration in the water results in organisms requiring a higher dissolved oxygen concentration to avoid oxygen related stress.

All aerobic organisms require dissolved oxygen, and much of the dissolved oxygen in aquatic ecosystems is used by bacteria and other saprophytic microorganisms that oxidize organic matter to carbon dioxide, water, and inorganic minerals. Chemoautotrophic microorganisms that oxidize reduced inorganic compounds to obtain energy, e.g., nitrifying and sulfide oxidizing bacteria, require an abundant supply of dissolved oxygen.

The amount of oxygen used by aquatic animals varies with species, size, temperature, time since feeding, physical activity, and other factors. Small organisms use more oxygen per unit of weight than larger ones of the same species. For example, 10-g channel catfish were reported to use about twice as much oxygen per unit weight as 500-g channel catfish. A 10 °C increase in water temperature within the range of temperature tolerance of an organism usually will double the oxygen consumption rate. Fish have been shown to use about twice as much oxygen 1 hour after feeding than did fish fasted overnight. Fish swimming against a current use oxygen at a greater rate as the current velocity increases (Andrews et al. 1973; Andrews and Matsuda 1975).

Average oxygen consumption rates for adult fish usually range from 200 to 500 mg oxygen/kg fish/hour (Boyd and Tucker 2014). There is less information on oxygen consumption by crustaceans, but it appears that rates are similar to those for fish. In natural waters where the standing crop of fish populations is typically less

Fig. 7.5 Effect of carbon dioxide concentration on saturation of hemoglobin with oxygen

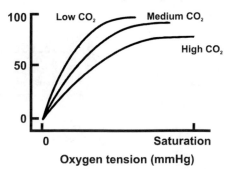

than 500 kg/ha, oxygen consumption by fish does not have a great effect on dissolved oxygen concentration (Ex. 7.14)

Ex. 7.15 *The amount of oxygen (in milligrams per liter) removed from a water body of 10,000 m³ by a fish standing crop of 500 kg will be estimated.*

Solution:

Assume that the fish respire at a rate of 300 mg/kg/hour over a 24-hour period. The oxygen use will be

$$O_2 \ used = 300 \ mg/kg/hr \times 500 \ kg \times 24 \ hr/d = 3,600,000 \ mg/d = 3,600 \ g/d$$

$$O_2 \ concentration = \frac{3,600 \ g/d}{10,000 \ m^3} = 0.360 \ g/m^3/d = 0.360 \ mg/L/d.$$

Fish held at high density in aquaria, holding tanks, or aquaculture production facilities cause a much larger demand for dissolved oxygen than found in Ex. 7.14.

Warmwater fish usually will survive for long periods at dissolved oxygen concentrations as low as 1.0 to 1.5 mg/L, while coldwater fish may survive at 2.5 to 3.5 mg/L dissolved oxygen. Nevertheless, fish and other aquatic animals are stressed, susceptible to disease, and grow slowly at low oxygen concentration. As a general rule, aquatic organisms do best when the dissolved oxygen concentration does not fall below 50% of saturation (Collins 1984; Boyd and Tucker 2014). In freshwater at sea level, 50% of saturation is around 5 mg/L at 15 °C and 4 mg/L at 26 °C. The effects of dissolved oxygen concentration on warmwater fish are summarized in Table 7.10. The effects on coldwater fish are similar, but occur at higher dissolved oxygen concentrations. Water quality criteria for aquatic ecosystems usually specify that dissolved oxygen concentrations be above 5 or 6 mg/L, and some may specify that dissolved oxygen concentrations should be at least 80 to 90% of saturation. Slightly lower dissolved oxygen standards may be listed for public water supplies, but few are below 4 mg/L. Even water for irrigation or livestock consumption should have at least 3 or 4 mg/L dissolved oxygen.

Carbon dioxide is not highly toxic to most aquatic animals, but fish have been shown to avoid carbon dioxide concentrations of 10 mg/L or more. Most species tolerate up to 60 mg/L of carbon dioxide provided there is plenty of dissolved oxygen, but high carbon dioxide concentrations have a narcotic effect on fish.

Table 7.10 Effects of dissolved oxygen concentration on warmwater fish

Dissolved oxygen (mg/L)	Effects
0–0.3	Small fish survive short exposure.
0.3–1.5	Lethal if exposure is prolonged for several hours.
1.5–5.0	Fish survive, but growth will be slow and fish will be more susceptible to disease.
5.0 – saturation	Desirable range
Above saturation	Possible gas bubble trauma if exposure prolonged.

Conclusions

Dissolved oxygen is the most important water quality variable in aquatic ecosystems. Molecular oxygen is the electron acceptor in aerobic metabolism, and all aerobic organisms must have an adequate supply of dissolved oxygen. When dissolved oxygen concentrations are low or dissolved oxygen is absent, decomposition of organic matter by anaerobic microorganisms releases reduced substances such as ammonia, nitrite, ferrous iron, hydrogen sulfide, and dissolved organic compounds into the water. In the absence of adequate dissolved oxygen, aerobic microorganisms do not function efficiently in oxidizing these reduced substances. The combination of low dissolved oxygen concentration and high concentrations of certain reduce, toxic metabolites cause drastic, negative impacts on the structure and function of aquatic ecosystems. "Clean water" species disappear and only those organisms that can tolerate highly polluted conditions thrive. Such ecosystems are of low biodiversity, unstable, and greatly impaired for most beneficial uses.

References

American Society of Civil Engineers (1992) Measurement of oxygen transfer in clean water, 2nd edn. ANSI/ASCE 2-91,. American Society of Civil Engineers, New York

Andrews JW, Matsuda Y (1975) The influence of various culture conditions on the oxygen consumption of channel catfish. Trans Am Fish Soc 104:322–327

Andrews JW, Murai T, Gibbons G (1973) The influence of dissolved oxygen on the growth of channel catfish. Trans Am Fish Soc 102:835–838

Benson BB, Krause D (1984) The concentration and isotopic fractionation of oxygen dissolved in freshwater and seawater in equilibrium with the atmosphere. Lim Ocean 29:620–632

Boyd CE, Teichert-Coddington D (1992) Relationship between wind speed and reaeration in small aquaculture ponds. Aqua Eng 11:121–131

Boyd CE, Tucker CS (2014) Handbook for aquaculture water quality. Craftmaster Printers, Auburn

Collins G (1984) Fish growth and lethality versus dissolved oxygen. In: Proceedings specialty conference on environmental engineering, ASCE, Los Angeles, June 25–27, 1984, pp 750–755

Colt J (1984) Computation of dissolved gas concentrations in water as functions of temperature, salinity, and pressure, Special Publication 14. American Fisheries Society, Bethesda

Colt J (2012) Dissolved gas concentration in water. Elsevier, New York

O'Connor DJ, Dobbins WE (1958) Mechanism of reaeration in natural streams. Trans Am Soc Civ Eng 123:641–666

Schroeder GL (1975) Nighttime material balance for oxygen in fish ponds receiving organic wastes. Bamidgeh 27:65–74

Welch HE (1968) Use of modified diurnal curves for the measurement of metabolism in standing water. Lim Ocean 13:679–687

Redox Potential

8

Abstract

Electrons are lost when a substance is oxidized and gained when a substance is reduced. Oxidations and reductions occur in couplets—known as half-cells—in which one substance, the oxidizing agent, accepts electrons from another substance, the reducing agent. The oxidizing agent is reduced and the reducing agent is oxidized. The flow of electrons between two half-cells can be measured as an electromotive force (in volts). The hydrogen half-cell ($H_2 \rightarrow 2H^+ + 2e^-$) has an electrical potential of 0.0 volt at 25 °C, 1 atm H_2 and 1 M H^+; it is said to have a standard electrode potential (E^0) of 0.0 volt. The flow of electrons to or from the hydrogen half-cell is the reference for determining E^0 values of other half-cells. The greater a positive E^0, the more oxidized a half-cell with respective to the hydrogen electrode; the opposite is true for a negative E^0. The redox potential (E or E_h) for non-standard conditions is measured with a calomel electrode or calculated with the Nernst equation. Water with measureable dissolved oxygen has $E_h = 0.50$ volt, and at oxygen saturation, $E_h = 0.56$ volt. The redox potential indicates whether given substances may exist in a particular environment and explains the sequence of reactions that occur as dissolved oxygen concentration declines in water bodies and sediments. The redox potential has many practical applications in analytical chemistry and industry, but it is not frequently measured in water quality investigations.

Introduction

The oxidation-reduction (redox) potential is a measurement of the direction and amount of electron transfer that occurs in oxidation-reduction reactions. The redox potential indicates whether a substance is a reducing agent and donates electrons to another substance or an oxidizing agent and accepts electrons from another substance. It also provides an indication of the relative oxidizing and reducing strengths of reactants.

C. E. Boyd, *Water Quality*, https://doi.org/10.1007/978-3-030-23335-8_8

In a redox reaction, electrons donated (lost) by the reducing agent must be accepted (gained) by the oxidizing agent. A redox reaction can be thought of as having two parts: electrons are released by the reducing agent; electrons released by the reducing agent are accepted by the oxidizing agent. The flow of electrons between the two parts of the reaction can be measured as an electrical current—the redox potential.

Hydrogen (H_2) is the base for comparing the relative oxidizing or reducing power of a substance. Hydrogen is assigned a redox potential of zero at standard conditions (1.0 MH^+; 1 atm pressure; 25 °C). Other substances can be compared to hydrogen, and their redox potentials can be compared to each other. Concentrations of substances, pH, and temperature influence redox potential, but redox can be adjusted for differences in these variables. Redox potential may be used to determine the endpoints of chemical reactions, predict whether or not two substances will react when brought together, ascertain if certain substances are likely to be present in the aquatic environment, and other purposes.

Oxidation-Reduction

Principles of the law of mass action, Gibbs free energy change of reaction, and the equilibrium constant can be applied to oxidation-reduction reactions. The driving force for oxidation-reduction reactions can be expressed in terms of a measurable electrical current. Consider the oxidation-reduction reaction

$$I_2 + H_2 = 2H^+ + 2I^-. \tag{8.1}$$

Iodine is reduced to iodide, and hydrogen gas is oxidized to hydrogen ion. The reaction can be divided into two parts, each called a half-cell, one showing the loss of electrons (e^-) resulting in oxidation and one showing the gain of electrons causing reduction:

$$H_2 = 2H^+ + 2e^- \tag{8.2}$$

$$I_2 + 2e^- = 2I^-. \tag{8.3}$$

The standard Gibbs free energy of reaction (ΔG^o) expressions for Eqs. 8.2 and 8.3, respectively, are:

$$\Delta G^o = 2\Delta G_f^0 H^+ + 2\Delta G_f^0 e^- - \Delta G_f^0 H_2$$

$$\Delta G^o = 2\Delta G_f^0 I^- - \Delta G_f^0 I_2 - 2\Delta G_f^0 e^-.$$

The $\Delta G_f^o e^-$ term cancels during addition of the right-hand sides of these expressions, and the sum is

$$2\Delta G_f^0 H^+ + 2\Delta G_f^0 I^- - \Delta G_f^0 H_2 - \Delta G_f^0 I_2.$$

The sum of the two expressions is the same as the expression for estimating ΔG° from Eq. 8.1.

Because the $\Delta G_f^0 e^-$ terms cancel when two half-cells are added, $\Delta G_f^0 e^-$ always is assigned a value of zero. Also, from Table 4.3, notice that $\Delta G_f^0 H^+ = \Delta G_f^0 H_2 = 0$. Thus, $\Delta G^\circ = 0$ for the hydrogen half-cell reaction (Eq. 8.2).

Standard Hydrogen Electrode

The hydrogen half-cell (Eq. 8.2) is often written as

$$\frac{1}{2} H_2 = H^+ + e^-. \tag{8.4}$$

The hydrogen half-cell is a useful expression. Any oxidation-reduction reaction can be written as two half-cell reactions with the hydrogen half-cell functioning as either the electron-donating or electron-accepting half-cell:

$$Fe^{3+} + e^- = Fe^{2+}, \qquad \frac{1}{2} H_2 = H^+ + e^- \tag{8.5}$$

$$Fe(s) = Fe^{2+} + 2e^-, \qquad 2H^+ + 2e^- = H_2(g). \tag{8.6}$$

In Eq. 8.5, hydrogen is the reductant because it donates electrons that reduce Fe^{3+} to Fe^{2+}. In Eq. 8.6, H^+ is the oxidant because it accepts electrons, and thereby oxidizes solid, metallic iron or Fe(s) to Fe^{2+}.

The flow of electrons between half-cells can be measured as an electrical current (Fig. 8.1). The chemical reaction involving reduction of I_2 by H_2 can be used to make a cell of two half-cells in which the flow of electrons can be measured. This cell consists of a solution 1 M in hydrogen ion and a solution 1 M in iodine. A platinum electrode coated with platinum black and bathed in hydrogen gas at 1 atm pressure is placed in the solution of 1 M hydrogen ion to form the hydrogen half-cell or hydrogen electrode. A shiny platinum electrode is placed in the iodine solution to form the other electrode. A platinum wire connected between the two electrodes allows free flow of electrons between the two half-cells. A salt bridge connected between the two solutions allows ions to migrate from one side to the other and maintain electrical neutrality. The flow of electrons is measured with a potentiometer. Electron flow in the cell shown in Fig. 8.1 is from the hydrogen electrode to the iodine solution. The iodine-iodide half-cell is initially more oxidized than the hydrogen electrode, and electrons move from the hydrogen side of the device to the iodine solution and reduce I_2 to I^-. The oxidation of H_2 to H^+ is the source of the electrons. Electrons continue to flow from the hydrogen electrode to the iodine-iodide half-cell until equilibrium is reached.

The potentiometer in Fig. 8.1 initially will read 0.62 volt—the standard electrode potential (E^0) for the iodine half-cell. The value of E^0 declines as the reaction proceeds, and at equilibrium $E^0 = 0$ volt. The standard electrode potential refers to the initial voltage that develops between the standard hydrogen electrode (half-cell) and any other half-cell under standard conditions (unit activities and 25 °C). The electrons transferred between the half-cells to drive the oxidation-reduction reaction to equilibrium do not always flow in the direction shown in Fig. 8.1. In some cases the standard hydrogen electrode may be more oxidized than the other half-cell, and electrons will flow toward the hydrogen electrode and the reduction will occur at the hydrogen electrode. It is necessary to give a sign to E^0. The positive sign usually is applied when electrons flow away from the hydrogen electrode and toward the other half-cell as done in Fig. 8.1. The positive sign means that the other half-cell is more oxidized than the hydrogen electrode. When the hydrogen electrode is more oxidized than the other half-cell, electrons flow toward the hydrogen electrode and a negative sign is applied to E^0. The sign is sometimes applied to E^0 in the opposite manner, and the sign of E^0 can cause considerable confusion unless specified.

Values of E^0 have been determined for many half-cells and selected ones are provided (Table 8.1). The E^0 of the hydrogen electrode (0 volt) is the reference or standard to which other half-cells are compared in Table 8.1. A half-cell with an E^0 greater than 0 volt is more oxidized than the hydrogen electrode, and one with an E^0 less than 0 volt is more reduced than the hydrogen electrode. The E^0 of any two half-cells may be compared; it is not necessary to compare only with the hydrogen electrode.

Reference to Table 8.1 can indicate the relative degree of oxidation (or reduction) of different substances. For example, ozone (O_3), hypochlorous acid (HOCl), and permanganate (MnO_4^-) are stronger oxidants than dissolved oxygen [$O_{2(aq)}$],

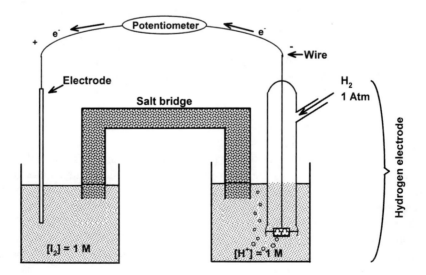

Fig. 8.1 The hydrogen half-cell or electrode connected to an iodine-iodide half-cell

Table 8.1 Standard electrode potentials at 25 °C

Reaction	E° (Volt)
$O_3(g) + 2H^+ + 2e^- = O_2(g) + H_2O$	+2.07
$Mn^{4+} + e^- = Mn^{3+}$	+1.65
$2HOCl + 2H^+ + 2e^- = Cl_2(aq) + 2H_2O$	+1.60
$MnO_4^- + 8H^+ + 5e^- = Mn^{2+} + 4H_2O$	+1.51
$Cl_2(aq) + 2e^- = 2Cl^-$	+1.39
$Cl_2(g) + 2e^- = 2Cl^-$	+1.36
$Cr_2O_7^{2-} + 14H^+ + 6e^- = 2Cr^{3+} + 7H_2O$	+1.33
$O_{2(aq)} + 4H^+ + 4e^- = 2H_2O$	+1.27
$2NO_3^- + 12H^+ + 10e^- = N_2(g) + 6H_2O$	+1.24
$MnO_2(s) + 4H^+ + 2e^- = Mn^{2+} + 2H_2O$	+1.23
$O_2(g) + 4H^+ + 4e^- = 2H_2O$	+1.23
$Fe(OH)_3(s) + 3H^+ + e^- = Fe^{2+} + 3H_2O$	+1.06
$NO_2^- + 8H^+ + 6e^- = NH_4^+ + 2H_2O$	+0.89
$NO_3^- + 10H^+ + 8e^- = NH_4^+ + 3H_2O$	+0.88
$NO_3^- + 2H^+ + 2e^- = NO_2^- + H_2O$	+0.84
$Fe^{3+} + e^- = Fe^{2+}$	+0.77
$I_2(aq) + 2e^- = 2I^-$	+0.62
$MnO_4 + 2H_2O + 3e^- = MnO_2(s) + 4OH^-$	+0.59
$O_2 + 2H_2O + 4e^- = 4OH^-$	+0.40
$SO_4^{2-} + 8H^+ + 6e = S(s) + 4H_2O$	+0.35
$SO_4^{2-} + 10H^+ + 8e^- = H_2S(g) + 4H_2O$	+0.34
$N_2(g) + 8H^+ + 6e^- = 2NH_4^+$	+0.28
$Hg_2Cl_2(s) + 2e^- = 2Hg(l) + 2Cl^-$	+0.27
$SO_4^{2-} + 9H^+ + 8e^- = HS^- + 4H_2O$	+0.24
$S_4O_6^{2-} + 2e^- = 2S_2O_3^{2-}$	+0.18
$S(s) + 2H^+ + 2e^- = H_2S(g)$	+0.17
$CO_2(g) + 8H^+ + 8e^- = CH_4(g) + 2H_2O$	+0.17
$H^+ + e^- = \frac{1}{2}H_2(g)$	0.00
$6CO_2(g) + 24H^+ + 24e^- = C_6H_{12}O_6 \text{ (glucose)} + 6H_2O$	−0.01
$SO_4^{2-} + 2H^+ + 2e^- = SO_3^{2-} + H_2O$	−0.04
$Fe^{2+} + 2e^- = Fe(s)$	−0.44

In this table, (g) = gas, (aq) = aqueous or dissolved, and (s) = solid or mineral form

because they have higher E^0 values than dissolved oxygen. But, for the same reason, dissolved oxygen is more oxidized than ferric iron (Fe^{3+}), nitrate (NO_3^-), and sulfate (SO_4^{2-})

Electrode potentials also can reveal if two or more specific substances can coexist. For example, dissolved oxygen and hydrogen sulfide do not coexist because oxygen is more highly oxidized than sulfide and its sources:

$$O_{2(aq)} + 4H^+ + 4e^- = 2H_2O \qquad\qquad E^0 = 1.27 \text{ volt} \qquad (8.7)$$

$$SO_4^{2-} + 10H^+ + 8e^- = H_2S(g) + 4H_2O \qquad E^0 = 0.34 \text{ volt} \qquad (8.8)$$

$$S(s) + 2H^+ + 2e^- = H_2S(g) \qquad\qquad E^0 = 0.17 \text{ volt.} \qquad (8.9)$$

If dissolved oxygen is present, hydrogen sulfide will be oxidized to sulfate. There are many other examples pertinent to water quality considerations. One is the occurrence of ferrous iron (Fe^{2+}) in water

$$Fe^{3+} + e^- = Fe^{2+} \qquad\qquad E^0 = 0.77. \qquad (8.10)$$

If the water contains dissolved oxygen ($E^0 = 1.27$ volt), the redox potential will be too high for Fe^{2+} to exist and only ferric iron (Fe^{3+}) can be present. Another example is nitrite—it seldom exists in appreciable concentration in aerobic water because its redox potential is less than that of dissolved oxygen.

Cell Voltage and Free-Energy Change

The relationship between voltage in cell reactions and the standard Gibbs free energy of reaction may be determined by the equation

$$\Delta G^0 = -nFE^0 \qquad (8.11)$$

where ΔG^0 = Gibbs standard free energy of reaction (kJ/mol), n = number of electrons transferred in the cell reaction, F = Faraday constant (96.485 kJ/volt-gram equivalent), and E^0 = voltage of the reaction in which all substances are at unit activity. The n and F values are multiplied by the voltage to convert the voltage to an energy equivalent. It follows that

$$\Delta G = -nFE \qquad (8.12)$$

where ΔG and E are for non-standard conditions.

Ex. 8.1 *The measured E^0 for the reaction in Eq. 8.1 is 0.62 volt (Table 8.1). It will be shown that Eq. 8.11 gives the same value of ΔG^0 as calculated by Eq. 4.8 using tabular values of ΔG_f^0.*

  Solution:
  From Eq. 8.11, ΔG^0 is

$$\Delta G^0 = -(2 \times 0.62 \text{ volt} \times 96.48 \text{ kJ/volt-gram equivalent}) = -119.6 \text{ kJ}.$$

The ΔG^0 may be calculated from ΔG_f^0 values with Eq. 4.8 as follows:

$$\Delta G^0 = 2\Delta G_f^0 H^+ + 2\Delta G_f^0 I^- - \Delta G_f^0 I_2 - \Delta G_f^0 H_2.$$

Using ΔG_f^0 values from Table 4.3 gives

$$\Delta G^o = 2(0) + 2(-51.56) - (16.43) - 0 = -119.6 \, kJ.$$

Thus, ΔG^o is the same whether calculated from the standard Gibbs free energy of reaction or from the standard electrode potential (Table 8.1).

Notice also, that the value of ΔG^o can be used in Eq. 8.11 to calculate E^o:

$$\Delta G^o = -nFE^o$$

$$-119.6 \, kJ = -(2)(96.485 \, kJ/volt\text{-}gram \; equivalent)(E^o)$$

$$E^o = \frac{119.6 \, kJ}{192.97 \, kJ/volt\text{-}gram \; equivalent}$$

$$E^o = 0.62 \, volt.$$

It was shown in Chap. 4 that the following equation may be used to estimate the free energy of reaction during a reaction until equilibrium is reached and $\Delta G = 0$ and $Q = K$:

$$\Delta G = \Delta G^o + RT \ln Q \tag{8.13}$$

where R = a form of the universal gas law (also called the ideal gas law) equation constant ($0.008314 \, kJ/mol/°A$) [$°A$ = absolute temperature ($273.15 + °C$)], and $\ln$ = natural logarithm, and Q = the reaction quotient.

Substituting Eqs. 8.11 and 8.12 into Eq. 8.13 gives.

$-nFe = -nFE^0 + RT \ln Q$, and by dividing through by $-nF$ we get

$$E = E^0 - \frac{RT}{nF} \ln Q. \tag{8.14}$$

Because $R = 0.008314 \, kJ/mol \cdot °A$; $T = 25 \, °C + 273.15°$; $F = 96.48 \, kJ/volt \cdot gram$ equivalent, and 2.303 converts from natural logarithms to common (base 10) logarithms, Eq. 8.14 can be written as

$$E = E^0 - \frac{0.0592}{n} \log Q. \tag{8.15}$$

Equations 8.14 and 8.15 are both forms of the Nernst equation; Eq. 8.15 is for 25 °C and Eq. 8.14 is for other temperatures. Just as with ΔG, when equilibrium is reached $E = 0.0$ volt and $Q = K$. The electrode potential E is the redox potential; it often is referred to as E_h.

Ex. 8.2 *The redox potential will be calculated using the Nernst equation for water at 25 °C with pH = 7 and a dissolved oxygen concentration of 8 mg/L.*
  Solution:

$$\frac{8\ mg/L\ O_2}{32,000\ mg/mol} = 10^{-3.60}\ M$$

$$(H^+) = 10^{-7}\ M\ at\ pH\ 7.$$

The appropriate reaction is

$$O_{2(aq)} + 4H^+ + 4e^- = 2H_2O$$

for which $E^0 = 1.27$ volt (Table 8.1)

$$E_h = E^0 - \frac{0.0592}{n}\log Q$$

where

$$\log Q = \frac{1}{(O_2)(H^+)^4} \qquad\qquad (H_2O\ is\ taken\ as\ unity).$$

so

$$E_h = 1.27 - \frac{0.0592}{4}\log\frac{1}{\left(10^{-3.6}\right)\left(10^{-28}\right)}$$

$$= 1.27 - \frac{0.0592}{4}\log\frac{1}{\left(10^{-31.6}\right)}$$

$$= 1.27 - (0.0148)(31.6) = 0.802\ volt.$$

  The pH has an effect on the redox potential, and E_h is often adjusted to its value at pH 7. To adjust E_h to pH 7, subtract 0.0592 volt for each pH unit below neutrality and add 0.0592 volt for each pH unit above neutrality. The origin of this correction factor is shown in Ex. 8.3. Many researchers adjust E_h values to pH 7 and report the redox potential using the symbol E_7 instead of E_h.

Ex. 8.3 *What is the E_h at pH 5, 6, 7, 8, and 9 for the reaction*

$$O_{2(aq)} + 4H^+ + 4e^- = 2H_2O$$

when dissolved oxygen is 8 mg/L ($10^{-3.60}M$) and water temperature is 25 °C?
    .

Solution:
The method of calculating E_h for the reaction was shown in Ex. 8.2:

$$E_h = 1.27 - \frac{0.0592}{4} \log \frac{1}{(10^{-3.6})(H^+)^4}.$$

$(H^+) = 10^{-5}, 10^{-6}, 10^{-7}, 10^{-8},$ and 10^{-9} M for pH 5, 6. 7, 8, and 9, respectively. Substituting into the preceding equation gives E_h values as follows: pH 5, 0.921 volt; pH 6, 0.862 volt; pH 7, 0.802 volt; pH 8, 0.743 volt; pH 9, 0.684 volt. Notice that E_h decreases by 0.059 volt for each unit increase in pH.

It can be seen from Eq. 8.14 that a change in temperature also causes Eh to change. However, Ex. 8.4 shows that the change caused by temperature (0.0016 volt/°C) is much less than caused by pH change. A pH change of 1 unit causes a pH to change about three-fold greater than does a 10 °C temperature change (compare Ex. 8.3 and Ex. 8.4). The factor 0.0016 volt/°C can be used to adjust E_h or E_7 to the standard temperature of 25 °C. This adjustment usually is not necessary, because modern redox instruments usually have a temperature compensator.

Ex. 8.4 What is the E_h at different temperatures for the reaction

$$O_{2(aq)} + 4H^+ + 4e^- = 2H_2O$$

at 8 mg/L dissolved oxygen and pH 7 for which log Q = 31.6?
Solution:
At 0 °C, the E_h is:

$$E_h = E° - \frac{(0.008314 \ kJ/mol)(273.15°A)}{(4)(96.48 \ kJ \ volt\text{-}gram \ equiv)} \times 2.303 \times 31.6$$

$$E_h = 0.842 \ volt.$$

Repeating the calculation for 0 °C for other temperatures gives:

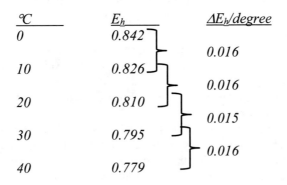

°C	E_h	ΔE_h/degree
0	0.842	
		0.016
10	0.826	
		0.016
20	0.810	
		0.015
30	0.795	
		0.016
40	0.779	

Practical Measurement of Redox Potential

Although the hydrogen electrode is a standard half-cell against which other half-cells are commonly compared, it is not used as the reference electrode in practical redox potential measurement. The most common reference electrode for redox measurement is the calomel (Hg_2Cl_2) electrode

$$Hg_2Cl_2 + 2e^- = 2Hg^+ + 2Cl^-. \qquad (8.16)$$

A KCl-saturated calomel electrode has $E^0 = 0.242$ volt at 25 °C. Standard electrode potentials for the popular KCl-saturated calomel electrode are 0.242 volt less than tabulated E^0 values of standard electrode potentials for the hydrogen electrode as shown in Ex. 8.5.

Ex. 8.5

E^o values from Table 8.1 will be adjusted to indicate potentials that would be obtained with a KCl-saturated calomel electrode.

$$(1) \quad O_{2(aq)} + 4H^+ + 4e^- = 2H_2O \qquad\qquad E^0 = 1.27 \ volt$$

$$(2) \quad Fe^{3+} + e^- = Fe^{2+} \qquad\qquad E^0 = 0.77 \ volt$$

$$(3) \quad NO_3^- + 2H^+ + 2e^- = NO_2^- + H_2O \qquad E^0 = 0.84 \ volt.$$

Solution:

Half-cell	E^0		Correction Factor	Calomel potential (volt)
(1)	1.27	–	0.242 =	1.028
(2)	0.77	–	0.242 =	0.528
(3)	0.84	–	0.242 =	0.598

As will be seen in other chapters, the redox potential is a useful concept in explaining many chemical and biological phenomena related to water quality. However, the practical use of redox potential in natural ecosystems is fraught with difficulty. In natural waters and sediments, redox potential is governed by dissolved oxygen concentration. Although oxygen is used up as it oxidizes reduced substances in the water, oxygen is continually replaced by diffusion of oxygen from the atmosphere or by oxygen produced in photosynthesis, and the redox potential usually remains fairly constant.

Measurement of redox potential is especially difficult in sediment. The redox potential often changes drastically across a few millimeters of sediment depth. The redox probe is rather large—about the size of a ballpoint pen—making it impossible to determine the precise depth in the sediment to which a redox reading applies. The act of inserting the probe in sediment also allows oxygenated water to follow the probe downward into the sediment changing the redox potential.

In the hypolimnion of stratified bodies of water, at the soil-water interface in unstratified water bodies, and in bottom soils and sediments, dissolved oxygen usually is at low concentration or absent, and reducing conditions develop. The typical pattern of redox potential in a water body containing dissolved oxygen throughout the water column is illustrated (Fig. 8.2). The redox potential is between 0.5 and 0.6 volt in the water column and in first centimeter of sediment. The redox potential then drops to less than 0.1 volt just below the sediment-water interface. The redox increases slightly at greater sediment depths possibly because the deeper sediment typically contains less organic matter.

The driving force causing a decrease in the redox potential is the consumption of oxygen by microbial respiration. When molecular oxygen is depleted, certain microorganisms utilize oxidized inorganic or organic substances as electron acceptors in metabolism. Redox reactions in reduced environments are very complicated, and while they occur spontaneously, microorganisms usually accelerate the process. There is a wide variety of compounds, concentrations of relatively reduced and relatively oxidized compounds vary greatly both spatially and temporally, and it is impossible to isolate specific components of the system. As a result, the best that can be done is to insert the redox probe into the desired place and obtain a reading. This reading can then be interpreted in terms of departure from the oxygen potential established by dissolved oxygen

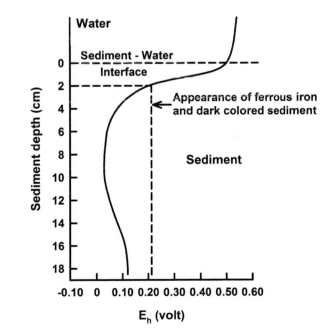

Fig. 8.2 Distribution of redox potential (E_h) in sediment with an oxidized surface

$$O_{2(aq)} + 4H^+ + 4e^- = 2H_2O. \tag{8.17}$$

At 25 °C and pH 7, the oxygen potential in well-oxygenated water should be about 0.802 volt (802 millivolts) against a standard hydrogen electrode (see Ex. 8.2). The oxygen potential of well-oxygenated surface water is about 0.560 volt (560 millivolt) when measured with a KCl-saturated calomel electrode. Redox potential drops in anaerobic waters or soils, and values below -0.250 volt have been observed in hypolimnetic waters and in sediments.

Commercially available devices for measuring redox potential do not all provide the same reading for the redox potential of well-oxygenated water or for a sample of reduced water or sediment. This makes it difficult to interpret redox potentials measured in natural waters and sediments. Nevertheless, a decrease in the redox potential below values obtained in oxygen-saturated water with a particular instrument indicates that reducing conditions are developing, and the reducing ability of the environment increases as the redox potential drops. Care must be taken not to introduce air or oxygenated water into the medium where redox potential is to be measured.

The presence of dissolved oxygen in water—even at low concentration—tends to stabilize the redox potential (Ex. 8.6). Thus, the oxygen potential (Eq. 8.17) is the standard for comparing redox potential among different places in a water body and its sediment.

Ex. 8.6 *The E_h of water at 25 °C and pH 7 will be calculated for dissolved oxygen concentrations of 1, 2, 4, and 8 mg/L.*

Solution:

The molar concentration of dissolved oxygen can be computed by dividing milligrams per liter values by 32,000 mg O_2/mol:1 mg/L $= 10^{-4.5}$ M; 2 mg/L $= 10^{-4.2}$ M; 4 mg/L $= 10^{-3.9}$ M; 8 mg/L $= 10^{-3.6}$ M. Use the molar concentrations of dissolved oxygen and a hydrogen ion concentration of 10^{-7} M (pH 7) to solve the expression

$$E_h = 1.27 - \frac{0.0592}{4} \log \frac{1}{(O_2)(H^+)^4}.$$

E_h values are as follows: 1 mg/L $= 0.789$ volt; 2 mg/L $= 0.793$ volt; 4 mg/L $= 0.798$ volt; 8 mg/L $= 0.802$ volt. The corresponding readings for a calomel electrode are 1 mg/L $= 0.547$ volt; 2 mg/L $= 0.551$ volt; 4 mg/L $= 0.556$ volt; 8 mg/L $= 0.560$ volt.

It can be seen from Ex. 8.6 that an amount of dissolved oxygen as low as 1 mg/L will maintain E_h near 0.8 volt (near 0.56 volt for a calomel electrode). Natural systems do not become strongly reducing until dissolved oxygen is entirely depleted.

Corrosion

The standard electrode potential provides a way of predicting metal corrosion. In corrosion, the metal is oxidized to its ionic form with release of electrons (the anode process), and the electrons are transferred to water, oxygen, or other oxidants (the cathode process) as illustrated in a simplified assessment of the corrosion of iron metal by aerobic water (Fig. 8.3). The overall reaction is

$$2Fe^0 + O_2 + 2H_2O \rightarrow 2Fe^{2+} + 4OH^-. \tag{8.18}$$

The anode process is

$$2Fe^0 \rightarrow 2Fe^{2+} + 4e^- \tag{8.19}$$

while the cathode process is

$$O_2 + 2H_2O + 4e^- \rightarrow 4OH^-. \tag{8.20}$$

In presence of oxygen, Fe^{2+} will precipitate as iron oxide on the surface of the corroding metal.

The potential for corrosion ($E_{corrosion}$) may be assessed by the difference between the E^0 of Eq. 8.20 and the E^0 of Eq. 8.19:

$$E_{corrosion} = E_C^0 - E_A^0$$

where subscripts C and A indicate cathode and anode processes, respectively.

$$E_{corrosion} = +0.40 - (-0.44) = +0.84 \text{ volt}.$$

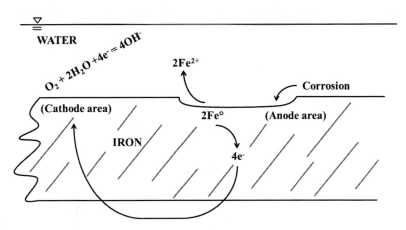

Fig. 8.3 Illustration of the process of metal corrosion

The positive value for $E_{corrosion}$ indicates that the reaction is possible and that corrosion will occur under the specific conditions. Of course, this simple approach does not indicate the degree of corrosion that may be expected.

The corrosive potential of water is increased by lower pH, and high concentrations of dissolved oxygen, carbon dioxide, and dissolved solids. Acidic waters have more hydrogen ion to facilitate the cathode reaction, oxygen reacts with hydrogen gas at the cathode speeding up the reaction, carbon dioxide reduces pH, and higher dissolved solids concentration increases electrical conductivity. Bacteria action and presence of oxidants such as nitrate or chlorine also can accelerate corrosion.

A common way of reducing corrosion of methods is to put a corrosion resistant coating on the metal to protect it from contact with the environment, e.g., zinc or tin coating on steel, resins, plastics, paints, or greases. Another way is to reduce the corrosiveness of the environment, e.g., removal of dissolved oxygen from water. There also are electrochemical means of lessening corrosion that will not be discussed.

Conclusions

The redox potential is directly related to dissolved oxygen concentration. When there is dissolved oxygen at concentrations above 1 or 2 mg/L, the redox potential will be high. The main value of the redox potential in aquatic ecology is for explaining how oxidations and reductions occur in sediment-water systems. The redox potential is difficult to measure, so it is not generally a useful variable in water quality criteria for aquatic ecosystems. Of course, concepts from electrochemistry discussed above in relation to the redox potential are the basis for understanding and controlling corrosion of metal structures and devices so essential to human society. Redox potential also has far-reaching importance in analytical chemistry and many industrial processes.

Carbon Dioxide, pH, and Alkalinity

9

Abstract

The pH or negative logarithm of the hydrogen ion concentration is a master variable in water quality because the hydrogen ion influences many reactions. Because dissolved carbon dioxide is acidic, rainwater that is saturated with this gas is naturally acidic—usually about pH 5.6. Limestone, calcium silicate, and feldspars in soils and other geological formations dissolve through the action of carbon dioxide to increase the concentration of bicarbonate in water and raise the pH. The total concentration of titratable bases—usually bicarbonate and carbonate—expressed in milligrams per liter of calcium carbonate is the total alkalinity. Total alkalinity typically is less than 50 mg/L in waters of humid areas with highly leached soils, but it is greater where soils are more fertile, limestone formations are present, or the climate is arid. Alkalinity increases the availability of inorganic carbon for photosynthesis in waters of low to moderate pH. Water bodies with moderate to high alkalinity are well-buffered against wide daily swings in pH resulting from net removal of carbon dioxide by photosynthesis during daytime and return of carbon dioxide to the water by respiratory process at night when there is no photosynthesis. The optimum pH range for most aquatic organisms is 6.5–8.5, and the acid and alkaline death points are around pH 4 and pH 11, respectively. Fish and other aquatic animals avoid high carbon dioxide concentration, but 20 mg/L or more can be tolerated if there is plenty of dissolved oxygen. The optimum alkalinity for aquatic life is between 50 and 150 mg/L.

Introduction

The pH is a master variable in water quality because many reactions are pH dependent. Normal waters contain both acids and bases and biological processes tend to increase either acidity or basicity. The interactions among these opposing acidic and basic substances and processes determine pH. Carbon dioxide is particularly influential in regulating the pH of natural water. This dissolved gas is acidic, and its concentration is in continual flux, because it is removed from water by aquatic plants for use in photosynthesis and released to the water by respiration of aquatic organisms. The alkalinity of water results primarily from bicarbonate and carbonate ions derived from the reaction of aqueous carbon dioxide with limestone and certain other minerals.

Alkalinity plays an important role in aquatic biology by increasing the availability of inorganic carbon for plants. It also tends to buffer water against excessive pH change. A basic knowledge of the relationships among pH, carbon dioxide, and alkalinity is essential to an understanding of water quality.

pH

The Concept

The hydrogen ion (H^+) is a naked nucleus, a proton with a high charge density. A proton cannot exist in water, because it is attracted to the negatively-charged side of water molecules creating an ion pair known as the hydronium ion (H_3O^+). The dissociation of water can be written as $H_2O + H_2O = H_3O^+ + OH^-$, but for simplicity H_3O^+ is replaced by H^+ and the dissociation of water expressed as

$$H_2O = H^+ + OH^-. \tag{9.1}$$

The pH is an index of the hydrogen ion concentration (H^+), and it is based on the mass action expression of Eq. 8.1

$$\frac{(H^+)(OH^-)}{(H_2O)} = K_w. \tag{9.2}$$

The equilibrium constant (K) for water is usually written as K_w. Water dissociates only slightly and can be considered unity in Eq. 9.2 giving

$$(H^+)(OH^-) = K_w. \tag{9.3}$$

Water dissociates into equal numbers of hydrogen and hydroxyl (OH^-) ions; hydrogen ion can be substituted into Eq. 9.3 to obtain

$$(H^+)(H^+) = K_w.$$

The hydrogen ion concentration is the square root of K_w

$$(H^+) = \sqrt{K_w}. \tag{9.4}$$

Of course, the hydroxyl ion concentration is related to hydrogen ion concentration

$$(OH^-) = \frac{K_w}{(H^+)}. \qquad (9.5)$$

In the early 1900s, the Danish chemist, S. P. L. Sørensen recommended taking the negative logarithm of hydrogen ion concentration to provide a convenient way of referring to small molar concentrations of this ion. The negative logarithm of the hydrogen ion concentration was called pH:

$$pH = -\log_{10}(H^+). \qquad (9.6)$$

The K_w for water at 25 °C is $10^{-14.00}$ (Table 9.1); the pH is

$$pH = -\log_{10}(10^{-7.00}) = -(-7.00) = 7.00.$$

Pure water is neutral—neither acidic nor basic in reaction—because it has equal concentrations of hydrogen ion and hydroxyl ion. The negative log concept can be applied to the hydroxyl ion concentration. The pOH is the negative log of the OH^- concentration. The pOH equals 14 – pH at 25 °C or pK_w – pH at other temperatures. The pOH, however, is not frequently used in discussions of water quality.

It is common practice to refer to a pH scale of 0–14 so that pH 7 is the midpoint. Acidic reaction intensifies as pH declines below 7.0; basic reaction intensifies as pH rises above 7.0. The pH of pure water is 7.00 only at 25 °C where $K_w = 10^{-14}$. The neutral point increases at lower temperature and decreases at higher temperature (Table 9.1). It is possible to have negative pH values and pH values above 14 as well. At 25 °C, a solution that is 2M ($10^{0.3}$ M) in hydrogen ion has a pH of -0.3; a 2M hydroxyl solution has a pH of 14.3.

Calculations

It is easy to estimate pH from solutions with hydrogen ion concentration such as 0.01M (10^{-2} M) or 0.000001M (10^{-6} M), because the negative logs of 10^{-2} and 10^{-5} M are 2 and 6, respectively. Hydrogen ion concentration usually is not an exact multiple of ten, and the conversion is more involved (Ex. 9.1).

Table 9.1 Ionization constants (K_w), molar concentrations of hydrogen ion (H^+), and pH of pure water at different temperatures. The molar concentrations of hydroxyl ion (OH^-) are the same as (H^+)

Water temperature (°C)	K_w	(H^+)	pH
0	$10^{-14.94}$	$10^{-7.47}$	7.47
5	$10^{-14.73}$	$10^{-7.36}$	7.36
10	$10^{-14.53}$	$10^{-7.26}$	7.26
15	$10^{-14.35}$	$10^{-7.18}$	7.18
20	$10^{-14.17}$	$10^{-7.08}$	7.08
25	$10^{-14.00}$	$10^{-7.00}$	7.00
30	$10^{-13.83}$	$10^{-6.92}$	6.92
35	$10^{-13.68}$	$10^{-6.84}$	6.84
40	$10^{-13.53}$	$10^{-6.76}$	6.76

Ex. 9.1 *The pH of water 0.002M in hydrogen ion concentration will be calculated.*
  Solution:

$$0.002M = 2 \times 10^{-3} \, M$$

$$pH = -log_{10}(H^+) = -\left(log_{10}2 + log_{10}10^{-3}\right) = -[0.301 + (-3)] = 2.70.$$

The pH scale is logarithmic; a solution of pH 6 has a ten times greater hydrogen ion concentration than does a water of pH 7. This brings us to the question of how to average pH values. The pH of a solution made by mixing 1 L each of solutions with pH 2, 4, and 6 is not 4; it is 2.47 as illustrated in Ex. 9.2.

Ex. 9.2 *The pH resulting from mixing 1-L each of solutions with pH values of 2, 4, and 6 will be calculated.*
  Solution:

pH	mol H+ in 1.0 L
2	0.01
4	0.0001
6	0.000001
Sum	0.010101 mol H+ in 3.0 L

$$Average = \frac{0.010101 \, mol \, H^+}{3 \, L} = 3.37 \times 10^{-3} \, mol \, H^+/L$$

$$pH = -log_{10}\left(3.37 \times 10^{-3}\right) = -[0.53 + (-3)] = 2.47.$$

Averaging pH directly would give an erroneous pH of 4.0.

The illustration of pH averaging (Ex. 9.2) appears to be mathematically correct and totally obvious. However, while the method is obligatory for calculating the resulting average pH of mixing solutions of different pH, the method is incorrect in most water quality applications (Boyd et al. 2011). The averaging of hydrogen ion concentration will not provide the correct average pH, because buffers present in natural waters have a greater effect on final pH than does dilution alone. When data sets are transformed to hydrogen ion to estimate average pH, extreme pH values will distort the average pH. Values of pH conform more closely to a normal distribution than do hydrogen ion concentrations, making pH values more acceptable for use in statistical analyses. Electrochemical measurements of pH and many biological responses to hydrogen ion concentration are described by the Nernst equation that states that the measured or observed response is linearly related to 10-fold changes in hydrogen ion concentration. Moreover, and possibly most important, is that in most applications, the relationships between pH and other variables are reported as the observed effect of measured pH—not H+ concentration—on the response of the other variables. The use of direct pH averaging is definitely correct in most water quality applications, so please be confident in yourself when rebuked for doing so.

pH of Natural Waters

The technique for measuring pH with a glass electrode was commercialized 1936, and the measurement of environmental pH is common. There have been millions of pH measurements in all kinds of natural waters, and values as low as 1 and as high as 13 have been reported.

Waters in humid areas with highly leached soils tend to have lower pH than waters in areas with limestone formations or those in semi-arid or arid regions. But, most natural waters have pH within the 6 to 9 range.

Waters that have high concentrations of humic substances tend to have a low pH because of acidic groups on the humic molecules as illustrated by the equation below in which R represents the organic (humic) moiety:

$$RCOOH = RCOO^- + H^+. \tag{9.7}$$

For example, waters in rivers flowing from South American jungles may have pH values of 4–4.5. Low pH from organic acids is not as harmful to aquatic organisms as low pH caused by strong, mineral acids. Moreover, species living in rivers flowing from jungle areas are more tolerant to low pH than organisms in other waters.

Carbon Dioxide

The atmosphere contains an average carbon dioxide concentration of roughly 0.040%, and raindrops become saturated with carbon dioxide while falling to the ground. The equilibrium concentrations of carbon dioxide in water of different temperatures and salinity were provided in Table 7.8.

Dissolved carbon dioxide reacts with water forming carbonic acid (H_2CO_3)

$$CO_2 + H_2O \rightleftharpoons H_2CO_3 \quad K = 10^{-2.75}. \tag{9.8}$$

Carbonic acid is a diprotic acid

$$H_2CO_3 \rightleftharpoons H^+ + HCO_3^- \quad K_1 = 10^{-3.6} \tag{9.9}$$

$$HCO_3^- \rightleftharpoons H^+ + CO_3^{2-} \quad K_2 = 10^{-10.33}. \tag{9.10}$$

The second dissociation may be ignored here, because, as explained later, it is not significant at pH less than 8.3.

Only a small fraction of dissolved carbon dioxide reacts with water. Using Eq. 9.8, it can be seen that the ratio (H_2CO_3):(CO_2) will be 0.00178:1:

$$\frac{(H_2CO_3)}{(CO_2)} = 10^{-2.75} = 0.00178.$$

The carbonic acid concentration at equilibrium will be only 0.18% of the carbon dioxide concentration. Although carbonic acid is a strong acid, little of it is present in water at equilibrium with atmospheric carbon dioxide. Moreover, common analytical procedures cannot distinguish carbonic acid from carbon dioxide.

A dilemma is avoided by considering the analytical carbon dioxide as a weak acid and deriving an apparent reaction and equilibrium constant for the acidic reaction of carbon dioxide in water. Multiplication of the equilibrium expressions of Eqs. 9.8 and 9.9 gives

$$\frac{(H_2CO_3)}{(CO_2)} \times \frac{(HCO_3^-)(H^+)}{(H_2CO_3)} = 10^{-2.75} \times 10^{-3.6}$$

$$\frac{(HCO_3^-)(H^+)}{(CO_2)} = 10^{-6.35}.$$

The apparent equilibrium expression derived above typically is used to explain the reaction of carbon dioxide with water as given below:

$$CO_2 + H_2O \rightleftharpoons H^+ + HCO_3^- \qquad K = 10^{-6.35}. \tag{9.11}$$

In Chap. 4, it was illustrated that the K for any reaction can be estimated from ΔG° using Eq. 4.11. The K for Eq. 9.11 estimated from ΔG° is $10^{-6.356}$—essentially the same value derived above by combining Eqs. 9.8 and 9.9 through multiplication.

The pH of rainwater not contaminated with acids stronger than carbon dioxide is around 5.6, and carbon dioxide normally will not lower the pH of water below 4.5. The two points will be illustrated in Exs. 9.3 and 9.4.

Ex. 9.3 *The pH of rainwater saturated with carbon dioxide at 25 °C will be estimated.*

Solution:
From Table 7.8, the carbon dioxide concentration in rainwater is 0.57 mg/L ($10^{-4.89}$M). Using Eq. 9.11,

$$\frac{(H^+)(HCO_3^-)}{(CO_2)} = 10^{-6.35}$$

$$but \quad (H^+) = (HCO_3^-)$$

$$and \quad (H^+)^2 = (10^{-6.35})(CO_2).$$

$$Thus, \quad (H^+)^2 = (10^{-6.35})(10^{-4.89}) = 10^{-11.24}$$

$$(H^+) = 10^{-5.62}$$

$$pH = 5.62.$$

Rainwater with a lower pH contains an acid stronger than carbon dioxide as discussed in Chap. 10.

Ex. 9.4 *Water from a well contains 100 mg/L ($10^{-2.64}$ M) carbon dioxide. The pH will be determined using Eq. 9.11.*
   Solution:
   As shown above in Ex. 9.3,

$$(H^+)^2 = (10^{-6.35})(10^{-2.64})(CO_2).$$

For a water with $10^{-2.64}$ M CO_2, we have $(H^+)^2 = (10^{-6.35})(10^{-2.64}) = 10^{-9.99}$,

$$(H^+) = 10^{-4.50}$$

$$pH = 4.5.$$

Although carbon dioxide concentrations up to 650 mg/L have been reported in natural water bodies (Stone et al. 2018), few natural waters contain more than 100 mg/L of carbon dioxide. It follows that a pH below 4.5 suggests that a water contains a stronger acid than carbon dioxide. In surface waters, the most common strong acid is sulfuric acid that originates from deposits of iron pyrite that may be exposed to the air and oxidize resulting in sulfuric acid (see Chap. 11). Rainwater may contain sulfuric and other strong acids as a result of air pollution.

Natural waters may contain more carbon dioxide than suggested by equilibrium concentrations provided in Table 7.8. This is because most waters contain more bicarbonate than that resulting from the reaction of carbon dioxide and water (Eq. 9.11). There is an equilibrium among H^+, HCO_3^-, and CO_2, and at a given temperature and pH, as the HCO_3^- concentration increases, the water can hold more CO_2 at equilibrium. For example, the equilibrium CO_2 concentrations at 25 °C and pH 7 in water with 61 mg/L (10^{-3} M) HCO_3^- would be 9.85, but it would be 27.8 mg/L in water with 122 mg/L ($10^{-2.55}$ M) HCO_3^-. In pure water, the equilibrium concentration at 25 °C is 0.57 mg/L (Table 7.8).

Total Alkalinity

The total alkalinity of water is defined as the total concentration of titratable bases expressed as calcium carbonate ($CaCO_3$). Calcium carbonate is the basis for expressing alkalinity, because it is commonly added to increase the pH and alkalinity of acidic water. Calcium carbonate also is the substance that precipitates from some waters when they are boiled as will be discussed in Chap. 10.

Examination of Eq. 9.11 reveals that bicarbonate results from the reaction of carbon dioxide with water. At equilibrium with atmospheric carbon dioxide, the concentrations of H^+ and HCO_3^- in unpolluted rainwater or other relatively pure water are both $10^{-5.62}$ M (Ex. 9.3). This molar concentration of HCO_3^- is equal to 0.144 mg/L. The ratio of equivalent weights of $CaCO_3$ to HCO_3^- is 50/61, and the total alkalinity concentration would be 0.118 mg/L as $CaCO_3$. At a carbon dioxide concentration of 100 mg/L (Ex. 9.4) the total alkalinity in pure water would be only 1.58 mg/L as $CaCO_3$. Such low concentrations are essentially undetectable analytically, and carbon dioxide is not a direct source of alkalinity. Moreover, $(H^+) = (HCO_3^-)$ in pure water, and the acidity of the water (see Chap. 11) would be the same as the alkalinity.

Carbon dioxide increases the dissolution of limestone and certain other minerals in soils and other geological formations imparting bicarbonate (and carbonate) to natural waters. Concentrations of total alkalinity usually range between 5 and 300 mg/L in freshwaters. Although alkalinity often results almost entirely from bicarbonate and carbonate, some waters—especially polluted ones—contain appreciable amounts of hydroxide, ammonia, phosphate, borate, or other bases that contribute alkalinity. The reactions of selected bases in neutralizing acidity are illustrated below:

$$OH^- + H^+ \rightleftharpoons H_2O$$

$$CO_3^{2-} + H^+ \rightleftharpoons HCO_3^-$$

$$HCO_3^- + H^+ \rightleftharpoons CO_2 + H_2O$$

$$NH_3 + H^+ \rightleftharpoons NH_4^+$$

$$PO_4^{3-} + H^+ = HPO_4^-$$

$$HPO_4^- + H^+ \rightleftharpoons H_2PO_4^-$$

$$H_2BO_4^- + H^+ \rightleftharpoons H_3BO_4$$

$$H_3SiO_4^- + H^+ \rightleftharpoons H_4SiO_4$$

$$RCOO^- + H^+ \rightleftharpoons RCOOH.$$

Bicarbonate and carbonate usually comprise most of the alkalinity in natural waters, but in some cases, other ions may contribute significantly to alkalinity. For example, Snoeyink and Jenkins (1980) presented an example from a major water supply to the Bay Area in California that originates in the Sierra Nevada Mountains where rock formations are rich in silicate minerals. The water has 20 mg/L alkalinity of which 4 mg/L is the result of silicic acid.

Alkalinity sometimes is described by the expression

$$\text{Alkalinity} = \left(HCO_3^-\right) + 2\left(CO_3^{2-}\right) + \left(OH^-\right) - \left(H^+\right) \qquad (9.12)$$

where the ions are entered in moles/L.

Ex. 9.5 *The alkalinity of a water sample with pH = 9 ($H^+ = 10^{-9}$ M; $OH^- = 10^{-5}$ M) that contains 61 mg/L HCO_3^- (10^{-3} m) and 30 mg/L CO_3^{2-} (5×10^{-4} M) will be calculated.*

Solution:

$Alkalinity = \left(10^{-3}\ M\ HCO_3^-\right) + 2\left(5 \times 10^{-4}\ M\ CO_3^{2-}\right) + \left(10^{-5}\ M\ OH^-\right) - \left(10^{-9}\ M\ H^+\right)$
$= (0.001\ M + 0.001\ M + 0.00001\ M) - (0.000000001\ M)$
$= 0.00201\ M.$

It would require 0.00201 M of H^+ to neutralize 0.00201 M alkalinity, and molarity and normality are the same for H^+.

Alkalinity is expressed as equivalent $CaCO_3$ and it takes 2 mol H^+ to neutralize 1 mol of $CaCO_3$, e.g., $CaCO_3 + 2H^+ = Ca^{2+} + CO_2 + H_2O$ (1 mol $CaCO_3$ = 2 equivalents $CaCO_3$).

And, 0.00201 eq H^+ × 50.04 g $CaCO_3$/eq = 0.10058 g/L (100.58 mg/L) alkalinity as $CaCO_3$.

Measurement of Alkalinity

The expression (Eq. 9.12) illustrates the concept of alkalinity, but it does not provide a practical way of estimating the alkalinity of natural waters. Total alkalinity is determined by titration of a water sample with standard sulfuric or hydrochloric acid (Eaton et al. 2005). The measurement traditionally was done by titration with 0.020 N acid to the methyl orange endpoint—the color of methyl orange changes from yellow to faint orange at pH 4.4. However, the pH in the water sample at the end of the titration is influenced by the amount of carbon dioxide produced in the reaction between acid and bicarbonate ($HCO_3^- + H^+ \rightarrow CO_2 + H_2O$), and samples of higher alkalinity will be at a lower pH when bicarbonate has been neutralized than will samples of lower alkalinity.

In the situation where all of the alkalinity in a sample is from bicarbonate, the maximum amount of carbon dioxide will be liberated per milligram per liter of alkalinity during titration. Assuming total alkalinity concentrations of 30, 150, and 500 mg/L and using the ratio $2HCO_3^-:CaCO_3$ (122:100), the carbon dioxide at the titration endpoint (assuming no loss by diffusion to the air) would be 36.6, 183, and 610 mg/L or $10^{-3.08}$, $10^{-4.36}$, and $10^{-4.10}$ M, respectively. With Eq. 9.11, it can be shown the potential pH at the endpoint would be 4.72, 4.36, and 4.10, for 30, 150, and 500 mg/L total alkalinity respectively. Of course, some carbon dioxide is lost to the air during titration. Taras et al. (1971) recommended the following endpoint pHs for different alkalinities: <30 mg/L, 5.1; 30–500 mg/L, 4.8; >500 mg/L, 4.5. Eaton et al. (2005) revised these recommendations to pHs of 4.9, 4.6, and 4.3 at alkalinities of 30, 150, 500 mg/L, respectively—roughly 0.2 pH units more than potential pHs calculated above.

Mixed bromocresol green-methyl red indicator can be used to detect pH 5.1 and 4.8, while methyl orange can be used to signal pH 4.5, but many analysts prefer to use a pH electrode to detect the endpoint. For routine titrations, pH 4.5 often is used as the endpoint for samples of all concentrations of total alkalinity.

The measurement and calculation of total alkalinity are illustrated in Ex. 9.6.

Ex. 9.6 A 100-mL water sample is titrated with 0.0210 N sulfuric acid to an endpoint of pH 4.5. The titration consumes 18.75 mL of sulfuric acid. The total alkalinity will be calculated.

Solution:

The milliequivalent quantity of alkalinity or $CaCO_3$ equals the milliequivalent amount of acid used in the titration:

$$(18.75 \ mL)(0.0210 \ N) = 0.394 \ meq \ CaCO_3. \qquad (Note : N = meq/mL)$$

The weight of equivalent $CaCO_3$ is

$$(0.394 \ meq)(50.04 \ mg \ CaCO_3/meq) = 19.72 \ mg \ CaCO_3.$$

To convert to milligrams per liter,

$$19.72 \ mg \ CaCO_3 \times \frac{1000 \ mL/L}{100 \ mL \ sample} = 197.2 \ mg/L.$$

In water samples with pH <8.3, the alkalinity is typically considered to be from bicarbonate alone. Samples with a pH >8.3 usually contain both carbonate and bicarbonate, but if the pH is especially high, bicarbonate may be absent. Such samples may contain carbonate only, carbonate and hydroxide, or hydroxide only. Phenolphthalein indicator is pink above pH 8.3 and colorless at lower pH and may be used in samples with pH >8.3 to separate the total alkalinity titration into two steps: determination of the volume of acid needed to decrease pH to 8.3 and continuation of the titration to pH 4.5 (or other selected endpoint) to determine the total volume of acid needed to neutralize all base in the sample. The first step

Table 9.2 Relationships of the three kinds of alkalinity to measurements by acid titration of total alkalinity (TA) and phenolphthalein alkalinity (PA)

Titration results	Kind of alkalinity		
	Bicarbonate	Carbonate	Hydroxide
PA $= 0$	TA	0	0
PA <0.5TA	TA $-$ 2PA	2PA	0
PA $= 0.5$TA	0	2PA (or TA)	0
PA >0.5 TA	0	2 (TA $-$ PA)	2PA $-$ TA
PA $=$ TA	0	0	TA (or PA)

provides an estimate of the phenolphthalein alkalinity, and of course, the result of the entire titration gives the total alkalinity. The phenolphthalein alkalinity (PA) total alkalinity (TA) may be used to calculate the three possible forms of alkalinity: bicarbonate, carbonate, and hydroxide.

There are five possible cases (Table 9.2) with respect to the relationship between phenolphthalein alkalinity and total alkalinity titration volumes:

1. PA $= 0$. There is only bicarbonate, and the titration is simply $HCO_3^- + H^+ = CO_2 + H_2O$.
2. PA <0.5 TA. There are two reactions. In the first step, the reaction is $CO_3^{2-} + H^+ = HCO_3^-$, and HCO_3^- produced from CO_3^{2-} mixes with HCO_3^- already in the sample. The second step is the neutralization of all of the bicarbonate. The carbonate alkalinity is 2PA and the bicarbonate alkalinity is TA $-$ 2PA.
3. PA $= 0.5$TA. A sample exhibiting this result contains only carbonate, and the second part of the titration involves the neutralization of HCO_3^- formed by conversion of carbonate to bicarbonate in the first step. The carbonate alkalinity $= $ 2PA or TA.
4. PA >0.5 TA. This result reveals the presence of hydroxide and the absence of bicarbonate. In the first step of the titration, the reactions are $OH^- + H^+ = H_2O$ and $CO_3^{2-} + H^+ = HCO_3^-$. The bicarbonate produced in the first step is neutralized in the second step. The hydroxide alkalinity is 2PA $-$ TA, while the carbonate alkalinity is 2(TA $-$ PA). There is no bicarbonate alkalinity.
5. PA $=$ TA. This type of sample contains only hydroxide alkalinity and will usually have pH $>$12. The pH will fall to 4.5 or less when the hydroxide is neutralized. The sample has no pH buffering capacity once hydroxide is neutralized, and the pH drop from around 10 to about 4.5 will be essentially vertical in the titration curve (Fig. 9.1).

The five alkalinity relationships are summarized in Table 9.2.

Sources of Alkalinity

The major source of alkalinity for many waters is limestone, and calcium carbonate commonly is used in the equation depicting the dissolution of limestone

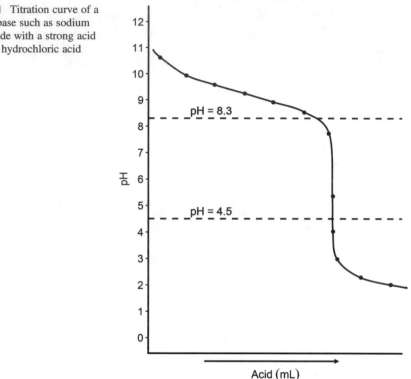

Fig. 9.1 Titration curve of a strong base such as sodium hydroxide with a strong acid such as hydrochloric acid

$$CaCO_3 = Ca^{2+} + CO_3^{2-}. \tag{9.13}$$

Several values of K_{sp} for Eq. 9.13 have been determined experimentally, but the most appropriate appears to be $10^{-8.3}$ (Akin and Lagerwerff 1965). This K_{sp} agrees well with the K_{sp} of $10^{-8.34}$ estimated from the ΔG^{o} of Eq. 9.13.

The equilibrium concentration of carbonate in a closed calcium carbonate-pure water system (free of carbon dioxide) calculated using K_{sp} $CaCO_3 = 10^{-8.3}$ is 4.25 mg/L (7.08 mg/L total alkalinity). This estimate is not exactly correct, because CO_3^{2-} reacts with H^+ forming bicarbonate (reverse reaction in Eq. 9.10). This allows more calcium carbonate to dissolve and increases OH^- concentration relative to H^+ concentration increasing the pH above neutral. Nevertheless, the total alkalinity concentration calculated by the K_{sp} of Eq. 9.13 will be far below the measured alkalinity in an open calcium carbonate-water system in equilibrium with atmospheric carbon dioxide.

Carbon dioxide reacts with limestone as illustrated below for both calcitic limestone ($CaCO_3$) and dolomitic limestone ($CaCO_3 \cdot MgCO_3$):

$$CaCO_3 + CO_2 + H_2O \rightleftharpoons Ca^{2+} + 2HCO_3^- \tag{9.14}$$

$$CaCO_3 \cdot MgCO_3 + 2CO_2 + 2H_2O \rightleftharpoons Ca^{2+} + Mg^{2+} + 4HCO_3^-. \qquad (9.15)$$

Most limestone is neither calcite nor dolomite, but a mixture in which the ratio of $CaCO_3{:}MgCO_3$ usually is greater than unity. Limestones are of low solubility, but their dissolution is increased greatly by the action of carbon dioxide.

The mass action expression for the solubility of calcium carbonate in the presence of carbon dioxide (see Eq. 9.14) is

$$\frac{\left(Ca^{2+}\right)\left(HCO_3^-\right)^2}{\left(CO_2\right)} = K$$

The same expression and the value of K may be derived by combining the inverse of the mass action form of Eq. 9.10 and the mass action forms of Eqs. 9.11 and 9.13 by multiplication as follows:

$$\frac{\left(HCO_3^-\right)}{\left(H^+\right)\left(CO_3^{2-}\right)} \times \frac{\left(H^+\right)\left(HCO_3^-\right)}{\left(CO_2\right)} \times \left(Ca^{2+}\right)\left(CO_3^{2-}\right) = \frac{1}{10^{-10.33}} \times 10^{-6.35} \times 10^{-8.3}.$$

In arranging the expression above, it was necessary to use the inverse of the mass action form of Eq. 9.10 in order to have (H^+) and $(CO_3{}^{2-})$ in the denominator. Upon multiplication, the terms reduce to

$$\frac{\left(Ca^{2+}\right)\left(HCO_3^-\right)^2}{\left(CO_2\right)} = 10^{-4.32}.$$

Revealing a K of $10^{-4.32}$ for Eq. 9.14 which agrees well with a K of $10^{-4.35}$ estimated using the ΔG^o of the equation. The K for Eq. 9.15 may be obtained as illustrated for Eq. 9.14.

The calcium and bicarbonate concentrations at any carbon dioxide concentration may be estimated using the mass action form of Eq. 9.14 as illustrated in Ex. 9.7 for the current atmospheric carbon dioxide level of about 400 ppm.

Ex. 9.7 *The concentration of calcium and bicarbonate at equilibrium between solid phase calcite and distilled water in equilibrium with atmospheric carbon dioxide at 25 °C will be calculated.*

Solution:
From Table 7.8, freshwater contains 0.57 mg/L ($10^{-4.89}$ M) carbon dioxide at equilibrium with atmospheric carbon dioxide concentration. From Eq. 9.14, each mole of $CaCO_3$ that dissolves results in one mole of Ca^{2+} and two moles of $HCO_3{}^-$. Thus, we may set $(Ca^{2+}) = X$ and $(HCO_3{}^-) = 2X$ and substitute these values into the mass action form of Eq. 9.14 to obtain

$$\frac{(X)(2X)^2}{10^{-4.89}} = 10^{-4.32}$$

and

$$4X^3 = 10^{-9.21} = 6.17 \times 10^{-10}$$

$$X^3 = 1.54 \times 10^{-10} = 10^{-9.81}$$

$$X = 10^{-3.27} M = 5.37 \times 10^{-4} M.$$

Thus, $Ca^{2+} = \left(5.37 \times 10^{-4} M\right)\left(40.08 \ g \ Ca^{2+}/mole\right) = 0.0215 g/L = 21.5 \ mg/L$

and $HCO_3^- = \left(5.37 \times 10^{-4} M\right)(2)\left(61 \ g \ HCO_3^-/mole\right) = 0.0655 \ g/L$
$$= 65.5 \ mg/L$$

or 53.6 *mg/L total alkalinity.*

The calcium and bicarbonate concentrations calculated in Ex. 9.7 agree quite well with concentrations of 22.4 mg/L calcium and 67.1 mg/L bicarbonate determined experimentally for a carbonate dioxide concentration of 0.04% of total atmospheric gases by Frear and Johnston (1929). The bicarbonate concentration calculated in Ex. 9.7 is equivalent to 54 mg/L total alkalinity. Of course, limestone seldom exists as pure calcium carbonate in form of calcite. The other forms have different solubilities, some are more soluble than calcite (Sá and Boyd 2017).

Examination of Eq. 9.14 reveals that half of the carbon in bicarbonate comes from calcium carbonate and half is from carbon dioxide. Twice as much soluble carbon results from each molecule of calcium carbonate that dissolves as occurs in a closed system (no carbon dioxide). Moreover, the removal of carbon dioxide from the water in the reaction with calcium carbonate allows more atmospheric carbon dioxide to enter the water. This effect is limited, because when equilibrium is reached, the dissolution of calcium carbonate stops. The upshot is that the action of carbon dioxide on calcium carbonate (or other alkalinity source) provides a means of increasing both the solubility of the calcium carbonate or other alkalinity source and the capture of atmospheric carbon dioxide.

Equation 9.14 also may be used to illustrate the effect of the increasing atmospheric carbon dioxide concentration on the solubility of calcite. In the 1960s, the atmospheric carbon dioxide concentration was about 320 ppm at the Mauna Loa Observatory, Hawaii, but today, the concentration measured at that observatory is slightly over 400 ppm. Using 320 and 400 ppm of atmospheric carbon dioxide, the dissolved carbon dioxide concentration at equilibrium in freshwater at 25 °C has increased from about 0.42 mg/L in the 1960s to 0.57 mg/L today. This has resulted in a corresponding increase in total alkalinity in a calcite-water-air system at 25 °C from 48 mg/L to about 54 mg/L—a 12.5% increase in alkalinity from a 25% increase in atmospheric carbon dioxide (Somridhivej and Boyd 2017). Of course, the solubilities of the other sources of alkalinity also increase with greater atmospheric carbon dioxide concentration.

The carbon dioxide concentration in water of meteoric origin tends to increase during infiltration through the soil and underlying formations. This results because hydrostatic pressure increases with depth allowing the water to hold more carbon dioxide at equilibrium. Moreover, there often is abundant carbon dioxide in soil because of respiration by plant roots and soil microorganisms. The solubility of minerals that are sources of alkalinity often will be considerably greater in the soil and underlying geological formations than at the land surface. Groundwater usually is higher in alkalinity than surface water.

In water bodies, decomposition of organic matter also produces carbon dioxide and waters may be supersaturated with carbon dioxide. This phenomenon can result in waters having greater total alkalinity concentrations than expected from calculations based on equilibrium carbon dioxide concentrations.

The dissolution of limestone and other sources of alkalinity will increase at greater concentrations of dissolved ions, because solubility products are based on activities rather than measured molar concentrations (Garrels and Christ 1965). Activities are estimated by multiplying an activity coefficient by the measured molar concentration. If the ionic strength of a solution increases, the activity coefficients of the ions in the solution decrease. Thus, more of the mineral must dissolve to attain equilibrium at greater ionic strength of the solution, i.e., the measured molar concentration of ions from the dissolving mineral must increase because the activity coefficients of these ions decrease. Calcite—like other minerals—dissolves to an increasing extent as ionic strength of the water increases.

Reactions involving formation of bicarbonate from carbonate are equilibrium reactions and a certain amount of carbon dioxide must be present to maintain a given amount of bicarbonate in solution. If the amount of carbon dioxide at equilibrium is increased or decreased, there will be a corresponding change in the concentration of bicarbonate. Additional carbon dioxide will result in the solution of more calcium carbonate and a greater alkalinity, while removal of carbon dioxide will result in precipitation of calcium carbonate and a lower alkalinity. When well water or spring water of high alkalinity and enriched with carbon dioxide contact the atmosphere, calcium carbonate often precipitates in response to equilibration with atmospheric carbon dioxide.

Texts on water quality and limnology often unintentionally leave the reader with the impression that limestone is the only source of alkalinity in water. Carbon dioxide also reacts with calcium silicate and other silicates such as the feldspars olivine and orthoclase shown below to give bicarbonate:

$$CaSiO_3 + 2CO_2 + 3H_2O \rightarrow Ca^{2+} + 2HCO_3^- + H_4SiO_4$$
$$\text{Calcium silicate} \tag{9.16}$$

$$Mg_2SiO_4 + 4CO_2 + 4H_2O \rightleftharpoons 2Mg^{2+} + 4HCO_3^- + H_4SiO_4$$
$$\text{Olivine} \tag{9.17}$$

$$2KAlSi_3O_8 + 2CO_2 + 11H_2O \rightleftharpoons Al_2Si_2O_5(OH)_4 + 4H_4SiO_4 + 2K^+ + 2HCO_3^-.$$
Orthoclase

$$(9.18)$$

Calcium silicate is a major source of alkalinity in natural waters (Ittekkot 2003). Weathering of feldspars also is an important source of alkalinity in areas with acidic soils that do not contain limestone or calcium silicate.

The Ks for Eqs. 9.16, 9.17, and 9.18 can be derived from their $\Delta G°$s. These K values will not be estimated because the effect of carbon dioxide on dissolution of mineral sources of alkalinity has been adequately demonstrated. The actual alkalinity of a water body should be determined, because the sources of alkalinity differ from one place to another and are often unknown making accurate calculation impossible. Natural waters often are not at equilibrium with atmospheric carbon dioxide or the minerals that contribute alkalinity. Alkalinity concentrations were calculated in examples above only to promote an understanding the processes that impart alkalinity to natural waters.

Equilibria

The aqueous dissolution products of calcium carbonate are calcium and bicarbonate (Eq. 9.14), but a very small amount of carbon dioxide and carbonate will be present. Carbon dioxide is present because it is in equilibrium with bicarbonate (Eq. 9.11) while carbonate results from dissociation of bicarbonate (Eq. 9.10). A bicarbonate solution at equilibrium contains carbon dioxide, bicarbonate, and carbonate. Carbon dioxide is acidic, and carbonate is basic because it hydrolyzes:

$$CO_3^{2-} + H_2O = HCO_3^- + OH^-.$$

$$(9.19)$$

The hydrolysis reaction also can be written as

$$CO_3^{2-} + H^+ = HCO_3^-.$$

$$(9.20)$$

Because the hydrogen ion comes from the dissociation of water, hydroxyl ion concentration rises to maintain the equilibrium constant. Equation 9.20 is the reverse reaction of Eq. 9.10. In Eq. 9.10, bicarbonate is an acid and the equilibrium constant for the forward reaction can be referred to as K_a. In the reverse reaction, carbonate acts as a base and the equilibrium constant for the basic reaction is called K_b. In such reactions, $K_aK_b = K_w$ or 10^{-14}. The equilibrium constant for the reverse reaction of Eq. 9.10 (or the forward reaction of Eq. 9.20) is therefore

$$K_b = 10^{-14} - 10^{-10.33} = 10^{-3.67}.$$

Thus, the strength of carbonate as a base ($K_b = 10^{-3.67}$) is greater than the strengths of either carbon dioxide ($K = 10^{-6.35}$) or bicarbonate ($K_a = 10^{-10.33}$) as acids. As a result, a bicarbonate solution at equilibrium will be basic.

We can determine the pH of a weak bicarbonate solution at equilibrium with calcium carbonate and atmospheric carbon dioxide by combining the mass action expressions for Eq. 9.11 and Eq. 9.10 through multiplication

$$\frac{(H^+)(HCO_3^-)}{(CO_2)} \times \frac{(H^+)(CO_3^{2-})}{(HCO_3^-)} = 10^{-6.35} \times 10^{-10.33}.$$

Because bicarbonate concentration is the same in both expressions and carbon dioxide and carbonate are of negligible concentration when bicarbonate is at maximum concentration, the overall expression reduces to

$$(H^+)^2 = 10^{-16.68}$$

$$(H^+) = 10^{-8.34}.$$

The calculation above shows that for practical purposes water above pH 8.3 does not contain free carbon dioxide, and that carbonate begins to appear at pH 8.3.

In order to see what happens when carbon dioxide is withdrawn from a bicarbonate solution at equilibrium, we can combine the reactions of bicarbonate acing as a base and as an acid:

$$HCO_3^- + H^+ = H_2O + CO_2 \qquad (as\ base\text{---}neutralizes\ H^+)$$

$$(+) \qquad HCO_3^- = H^+ + CO_3^{2-} \qquad (as\ acid\text{---}releases\ H^+$$

to obtain

$$2HCO_3^- = CO_2 + CO_3^{2-} + H_2O. \qquad (9.21)$$

When carbon dioxide is removed, carbonate will increase, and hydrolysis of carbonate will cause pH to increase (Eqs. 9.19 and 9.20). Most natural waters contain calcium, and the pH rise will be moderated because of precipitation of calcium carbonate in response to the increasing carbonate concentration.

During photosynthesis, aquatic plants remove carbon dioxide to cause the pH to rise. Free carbon dioxide is depleted once pH rises above 8.3, but most aquatic plants can use bicarbonate as a carbon source (Korb et al. 1997; Cavalli et al. 2012). The removal of carbon dioxide from bicarbonate by aquatic plants involves the enzyme carbonic anhydrase, but the exact mechanism has not been elucidated entirely. The overall affect is illustrated in Eq. 9.21, and it shows that bicarbonate concentration will diminish while carbonate concentration will increase. Carbonate accumulating in the water will hydrolyze to produce bicarbonate and hydroxyl ion (Eq. 9.19), but two bicarbonates are removed for each carbonate formed (Eq. 9.21). Moreover, the hydrolysis reaction only converts a portion of the carbonate to bicarbonate. In waters with pH > 8.3, carbonate and hydroxyl ion concentrations increase in water as photosynthesis proceeds, and the pH rises. Of course, depending upon the calcium concentration, the rise in pH will be limited by the precipitation of calcium carbonate.

Aquatic plants that tolerate very high pH continue using bicarbonate—including that formed by the hydrolysis of carbonate—resulting in accumulation of hydroxyl ion in the water and extremely high pH (Ruttner 1963). This effect is particularly common in waters of low calcium concentration where the anions are balanced mainly by potassium and magnesium (Mandal and Boyd 1980).

Some aquatic plants accumulate calcium carbonate on their surfaces as a result of using bicarbonate as a carbon source in photosynthesis. An excellent example is species of the macroalgal genus *Chara*—called stoneworts because of their tendency to become encrusted by calcium carbonate. Samples of *Chara* have been reported to contain as much as 20% of dry weight as calcium (Boyd and Lawrence 1966). Most macroalgae contain 1 or 2% of dry weight as calcium, and the *Chara* samples likely were encrusted with an amount of calcium carbonate equal to about half of the dry weight.

The precipitation of calcium carbonate as the result of photosynthetic removal of carbon dioxide by phytoplankton often causes a phenomenon known as whitening of lakes, streams, or ponds (Thompson et al. 1997). Calcium carbonate forms a fine precipitate that does not settle rapidly, and the water takes on a milky or whitish appearance. The calcium carbonate precipitate dissolves at night when carbon dioxide concentration increases, but in some waters, calcium carbonate precipitates on the bottom and becomes mixed with the sediment. The resulting sediment of high calcium carbonate concentration is called marl (Pentecost 2009).

The interdependence of pH, carbon dioxide, bicarbonate, and carbonate is illustrated in Fig. 9.2. The graph shows that below about pH 5, carbon dioxide is the only significant species of inorganic carbon. Above pH 5, the proportion of bicarbonate increases relative to carbon dioxide until bicarbonate becomes the only significant species at about pH 8.3. Above pH 8.3, carbonate appears and it increases in importance relative to bicarbonate if pH continues to rise.

The amount of inorganic carbon available to aquatic plants depends upon pH and alkalinity. At the same alkalinity, as pH rises, the amount of carbon dioxide decreases, but at the same pH, as alkalinity rises, carbon dioxide concentration increases. Of course, aquatic plants use bicarbonate and even carbonate as carbon sources. Thus, it is difficult to estimate the actual amount of inorganic carbon available to plants in a particular water body. Saunders et al. (1962) provided a nomograph of converting total alkalinity (mg/L) to available carbon (mg/L) a portion of which is provided in Table 9.3. This nomograph probably provides a reasonable estimate of available inorganic carbon.

Buffering

Buffers are substances that allow water to resist pH change. A buffer consists of a mixture of a weak acid and its conjugate base (salt) or a weak base and its conjugate acid. An acidic buffer can be made from acetic acid and its conjugate base sodium acetate, while an alkaline buffer can be made from ammonium hydroxide and its conjugate acid ammonium chloride.

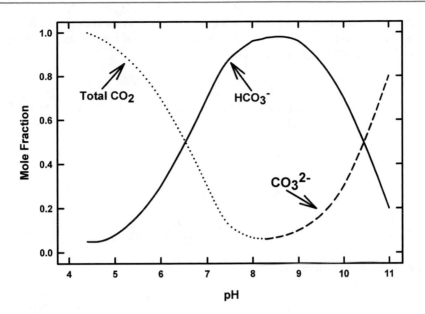

Fig. 9.2 Effects of pH on the relative proportions of total CO_2, HCO_3^-, and CO_3^{2-}. The mole fraction of a component is its decimal fraction of all the moles present

Table 9.3 Factors for converting total alkalinity (mg/L as $CaCO_3$) to carbon available to aquatic plants (mg/L as C). Multiply factors by total alkalinity

pH	Temperature (°C)					
	5	10	15	20	25	30[a]
5.0	8.19	7.16	6.55	6.00	5.61	5.20
5.5	2.75	2.43	2.24	2.06	1.94	1.84
6.0	1.03	0.93	0.87	0.82	0.78	0.73
6.5	0.49	0.46	0.44	0.42	0.41	0.40
7.0	0.32	0.31	0.30	0.30	0.29	0.29
7.5	0.26	0.26	0.26	0.26	0.26	0.26
8.0	0.25	0.25	0.25	0.24	0.24	0.24
8.5	0.24	0.24	0.24	0.24	0.24	0.24
9.0	0.23	0.23	0.23	0.23	0.23	0.23

[a]Estimated by extrapolation

A buffer made of acetic acid (CH_3COOH) and sodium acetate (CH_3COONa) will be used for illustration. The initial pH is determined by the dissociation of acetic acid

$$CH_3COOH = CH_3COO^- + H^+ \qquad K = 10^{-4.74}. \tag{9.22}$$

If more H^+ is added, it combines with CH_3COO^- to form CH_3COOH and resist pH change. If OH^- is added, it reacts with H^+ to form H_2O, CH_3COOH dissociates providing CH_3COO^-, and pH remains fairly constant. Sodium acetate is a source of CH_3COO^- in excess of that possible from the dissociation of CH_3COOH alone to increase buffering capacity.

The pH of a buffer can be calculated by aid of an equation derived from the equilibrium expression of a weak acid (HA) in a solution containing its conjugate base (A^-) as illustrated below for a weak acid:

$$HA = H^+ + A^- \qquad (9.23)$$

where HA = a weak acid such as acetic acid and A^- = the conjugate base of the weak acid such as sodium acetate. The equilibrium constant (K_a) for the weak acid is

$$K_a = \frac{(H^+)(A^-)}{(HA)}. \qquad (9.24)$$

Taking the negative $\log_{10}$ of both sides of Eq. 9.24 gives

$$-\log_{10}K_a = -\log_{10}\left[\frac{(H^+)(A^-)}{(HA)}\right]$$

which may be rearranged as

$$-\log K_a = -\log(H^+) - \log\left[\frac{(A^-)}{(HA)}\right]. \qquad (9.25)$$

By substituting the equivalents pK for $-\log K_a$ and pH for $-\log$ pH, Eq. 9.25 becomes

$$pK_a = pH - \log\left[\frac{(A^-)}{(HA)}\right]$$

$$or \quad pK_a + \log\left[\frac{(A^-)}{(HA)}\right] = pH$$

which may be rearranged to give

$$pH = pK_a + \log_{10}\left[\frac{(A^-)}{(HA)}\right]. \qquad (9.26)$$

Equation 9.26 is known as the Henderson-Hasselbalch equation. It was named after the American biochemist, L. J. Henderson, who first described the relationship of equilibrium constants of buffers to their pH values, and the Danish biochemist, K. A. Hasselbalch, who put Henderson's equation into logarithmic form.

The Henderson-Hasselbalch equation reveals that the weak acid and its conjugate base buffer the solution at a pH near the pK_a of the weak acid. Moreover, the pH of the solution will be determined by the ratio of the concentration of the conjugate base:the concentration of the weak acid. When $(A^-) = (HA)$ in Eq. 9.26, the ratio $(A^-)/(HA) = 1.0$, log $1.0 = 0$, and pH = pK_a. When H^+ is added HA is formed, but more A^- is available from the conjugate base to minimize pH change. Adding OH^- results in removal of H^+, i.e., $OH^- + H^+ = H_2O$, but more H^+ is available from the weak acid to minimize pH change.

The pH of a buffer changes as H^+ or OH^- is added, but the changes are relatively small as shown in Ex. 9.8 for addition of OH^- to a buffer.

Ex. 9.8 *The pH change in a buffer ($pK_a = 7.0$) that is 0.1 M in HA and A^- will be calculated for the addition of 0.01 mol/L and 0.05 mol/L of OH^-.*

Solution:

For simplicity, it will be assumed that there is no change in buffer volume as a result of the OH^- addition. The addition of 0.01 mol/L OH^- will reduce (AH) to 0.09 M and increase (A^-) to 0.11 M, while the 0.05 mole/L increase in OH^- will result in concentrations of 0.15 M A^- and 0.05 M AH.

The pH values will be

0.01 mol/L OH^- added:

$$pH = 7.0 + log\left[\frac{0.11}{0.09}\right]$$

$$pH = 7.0 + log1.222 = 7.0 + 0.087 = 7.09.$$

0.05 mole/L OH^- added:

$$pH = 7.0 + log\left[\frac{0.15}{0.05}\right]$$

$$pH = 7.0 + log3 = 7.0 + 0.48 = 7.48.$$

It can be seen from Ex. 9.8 that the addition of hydroxyl ion changed the pH of the buffer. However, the increase in pH was much less than would have occurred had the same quantity of OH^- been added to pure water at pH 7 as illustrated in Ex. 9.9.

Ex. 9.9 *The effect of OH^- additions of 0.01 mol/L and 0.05 mol/L on the pH of pure water initially at pH 7 will be calculated.*

Solution:

0.01 mol/L OH^- added:

$$10^{-7}\,OH^- + 10^{-2}\,mol/L\,OH^- = 0.0000001\,M + 0.01\,M \approx 10^{-2}\,M$$

$$pOH = -log\left(10^{-2}\right) = -log\left(^{-2}\right) = 2$$

$$pH = 14 - 2 = 12.$$

0.05 mol/L OH^- added:

$$10^{-7}\,OH^- + 0.05\,mol/L\,OH^- = 10^{-1.30}\,M\,OH^-$$

$$(H^+) = 10^{-12.7} M$$

$$pH = 12.7.$$

Carbon dioxide, bicarbonate, and carbonate buffer waters against changes in pH. This can be readily seen by comparing the shape of the titration curves of sodium hydroxide solution (no buffering capacity) with the titration of a natural water sample containing alkalinity. The pH drops abruptly in the sodium hydroxide solution (Fig. 9.1), but gradually in the water sample (Fig. 9.3). Water with low alkalinity will exhibit a greater pH fluctuation during a 24-hour period as a result of fluctuations in carbon dioxide concentration caused by photosynthesis and respiration than will water of greater alkalinity (Fig. 9.4).

At pH below 8.3, the buffering action of alkalinity results because added hydrogen ion reacts with bicarbonate to form carbon dioxide and water so that the pH changes only slightly. A small addition of hydroxyl ion will reduce the hydrogen ion concentration, but carbon dioxide and water react to form more hydrogen ion, thereby minimizing change in pH. The buffer system in natural water for pH below 8.3 may be expressed in terms of the Henderson-Hasselbalch equation as follows:

$$pH = 6.35 + \log_{10} \frac{(HCO_3^-)}{(CO_2)} \qquad (9.27)$$

where 6.35 is the negative logarithm—or pK—of the reaction constant for Eq. 9.11. Notice that in terms of buffers, carbon dioxide is the acid and the bicarbonate ion is the salt or conjugate base.

Fig. 9.3 Titration of a sample of natural water with hydrochloric acid

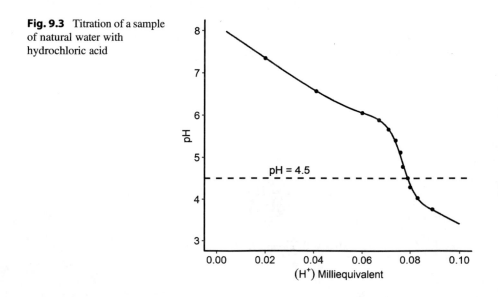

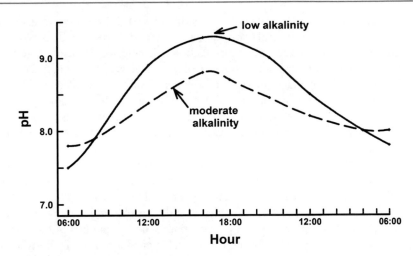

Fig. 9.4 Daily pH changes in waters of low or moderate alkalinity in response to a high rate of phytoplankton photosynthesis

At pH 8.34 and above, when hydrogen ion is added, it will react with carbonate ion to form bicarbonate. Adding of hydroxide results in a reaction with bicarbonate to form carbonate and water. Putting Eq. 9.10 into the Henderson-Hasselbalch equation form gives

$$pH = 10.33 + \log_{10} \frac{(CO_3^{2-})}{(HCO_3^-)} \qquad (9.28)$$

Alkalinity and pH of Natural Waters

Total alkalinity concentrations for natural waters may range from 0 mg/L to more than 500 mg/L. One total alkalinity scheme for classifying concentration is given below:

Less than 10 mg/L   —   Very low
10–50 mg/L   —   Low
50–150 mg/L   —   Moderate
150–300 mg/L   —   High
More than 300 mg/L   —   Very high

Waters of moderate to high total alkalinity often are associated with limestone deposits in watershed soils. For example, total alkalinities of waters from different physiographic regions of the Coastal Plain region of Alabama and Mississippi ranged from 2 to 200 mg/L. The lower values consistently were associated with sandy soils and the higher values were for areas where soils contain free calcium carbonate (Boyd and Walley 1975). Livingstone (1963) did not present alkalinity data in his survey of the chemical composition of the world's rivers and lakes, but his bicarbonate data allow inference about alkalinity. Livingstone's data suggest that areas with weakly-developed or highly-leached soils have very low or low alkalinity, humid regions with abundant limestone in watershed soils have moderate alkalinity, and areas with semi-arid or arid climate have moderate to high alkalinity waters. Seawater has an average total alkalinity of 116 mg/L. Alkalinity is very high in closed basin lakes. Waters from places with organic soils tend to have very low or low alkalinity.

The alkalinity of water is closely related to pH. In general, pH tends to increase as alkalinity increases. Waters that do not contain measurable alkalinity have pH values below 5, those with very low or low alkalinity tend to have pH values between 5 and 7, pH usually is between 7 and 8.5 in moderate and high alkalinity waters, and waters of very high alkalinity may have pH of 9 or more. The average pH of the ocean is about 8.1. The influence of photosynthesis on pH is greater in low alkalinity waters because of their low buffering capacity. In waters of very low to moderate alkalinity, a dense phytoplankton bloom may cause afternoon pH to rise above 9. At night, pH will usually decline to a much lower value in such waters.

Along coastal plains in several countries including the United States, there are areas where water percolates through calcium carbonates deposit and accumulates moderate to high concentrations of calcium and bicarbonate. This water then enters an aquifer that contains a lot of sodium from having been saturated with seawater in the geological past. Calcium in the water recharging the aquifer is exchanged for sodium in the geological matrix of the formation. The water removed from the aquifer in wells will be high in alkalinity and sodium, but low in calcium. Hem (1970) referred to this process as natural softening of groundwater. When such water is used to fill fish ponds, an extremely high pH that may be toxic to fish can result from the effects of photosynthesis (Mandal and Boyd 1980).

Importance of Alkalinity System

pH

The pH of water is important in aquatic ecosystems, because it affects aquatic life. Gill tissue is the primary target organ in fish affected by excessively low pH. When fish are exposed to low pH, the amount of mucus on gill surfaces increases. Excess mucus interferes with the exchange of respiratory gases and ions across the gill. Therefore, failure in blood acid-base balance resulting in respiratory stress and decreasing blood concentrations of sodium chloride which cause osmotic disturbance are the dominant physiologic symptoms of acid stress. Of course, at low pH, aluminum ion concentration increases in water, and sometimes, toxic effects of aluminum may occur in addition to pH effects.

Gill damage in fish also can occur in alkaline solutions (high pH). Mucus cells at the base of the gill filaments become hypertrophic and the gill epithelium separates

Table 9.4 Effects of pH on fish and other aquatic life

pH	Effects
4	Acid death point
4-5	No reproduction[a]
4-6.5	Slow growth of many species[a]
6.5-9	Optimum range
9–11	Slow growth and adverse reproductive effects
11	Alkaline death point

[a]Some fish in rivers flowing from jungles do very well at low pH

from the pilaster cells. Gill damage contributes to problems with respiration and blood acid-base balance. Damage to the lens and cornea of the eye also occurs in fish at high pH (Boyd and Tucker 2014).

The relationship of pH to aquatic animals is summarized in Table 9.4. The acid and alkaline death points are approximately pH 4 and pH 11. Waters with pH values ranging from about 6.5 to 9 are most suitable for aquatic life. In fish and some other aquatic animals, reproduction diminishes as pH declines below 6.5. Most authorities suggest that aquatic ecosystems should be protected from acidic or basic pollutants so that pH remains between 6.5 and 8.5 or 9.0. Of course, many waters may naturally be lower in pH than 6.5 and a few may be higher than 9.0 in pH.

Drinking waters should have pH of 6.5 to 8.5. Also, waters with excessively low or high pH will be corrosive. Where water supplies are acidic, lime (calcium oxide, or calcium hydroxide) may be used to increase pH and minimize pipe corrosion.

Carbon Dioxide

High concentrations of carbon dioxide have a narcotic effect on fish and even higher concentrations may cause death. Environmental concentrations are seldom high enough to elicit narcotic effects or death; the usual effect is a respiratory one. Carbon dioxide must leave a fish or invertebrate by diffusion from gills, and high external concentrations in the surrounding water decrease the rate of loss of carbon dioxide. Thus, carbon dioxide accumulates in the blood and depresses blood pH causing detrimental effects. More importantly, high carbon dioxide concentrations interfere with the loading of hemoglobin in fish blood with oxygen. This results in an elevation of the minimum tolerable dissolved oxygen concentration. In this regard, when dissolved oxygen concentrations in natural waters are low, carbon dioxide concentrations almost always are high.

Fish can sense small differences in free carbon dioxide concentrations and apparently attempt to avoid areas with high carbon dioxide concentrations. Nevertheless, 10 mg/L or more of carbon dioxide may be tolerated provided dissolved oxygen concentration is high. Most species will survive in waters containing up to 60 mg/L of free carbon dioxide. Water supporting good fish populations normally contained less than 5 mg/L of free carbon dioxide (Ellis 1937). In low alkalinity, eutrophic waters, free carbon dioxide typically fluctuates from 0 mg/L in the afternoon to 5 or 10 mg/L at daybreak with no obvious ill effect on aquatic organisms.

Carbon dioxide is necessary in photosynthesis, but its concentration is closely related to pH and total alkalinity. At the same pH, a water with a greater alkalinity will contain more available carbon for photosynthesis than a water with lesser alkalinity. But, at the same alkalinity, the availability of free carbon dioxide (gaseous form) will decrease as pH increases.

Alkalinity

The alkalinity of water is an important variable as productivity is related to alkalinity because of the relationship between alkalinity and carbon availability. Waters with total alkalinity values of 0 to 50 mg/L usually are less productive than those with total alkalinity concentrations of 50 to 200 mg/L (Moyle 1946). Productivity tends to decrease at higher alkalinities. Liming sometimes may be used to increase the alkalinity of water.

Alkalinity has a great influence on the use of water. Waters with a high carbonate hardness (high alkalinity and high hardness) will be problematic in causing deposits when heated. Total alkalinities up to 400–500 mg/L are permissible in public water supplies.

Conclusions

The relationships among carbon dioxide, bicarbonate, and carbonate concentrations control the pH of most natural waters. Carbon dioxide has an acidic reaction in water. Plants normally remove more carbon dioxide from water than is returned by respiration during daylight, and pH rises. Carbon dioxide becomes depleted at pH 8.3, but many aquatic plants can use bicarbonate as an inorganic carbon source for photosynthesis. Plants use one-half of the carbon in bicarbonate, and the other half is excreted into the water as carbonate. The carbonate hydrolyzes causing pH to continue rising. Photosynthesis stops at night and carbon dioxide from respiration results in a decline in pH.

Dissolution of limestone, calcium silicate, and feldspars is the source of alkalinity in natural waters, and dissolved carbon dioxide enhances the solubilities of these minerals. Alkalinity usually consists primarily of bicarbonate which is in equilibrium with carbon dioxide and hydrogen ion. Alkalinity is a reserve of inorganic carbon, and the CO_2-HCO_3^--CO_3^{2-} system acts to buffers water. The buffering capacity increases with greater alkalinity.

References

Akin GW, Lagerwerff JV (1965) Calcium carbonate equilibria in aqueous solution open to the air. I. The solubility of calcite in relation to ionic strength. Geochimica et Cosmochimica 29:343–352

Boyd CE, Lawrence JM (1966) The mineral composition of several freshwater algae. Proc Ann Conf SE Assoc Game Fish Comm 20:413–424

Boyd CE, Tucker CS (2014) Handbook for aquaculture water quality. Craftmaster Printers, Auburn

Boyd CE, Walley WW (1975) Total alkalinity and hardness of surface waters in Alabama and Mississippi. Bulletin 465, Alabama Agricultural Experiment Station, Auburn University.

Boyd CE, Tucker CS, Viriyatum R (2011) Interpretation of pH, acidity and alkalinity in aquaculture and fisheries. N Am J Aqua 73:403–408

Cavalli G, Riis T, Baattrup-Pedersen A (2012) Bicarbonate use in three aquatic plants. Aqua Bot 98:57–60

Eaton AD, Clesceri LS, Rice EW, Greenburg AE (eds) (2005) Standard methods for the examination of water and wastewater. American Public Health Association, Washington, DC

Ellis MM (1937) Detection and measurement of stream pollution. U.S. Bur Fish Bull 22:367–437

Frear CL, Johnston J (1929) The solubility of calcium carbonate (calcite) in certain aqueous solutions at 25 °C. J Am Chem Soc 51:2082–2093

Garrels RM, Christ CL (1965) Solutions, minerals, and equilibria. Freeman, Cooper, and Company, San Francisco

Hem JD (1970) Study and interpretation of the chemical characteristics of natural water. Water-Supply Paper 1473, United States Geological Survey, United States Government Printing Office, Washington, DC.

Ittekkot V (2003) A new story from the Ol' Man River. Science 301:56–58

Korb RE, Saville PJ, Johnston AM, Raven JA (1997) Sources of inorganic carbon for photosynthesis by three species of marine diatom. J Phycol 33:433–440

Livingstone DA (1963) Chemical composition of rivers and lakes. Professional Paper 440-G, United States Government Printing Office, Washington, DC.

Mandal BK, Boyd CE (1980) The reduction of pH in water of high total alkalinity and low total hardness. Prog Fish-Cult 42:183–185

Moyle JB (1946) Some indices of lake productivity. Trans Am Fish Soc 76:322–334

Pentecost A (2009) The marl lakes of the British Isles. Freshwater Rev 2:167–197

Ruttner F (1963) Fundamentals of limnology. University of Toronto Press, Toronto

Sá MVC, Boyd CE (2017) Variability in the solubility of agricultural limestone from different sources and its pertinenece for aquaculture. Aqua Res 48:4,292–4,299

Saunders GW, Trama FB, Bachmann RW (1962) Evaluation of a modified C-14 technique for shipboard estimation of photosynthesis in large lakes. Publication No. 8, Great Lakes research Division, University of Michigan, Ann Arbor.

Snoeyink VL, Jenkins D (1980) Water chemistry. John wiley and Sons, New York

Somridhivej B, Boyd CE (2017) likely effects of the increasing alkalinity of inland waters on aquaculture. J World Aqua Soc 48:496–502

Stone DM, Young KL, Mattes WP, Cantrell MA (2018) Abiotic controls of invasive nonnative fishes in the Little Colorado River, Arizona. Am Midl Nat 180:119–142

Taras MJ, Greenberg AE, Hoak RD, Rand MC (1971) Standard methods for the examination of water and wastewater, 13th edn. American Public Health Association, Washington, DC

Thompson JB, Schultzepam S, Beveridge TJ, DesMarais DJ (1997) Whiting events: biogenic origin due to the photosynthetic activity of cyanobacterial picoplankton. Lim Ocean 42:133–141

Total Hardness

10

Abstract

The total hardness of water results from divalent cations—mainly from calcium and magnesium—expressed as equivalent calcium carbonate. The total hardness equivalence of 1 mg/L calcium is 2.5 mg/L, while 1 mg/L magnesium equates to 4.12 mg/L. Hardness and alkalinity often are similar in concentration in waters of humid regions, but hardness frequently exceeds alkalinity in waters of arid regions. Hardness generally is less important than alkalinity as a biological factor, but it is quite important in water supply and use. High concentrations of calcium and magnesium in water containing appreciable alkalinity lead to scale formation when the water is heated or its pH increases. This leads to clogging of water pipes and scale accumulation in boilers and on heat exchangers. The Langelier Saturation Index often is used to determine if water has potential to cause scaling. Divalent ions also precipitate soap increasing soap use for domestic purposes and in commercial laundries. The traditional method for softening water is to precipitate calcium as calcium carbonate and magnesium as magnesium hydroxide by the lime-soda ash process. Water also may be softened by passing it through a cation exchange medium such as zeolite.

Introduction

The rather strange designations of waters as soft or hard apparently originated from the observation that waters from some sources imparted deposits in pipes and in vessels in which they were boiled. Waters which caused the deposits were called hard as opposed to soft waters that did not produce such deposits. Hard waters also leave spots on drinking glasses, plumbing fixtures, and other objects after drying. Moreover, hard waters do not produce copious lather with soap, a failure resulting from precipitation of soap by minerals in the water and formation of the familiar bathtub ring.

The degree of hardness in water obviously is important to water use. Hard waters can clog pipes, form scales in boilers, waste soap, and cause unwanted spots and stains on various fixtures and household items. Hardness also is closely related to alkalinity and the amount of hardness and the ratio of hardness to alkalinity influences other water quality variables and biological processes in natural waters.

This chapter describes hardness, its measurements, and its effect on water quality and water use.

Definition and Sources

Total hardness is defined as the total concentration of divalent cations in water expressed as equivalent calcium carbonate. If a water contains 10 mg/L calcium, the resulting hardness is 25 mg/L as $CaCO_3$, because calcium has an atomic weight of 40 as opposed to a molecular weight of 100 for calcium carbonate. The ratio 100/40 multiplied by calcium concentration gives hardness as equivalent calcium carbonate.

The divalent cations—calcium, magnesium, strontium, ferrous iron, and manganous manganese—cause hardness in water. Surface waters usually are oxygenated and will not contain divalent iron or manganese. Few inland waters contain more than 1 or 2 mg/L of strontium, and calcium and magnesium are the major sources of hardness in nearly all surface waters including the ocean. Groundwaters usually are anaerobic and contain reduced iron and manganese (Fe^{2+} and Mn^{2+}) which contribute hardness, but hardness from iron and manganese is lost when these two ions precipitate when water removed from aquifers becomes oxygenated.

As with alkalinity, the dissolution of limestone, calcium silicate, and feldspars is a source of hardness. As discussed in Chap. 9, carbon dioxide in rainwater accelerates the dissolution of limestone, calcium silicate, and feldspars, and this capacity increases as water infiltrates downward dissolving carbon dioxide released by root and microbial respiration. Some aquifers are actually contained in underground limestone formations, and groundwater from such aquifers usually has high hardness. The acquisition of hardness (calcium and magnesium) by the dissolution of dolomitic limestone, calcium silicate, and feldspars is illustrated in Eqs. 9.14, 9.15, 9.16, 9.17, and 9.18. In addition, when gypsum ($CaSO_4 \cdot 2H_2O$) and certain magnesium salts common in soils of arid regions dissolve, they impart calcium and magnesium to the water. In arid regions, ions in water are concentrated by evaporation increasing hardness.

Because hardness often is derived from the dissolution of limestone or calcium silicate, concentrations of calcium and magnesium are nearly equal in chemical equivalence to concentrations of bicarbonate and carbonate in many natural waters. However, in areas with acidic soils, neutralization of alkalinity may result in greater hardness than alkalinity. In arid regions, concentration of ions by evaporation causes precipitation of alkalinity as calcium carbonate causing alkalinity to decline relative to hardness. Alkalinity, which also is expressed as calcium carbonate equivalent, sometimes exceeds hardness in groundwaters that have been naturally softened (see Chap. 9).

Determination of Hardness

As mentioned above, hardness is reported as its calcium carbonate equivalent, and each milligram per liter of calcium is equal to 2.5 mg/L of hardness as calcium carbonate. Similar calculations reveal that the hardness factors are 4.12 for magnesium, 1.82 for manganese, 1.79 for iron, and 1.14 for strontium. If the ionic composition of a water is known, the total hardness may be calculated from the concentrations of divalent ions as illustrated in the next example.

Ex. 10.1 *A water sample contains 50 mg/L of calcium, 5 mg/L of magnesium, and 0.5 mg/L of strontium. The total hardness will be calculated.*
   Solution:
   50 mg Ca^{2+}/L × 2.50 = 125 mg/L of $CaCO_3$.
   5 mg Mg^{2+}/L × 4.12 = 20.6 mg/L of $CaCO_3$.
   0.5 Mg Sr^{2+}/L × 1.14 = 0.57 Mg/L of CaCO3.
   Total hardness = (125 + 20.6 + 0.57) = 146.17 mg/L as $CaCO_3$.

Rather than calculating hardness from divalent cation concentrations, it is more common to measure the concentration of total hardness by titration of samples with the chelating agent ethylenediaminetetraacetic acid (EDTA) as described by Eaton et al. (2005). This titrating agent forms stable complexes with divalent cations as illustrated for calcium:

$$Ca^{2+} + EDTA = Ca \cdot EDTA. \tag{10.1}$$

Each molecule of EDTA complexes one divalent metal ion, and the endpoint is detected with the indicator eriochrome black T. The indicator combines with a small amount of calcium in the sample and holds it rather tightly creating a wine-red color. When all Ca^{2+} in the water sample has been chelated, the EDTA strips the Ca^{2+} from the indicator and the solution turns blue. The endpoint is very sharp and distinct. Each mole of EDTA equals 1 mol of $CaCO_3$ equivalence. Example 10.2 shows the calculation of hardness from EDTA titration.

Ex. 10.2 *Titration of a 100-mL water sample requires 12.55 mL of 0.0095 M EDTA solution. The total hardness will be calculated.*
   Solution:
   Each mole of EDTA equals one mole of $CaCO_3$, and

$$(12.55\ mL)(0.0095\ M) = 0.119\ mM\ CaCO_3$$

$$(0.119\ mM)\left(100.08\ mg\frac{CaCO_3}{mM}\right) = 11.91\ mg\ CaCO_3$$

$$11.91 \ mg \ CaCO_3 \times \frac{1,000 \ mL/L}{100 \ mL} = 119 \ mg/L \ total \ hardness \ as \ equivalent \ CaCO_3.$$

The steps in the solution shown in Ex. 10.2 can be combined into an equation for estimating hardness from EDTA titration:

$$\text{Total hardness (mg/L } CaCO_3) = \frac{(M)(V)(100,080)}{S} \qquad (10.2)$$

where M = molarity of EDTA, V = volume of EDTA (mL), and S = sample volume (mL).

Hardness Concentrations

The hardness of inland waters may be around 5 to 75 mg/L in humid areas with highly-leached, acidic soil, 150 to 300 mg/L in humid areas with calcareous soil, and 1000 mg/L or more in arid regions. Average ocean water contains about 1350 mg/L magnesium, 400 mg/L calcium, and 8 mg/L strontium (Table 5.8); its calculated hardness is about 6571 mg/L. Hardness of water for water supply purposes often is classified as follows:

Less than 50 mg/L   —   Soft
50–150 mg/L   —   Moderately hard
150–300 mg/L   —   Hard
More than 300 mg/L   —   Very Hard

This is similar to the classification given in Chap. 9 for total alkalinity.

Problems with Hardness

Common soap, sodium stearate, dissolves into stearate ions and sodium ions, but in hard water, stearate ions are precipitated as calcium stearate

$$2C_{17}H_{35}COO^- + Ca^{2+} \rightarrow (C_{17}H_{35}COO)_2Ca \downarrow . \qquad (10.3)$$

As a result, lather formation is limited, and the calcium stearate precipitates. This wastes soap and is particularly troublesome in commercial laundries. It also can be difficult to remove the precipitate from household items washed in hard water.

A scale often forms in containers in which hard waters are boiled. This is caused by loss of carbon dioxide to the air by boiling and the precipitation of calcium carbonate

$$Ca^{2+} + 2HCO_3^- \xrightarrow[\Delta]{} CaCO_3 \downarrow + CO_2 \uparrow + H_2O. \qquad (10.4)$$

This phenomenon results in the accumulation of a scale in household boiling devices and in large commercial boilers. These precipitates often are called boiler scale. Scale also can form in water pipes—especially hot water pipes—partially clogging them and lessening flow. Heat exchangers can be rendered less effective by calcium carbonate forming on their surfaces.

Types of Hardness

The hardness of water sometimes is separated into calcium hardness and magnesium hardness. This is done by subtracting calcium hardness from total hardness to give magnesium hardness. Calcium hardness can be estimated from calcium concentration or measured directly by EDTA titration using murexide indicator in a sample from which the magnesium has been removed by precipitation at high pH (Eaton et al. 2005). The murexide indicator is stable at the high pH necessary to precipitate magnesium, but it functions in the same manner as eriochrome black T.

Hardness also can be separated into carbonate and noncarbonate hardness. In waters where total alkalinity is less than total hardness, carbonate hardness is equal to total alkalinity. This results because there are more milliequivalents of hardness cations than of alkalinity anions (bicarbonate and carbonate). In waters where total alkalinity is equal to or greater than total hardness, carbonate hardness equals total hardness. In such a water, the milliequivalents of alkalinity anions is equal to or exceeds the milliequivalents of hardness cations (calcium and magnesium). Examples of waters with total alkalinity less than total hardness, and waters with total alkalinity equal or greater than total hardness are provided in Table 10.1. Where total alkalinity is less than total hardness, there will be a large amount of sulfate and chloride relative to carbonate and bicarbonate. In the water where total alkalinity is greater than total hardness, there will be a large amount of sodium and potassium relative to calcium and magnesium.

Table 10.1 Hardness fractions and ionic concentrations in waters with and without non-carbonated hardness

Variable	Sample A	Sample B
Total alkalinity (mg $CaCO_3$/L)	52.2	162.5
Total hardness (mg $CaCO_3$/L)	142.5	43.1
Calcium hardness (mg $CaCO_3$/L)	52.5	43.1
Non-carbonate hardness (mg $CaCO_3$/L)	90.0	0.0
$HCO_3^- + CO_3^{2-}$ (meq/L)	1.05	3.25
$SO_4^{2-} + Cl^-$ (meq/L)	1.81	0.25
$Ca^{2+} + Mg^{2+}$ (meq/L)	2.23	0.86
$Na^+ + K^+$ (meq/L)	0.63	2.54

If a water containing carbonate hardness is heated, carbon dioxide is driven off and calcium carbonate, magnesium carbonate or both precipitate as was shown for calcium carbonate precipitation in Eq. 10.4. Sulfate, chloride, sodium, and potassium are not precipitated by boiling, but the carbonate hardness can be removed from water by precipitation following boiling for a sufficient time. Because it can be removed by heating, carbonate hardness is called temporary hardness. Of course, boiling will not remove all of the hardness or alkalinity from a water with no noncarbonate hardness because calcium and magnesium carbonates have a degree of solubility.

Carbonate hardness is responsible for boiler scale. Hardness remaining in water after boiling is the permanent hardness. Because bicarbonate and carbonate precipitate with part of the calcium and magnesium during boiling, the permanent hardness also is known as noncarbonate hardness, and

$$\text{Noncarbonate hardness} = \text{total hardness} - \text{carbonate hardness.} \qquad (10.5)$$

Biological Significance of Hardness

The biological productivity of natural freshwaters increases with greater hardness up to a concentration of 150–200 mg/L (Moyle 1946, 1956). Hardness *per se* has less biological significance than does alkalinity. Productivity depends upon the availability of carbon dioxide, nitrogen, phosphorus, and other nutrients, a suitable pH range, and many other factors. As a rule, alkalinity tends to increase along with concentration of other dissolved ions where watershed soils are fertile but not highly acidic. Hardness and alkalinity often increase in proportion to each other, and in many waters of humid regions, they may be roughly equal. Nevertheless, for analytical reasons, it was easier in the past to measure total hardness than to determine total alkalinity. Hardness became a common index of productivity. Today, there is no reason to follow this tradition, and indices of productivity in aquatic ecosystems should be based on total alkalinity or other variables.

Water Softening

Two methods have traditionally been used for softening water—the lime-soda ash process and the zeolite process (Sawyer and McCarty 1978) . The lime-soda ash procedure involves treating water with calcium hydroxide (lime) to remove calcium carbonate and to convert magnesium bicarbonate to magnesium carbonate as shown in the following reactions:

$$Ca(OH)_2 + CO_2 = CaCO_3 \downarrow + H_2O \qquad (10.6)$$

$$Ca(OH)_2 + Ca(HCO_3)_2 = 2CaCO_3 \downarrow + 2H_2O \qquad (10.7)$$

$$Ca(OH)_2 + Mg(HCO_3)_2 = MgCO_3 + CaCO_3 \downarrow + 2H_2O. \qquad (10.8)$$

Magnesium carbonate is soluble, and it will react with lime to precipitate magnesium hydroxide as shown in the following equation:

$$Ca(OH)_2 + MgCO_3 = CaCO_3 \downarrow + Mg(OH)_2 \downarrow . \qquad (10.9)$$

Lime also removes magnesium associated noncarbonate hardness by the reaction below:

$$Ca(OH)_2 + MgSO_4 \rightarrow CaSO_4 \downarrow + Mg(OH)_2 \downarrow . \qquad (10.10)$$

Sodium carbonate (soda ash) is then added to remove the calcium associated noncarbonate hardness by the reaction

$$Na_2CO_3 + CaSO_4 \rightarrow Na_2SO_4 + CaCO_3 \downarrow . \qquad (10.11)$$

The water contains excess lime after treatment and has a high pH. Carbon dioxide is added to remove the lime and reduce the pH to an acceptable level. The total hardness concentration following lime-soda ash treatment is 50–80 mg/L.

Zeolite is an aluminosilicate mineral of high cation exchange capacity that is used as an ion exchanger. Natural zeolite is of volcanic origin, and deposits of this mineral that occur in many locations are mined as a source of zeolites for water softening or other purposes. Synthetic zeolites also are widely used for the same purposes. In an industrial process, the cation exchange sites on zeolites for softening water are saturated with sodium by exposure to a sodium chloride solution. This zeolite is contained in a column or bed through which hard water is passed, and calcium and magnesium ions in the water are exchanged with sodium on the zeolite resulting in softening of the water as illustrated below:

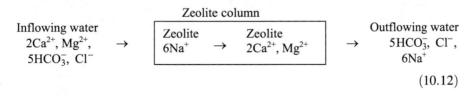

$$(10.12)$$

When the zeolite bed has exchanged so much sodium for calcium and magnesium that it becomes ineffective, its water softening capacity is regenerated by backwashing the zeolite column or bed with a concentrated sodium chloride solution.

Calcium Carbonate Saturation

The degree of calcium carbonate saturation of water affects the likelihood of calcium carbonate precipitation or scaling by a water during use. Several indices of scaling potential have been formulated, and one of the most popular is the Langelier Saturation Index (Langelier 1936). The full explanation of this index is beyond the scope of this book but the equation is

$$LSI = pH - pH_{sat} \qquad (10.13)$$

where LSI = Langelier Saturation Index; pH = pH of water; pH_{sat} = pH at calcium carbonate saturation of the water.

$$pH_{sat} = (9.3 + A + B) - (C + D) \qquad (10.14)$$

where A = (log TDS – 1)/10, B = $[-13.12 \times log\ (°C + 273)] + 34.55$, C = log $(Ca^{2+} \times 2.5) - 0.4$, and D = log (alkalinity as $CaCO_3$).

For waters with TDS concentration < 500 mg/L, a simplified equation (Gebbie 2000) may be used for pHsat:

$$pH_{sat} = 11.5 - log\ (Ca^{2+}) - log\ (Alkalinity). \qquad (10.15)$$

At LSI = 0, the water is saturated with calcium carbonate. Waters with a negative LSI are undersaturated with calcium carbonate and potentially corrosive (see Chap. 8), because the precipitation of calcium carbonate over a surface protects it from corrosion. When the LSI is positive, the water is not corrosive, but it can cause scaling that can be problematic. Calculation of the LSI is illustrated in Ex. 10.3.

Ex. 10.3 *Calculation of the Langelier Saturation Index for a water at 20 °C with pH 7.7, 20 mg/L Ca^{2+}, 75 mg/L total alkalinity, and 375 mg/L total dissolved solids.*
Solution:
The pH_{sat} value is calculated with Eq. 10.14.
A = (log 375–1)/10 = 0.16.
B = [–13.12 × log (20 + 273)] + 34.55 = 2.18.
C = log (20 × 2.5) – 0.4 = 1.30.
D = log (75) = 1.88

$$pH_{sat} = (9.3 + 0.16 + 2.18) - (1.30 + 1.88) = 8.46.$$

The LSI is computed using Eq. 10.13

$$LSI = 7.7 - 8.46 = -0.76.$$

The simplified Gebbie equation (Eq. 10.15) can be used instead of Eq. 10.14 because TDS is < 500 mg/L. The result is

$$pH_{sat} = 11.5 - \log 20 - \log 75 = 8.32$$

and

$$LSI = 7.7 - 8.32 = -0.62.$$

The calculation by either procedure suggests that the water would be corrosive, but it would not cause scaling.

Suppose that the water referred to in Ex. 10.3 is in a lake that develops a dense plankton bloom in summer and pH rises to 8.6 and the water temperature increases to 28 °C. The LSI would be +0.29 by the Langelier method for estimating pH_{sat} (Eq. 10.14) and +0.28 by the Gebbie (2000) shortcut method for estimating pH_{sat} (Eq. 10.15). At the higher pH and water temperature, the water would have potential to cause scaling, particularly in the afternoon when pH is high because of photosynthesis as discussed in Chap. 9.

Conclusions

Hardness is an important factor regarding the use of water for many purposes. Hardness of water is manifest most commonly by the amount of soap needed to produce suds. Hardness might be called the soap-wasting property of water, because no suds will be produced in a hard water until the minerals causing the hardness have been removed from the water by combining with the soap. The material that is removed by the soap is evident as an insoluble scum—familiar ring on the bathtub— that forms during bathing in some waters.

Waters with a hardness of less than 50 mg/L are considered soft. A hardness of 50–150 mg/L is not objectionable for most purposes, but the amount of soap needed increases with hardness. Laundries or other industries using large quantities of soap generally find it profitable to lower hardness concentrations to about 50 mg/L. Water having 100–150 mg/L hardness will deposit considerable scale in steam boilers. Hardness of more than 150 mg/L is decidedly noticeable. At levels of 200–300 mg/L or higher, it is common practice to soften water for household uses. Where municipal water supplies are softened, the hardness is reduced to about 85 mg/L. Further softening of a whole public water supply is not considered economical.

Where scale forms when water is heated, calcium carbonate scale deposits first, because it is more insoluble than magnesium carbonate. In the absence of carbon dioxide, water will carry only about 14 mg/L calcium carbonate in solution. Under the same conditions, the solubility of magnesium carbonate is more than five times as great, or about 80 mg/L.

References

Eaton AD, Clesceri LS, Rice EW, Greenburg AE (eds) (2005) Standard methods for the examina-
    tion of water and wastewater. American Public Health Association, Washington
Gebbie P (2000) Water stability—what does it mean and how do you measure it? In: Proceedings of
    63rd Annual Water Industry Engineers and Operators Conference, Warrnambool, pp 50–58.
    (http://wioa.org.au/conference_papers/2000/pdf/paper7.pdf).
Langelier WF (1936) The analytical control of anti-corrosion water treatment. J Am Water Works
    Assoc 28:1500–1521
Moyle JB (1946) Some indices of lake productivity. Trans Am Fish Soc 76:322–334
Moyle JB (1956) Relationships between the chemistry of Minnesota surface waters and wildlife
    management. J Wildlife Man 20:303–320
Sawyer CN, McCarty PL (1978) Chemistry for environmental engineering. McGraw-Hill,
    New York

Acidity

11

Abstract

Acidity in water bodies with pH of about 4 up to 8.3 is caused by carbon dioxide and dissolved humic substances. Waters with pH <4.0 usually contain sulfuric acid or possibly another strong acid, but they still contain carbon dioxide and possibly humic compounds. Low pH is associated with low productivity and biodiversity. Acidity is measured by titration with standard sodium hydroxide and reported in milligrams per liter of calcium carbonate. Low pH and elevated acidity result naturally from oxidation of sulfides in soil or sediment. Oxidation of pyritic material exposed by mining and oxidation of sulfur and nitrogen compounds released into the atmosphere by combustion of fuels also are sources of acidity to surface waters. The increasing solubility of carbon dioxide as a result of higher atmospheric carbon dioxide concentration does not have a large effect on pH of rainfall and freshwater bodies. However, it is causing ocean acidification and interfering with biological calcium carbonate precipitation by calcifying organisms. Liming materials typically are used to counteract acidity.

Introduction

Freshwater bodies usually have pHs between 6 and 9, but some have lower pHs that restrict the type and abundance of biota. Acidic waters can be put into three very general classes: low alkalinity and early morning pHs of 5–6; no alkalinity and pHs of 4–5; highly acidic with pH <4. Carbon dioxide is the main source of acidity in waters with low but measurable alkalinity. In waters with pH between 4 and 5 there is no alkalinity, and carbon dioxide and humic substances contribute to acidity. Humic substances have been reported to depress pH slightly below 4, but most waters with pH <4 contain an acid stronger than either carbon dioxide or humic compounds. Carbon dioxide and usually humic substances also will be present in water of pH <4. The most common cause of pH <4.0 in natural waters is the presence of a strong acid (often sulfuric acid) of natural or anthropogenic origin.

© Springer Nature Switzerland AG 2020

C. E. Boyd, *Water Quality*, https://doi.org/10.1007/978-3-030-23335-8_11

Acidity is not commonly measured in water quality investigations, but some waters are impacted by low pH and measurements of acidity are necessary for assessing water quality conditions. This chapter explains the concept of acidity, its measurement, its effects, and its mitigation.

The Concept of Acidity in Water

The pH is an index of the hydrogen ion concentration in water, but hydrogen ion concentration and pH are not the same as acidity. The acidity of water is defined by Eaton et al. (2005) as a capacity factor; the amount of strong base necessary to neutralize the hydrogen ion concentration in a given volume of water. In other words, acidity can be described as the total concentration of titratable acids in water. By this description, acidity can be considered the opposite of alkalinity, because alkalinity is the total concentration of titratable bases in water. This description could be taken to imply that acidity is negative alkalinity—not a bad analogy in many respects. The concept of acidity is perplexing, because most natural waters contain both acidity and alkalinity. This results because carbon dioxide is an acid, and is present in water at pH <8.3, while water contains bicarbonate as its main source of alkalinity in waters with pH >5.6. The practical endpoint pH for alkalinity titrations is lower than 5.6 as the result of accumulation of carbon dioxide produced in the titration vessel by reaction of acid with bicarbonate (see Chap. 9).

The concept of water with pH between 5.6 and 8.3 containing both acidity and alkalinity seems to contrast with the usual concept that pH <7 is an acidic condition and pH >7 is an alkaline condition. Actually both concepts are true; a water with pH between 5.6 and 7 is acidic, but it contains alkaline substances, while a water of pH 7–8.3 is alkaline, but it contains acidic substances.

Acidity in most natural waters is caused by carbon dioxide which acts as a weak acid, and humic substances which also are weak acids. Carbon dioxide concentration is normally not high enough to depress pH below 4.5; humic substances may depress pH as low as 3.7 or 3.8. Waters of lower pH contain strong acids. A water sample with pH ≥ 4.5 usually is said to contain only carbon dioxide acidity. A water with pH <4.5 is considered to contain an acid stronger than carbon dioxide, but it also contains acidity from carbon dioxide. Waters may have only carbon dioxide acidity, but most waters also have acidity from humic substances.

A weak acid does not dissociate appreciably and acids with $K_a > 10^{-2}$ often are designated to be weak acids. The reader should note that K_a 10^{-2} is greater than K_a of 10^{-3}. A strong acid has a greater K_a and is assumed to dissociate completely (Table 11.1). The K_a usually is not reported for a strong acid, because it is a large value. The pH of a 0.01 M solution of acetic acid is 3.39 as compared to 2.0 for 0.01 M hydrochloric acid (a strong acid). The ratio, acetate ion to undissociated acetic acid, in a 0.01 M solution of acetic acid is:

$$\frac{(CHCOO^-)}{(CH_3COOH)} = \frac{K_a}{(H^+)} = \frac{10^{-4.74}}{10^{-3.39}} = 10^{-1.35} = 0.045.$$

Acetic acid in a 0.01 M solution is about 4.3% dissociated [(0.045/1 + 0.045)100], but a strong acid such as HCl in a 0.01 M solution is completely dissociated.

Strong acids are inorganic and often called mineral acids, but there also are weak inorganic acids, e.g., carbon dioxide, boric acid, and silicic acid (Table 11.1). Strong acids include nitric, sulfuric, and hydrochloric acids are used so commonly in chemistry laboratories as to be referred to as bench acids. The molecular weight, number of steps in its dissociation, and its molarity in concentrated solution does not determine whether an acid is weak or strong. The designation of weak or strong depends upon the extent to which an acid dissociates. It requires the same number of milliequivalents of hydroxyl ion to neutralize 50 mL of 0.1 N boric acid as it does to neutralize the same volume of 0.1 N sulfuric acid.

The reason carbon dioxide is a weak acid is explained in Chap. 9, and its apparent reaction in water $CO_2 + H_2O = HCO_3^- + H^+$ has $K_a = 10^{-6.35}$. Dissolved humic substances are weak acids with K_a values around from 10^{-4} to 10^{-8} (Steelink 2002). Fulvic acid is light yellow to yellow brown in color, while humic acid is dark brown to grey-black in color. In soils for which most studies of humic substances have been made, molecular weights of fulvic acids are in the range of 1000–10,000 g/mole,

Table 11.1 Properties of some common weak and strong acids

Acid	Dissociation	K_a	Typical normality of bench acid	pH of 0.1 N solution
Carbon dioxide	$CO_2 + H_2O - H^+ + HCO_3^-$	$10^{-6.35}$		
Acetic acid	$CH_3COOH = H^+ + CH_3COO^-$	$10^{-4.74}$	17.4	2.9
Fulvic acids	–	$\approx 10^{-3.20}$		
Humic acids	–	$\approx 10^{-3.80}$ to $10^{-4.9}$		
Tannic acid	–	$\approx 10^{-5}$ to 10^{-6}		
Phosphoric	$H_3PO_4 = H^+ + H_2PO_4^-$	$10^{-2.13}$	14.8	1.5
	$H_2PO_4^- = H^+ + HPO_4^{2-}$	$10^{-7.21}$	–	
	$H_2PO_4^{2-} = H^+ + PO_4^{3-}$	$10^{-12.36}$	–	
Boric acid	$H_3BO_3 = H^+ + H_2BO_3^-$	$10^{-9.14}$	–	5.2
	$H_2BO_3^- = H^+ + HBO_3^{2-}$	$10^{-12.74}$	–	
	$HBO_3^{2-} = H^+ + BO_3^{3-}$	$10^{-13.8}$	–	
Hydrochloric acid	$HCl = H^+ + Cl^-$	Very large	11.6	1.1
Nitric acid	$HNO_3 = H^+ + NO^{3-}$	Very large	15.6	1.0
Sulfuric acid	$H_2SO_4 = H^+ + HSO_4^-$	Very large	36	1.2
	$HSO_4^- = H^+ + SO_4^{2-}$	$10^{-1.89}$	–	

while humic acids have molecular weights of 10,000–100,000 g/mol. Complex molecules of humic substances are organic polymers with functional groups such as carboxyl, phenolic, ester, lactone, hydroxyl, ether, and quinone attached to a framework of organic polymers (Steelink 1963, 2002). Humic acids such as the one illustrated in Fig. 11.1 consist of aromatic groups linked by aliphatic groups, and functional groups are attached to the aromatic and aliphatic groups. A fulvic acid is presented in Fig. 11.2, and it does not have the linkages to aliphatic groups found in humic acids. Fulvic acids also have more hydroxyl and carboxyl groups in proportion to molecular weight than do humic acids.

The C/O weight ratio is about 1.0 in fulvic acid, but it is around 1.8 in humic acid. The oxygen-containing functional groups, carboxyl (COOH) and phenolic (OH), dissociate to release hydrogen ion into water:

$$R - COOH = R - COO^- + H^+$$

$$R - OH = R - O^- + H^+.$$

Shinozuka et al. (2004) found that carboxyl groups were present in fulvic acids at an average of 5.34 mmol/g, but at an average of only 3.08 mmol/g in humic acids. Fulvic acids are more acidic than humic acids, because they have more carboxyl groups and presumably more phenolic groups also. Both fulvic and humic acids are found in natural water, but their molecular weights in water tend to be smaller than in soil. Beckett et al. (1987) found an average molecular weight of 1910 g/mol for fulvic acids in stream water; the average molecular weight of humic acids was 4390 g/mol.

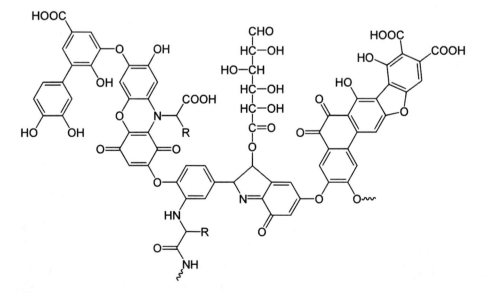

Fig. 11.1 Structural formula of a typical humic acid. (Photograph courtesy of Creative Commons)

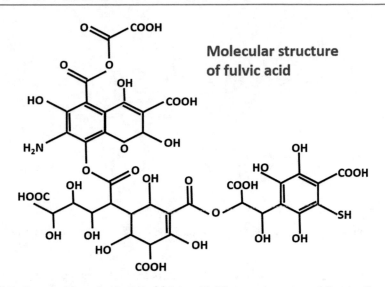

Fig. 11.2 Structural formula of a typical fulvic acid. (Photograph courtesy of Creative Commons)

Tannic acid is a humic substance of particular interest, because it is commonly present in plants and is highly water soluble. Some aquatic plants are particularly high in concentration of tannic acid. Boyd (1968) reported that 12 species of submerged, vascular plants contained an average tannin concentration of $4.6 \pm 4.63\%$ (range, 0.8–15.6% dry weight), while 20 species of emergent vascular plants had an average of $6.1 \pm 4.6\%$ (range, 0.8–15.6% dry weight). The submerged species *Myriophyllum brasiliense* and *Cabomba caroliniana* and the emergent species *Jussiaea peruviana, J. diffusa, Brasenia schreberi,* and *Nymphaea adorata* contained more than 10% tannic acid. Macrophyte algae had tannin concentrations less than 1% (dry matter basis).

Tannic acid is a polyphenolic molecule with a large number of phenolic hydroxyl groups but no carboxyl groups (Fig. 11.3). Tannins are classified as hydrolysable or condensed. The hydrolysable form consists of polymers of gallic acid linked by carbohydrate residues while the condensed form is made of polymers of flavoids (Geissman and Crout 1969). The phenolic groups impart a weak acidity. Purified tannic acid has a molecular weight of 1701.2 g/mol and an empirical formula of $C_{76}H_{52}O_{46}$. The K_a of tannic acid is $\approx 10^{-10}$, so it is not as acidic as most flavic and humic acids that have both carboxyl and phenolic groups. Lignosulfonic acid is a breakdown product of plant lignins, and it also is a weak acid. Lawrence (1980) analyzed waters from six small streams in southern Ontario, Canada, and found an average of 4.0 mg/L fulvic acid (46% C), 1.0 mg/L tannic acid (54% C), and 5.6 mg/L lignosulfonic acid (49% C). Thus, the three humic substances contributed about 5.12 mg/L of the measured dissolved organic carbon concentration of 6.78 mg/L. Much of the remainder of the dissolved organic carbon was likely humic acid.

Waters in some dystrophic lakes may contain over 100 mg/L of humic substances in solution. The pH of such waters usually is 4–6, but pH as low as 3.7 has been

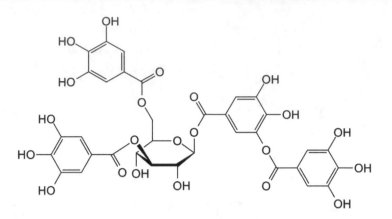

Fig. 11.3 Structural formula of tannic acid. (Photograph courtesy of Creative Commons)

recorded (Klavins et al. 2003). Dystrophic waters are highly stained a dark color by humic substances and they contain little or no alkalinity.

Acid-Sulfate Acidity in Soil and Sediment

The most common strong acid in the environment is sulfuric acid which may be of natural origin or from anthropogenic sources. There are areas where soils and other geologic formations contain large concentrations of metal sulfides and iron sulfide in particular. Sulfide-rich soils are common in existing or former coastal swamps such as salt marshes and mangrove ecosystems. Sediment in such areas contains abundant organic matter from wetland plants, iron-bearing soil particles from river inflow, and sulfate from seawater. Decomposition of organic matter from wetland plants creates anaerobic conditions in the sediments of these coastal wetlands. Anaerobic bacteria use oxygen from iron oxides and sulfate to oxidize organic matter in the sediment. Ferrous iron and hydrogen sulfide, metabolites of microbial respiration, accumulate in sediment pore water. This condition is conducive for iron sulfide production as illustrated in the following unbalanced equations:

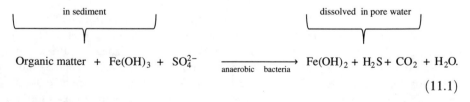

$$\text{Organic matter} + \text{Fe(OH)}_3 + \text{SO}_4^{2-} \xrightarrow[\text{anaerobic bacteria}]{} \text{Fe(OH)}_2 + \text{H}_2\text{S} + \text{CO}_2 + \text{H}_2\text{O}.$$

$$(11.1)$$

The reduced products of anaerobic microbial respiration in the solution phase of pore water react forming iron sulfide

$$Fe(OH)_2 + H_2S \rightarrow FeS + 2H_2O. \tag{11.2}$$

Ferrous sulfide is readily transformed to iron pyrite (FeS_2) in the presence of hydrogen sulfide that may be abundant in anaerobic environments (Rickard and Luther 1997). The reaction is:

$$FeS + H_2S \rightarrow FeS_2 + H_2 \uparrow. \tag{11.3}$$

Notice that in Eq. 11.3, sulfur in FeS_2 should have a valence of -1, a valence not usually assigned to sulfur. Schippers (2004) and Borda (2006) assigned a valence of -1 to S in iron pyrite, but the -1 valence causes a dilemma with others (Nesbitt et al. 1998). Those who do not agree with a -1 valence for sulfur solve this dilemma by referring to the S_2 in FeS_2 as the persulfide ion with a valence of -2.

Some soils in coastal areas of South Carolina in the southern United States were reported to contain up to 5.5% sulfide-sulfur (Fleming and Alexander 1961). Coastal areas with soil high in sulfide-sulfur concentration are common in Indonesia, the Philippines, Thailand, Malaysia, Australia, and many other countries. Metal sulfides also form in marine sediment not associated with wetlands or warm climates. The first description of acid-sulfate soils was made in the eighteenth century for certain soils in the Netherlands by Carl Linneaus, the Swedish botanist of plant and animal taxonomy fame (Fanning 2006).

In addition to iron sulfide, copper sulfide (CuS), zinc sulfide (ZnS), manganous sulfide (MnS), and other metal sulfides form in anaerobic environments (Rickard and Luther 2006). Iron pyrite and other metal sulfides are common constituents of coal and some other mineral deposits. Mining often exposes overburden of high sulfide-sulfur content and creates spoil piles that contain sulfide-sulfur.

Soils containing sulfur concentrations of 0.75% or more are classified as potential acid-sulfate soils (Soil Survey Staff 1994). While soils containing metal sulfides are anaerobic, these compounds are insoluble and have little influence on soil acidity. When potential acid-sulfate (PAS) soils are exposed to the air or oxygenated water, oxidation occurs and sulfuric acid is produced. Potential acid-sulfate soils are then called active acid-sulfate soils. The rate of oxidation of sulfide is much greater in soils exposed to the atmosphere than in sediment exposed to oxygenated water. The oxidation of iron pyrite is a complex reaction occurring in several ways depending upon conditions in pyritic sediment or PAS soil (Chandra and Gerson 2011). According to Sorensen et al. (1980) iron pyrite reacts with oxygen to produce ferrous sulfate and sulfuric acid. Presence of sulfuric acid allows further oxidation of ferrous sulfate to ferric sulfate. The presence of ferric sulfate also permits oxidation of iron pyrite to sulfuric acid. These reactions are shown below:

$$FeS_2 + H_2O + 3.5\,O_2 \rightarrow FeSO_4 + H_2SO_4 \tag{11.4}$$

$$2FeSO_4 + 0.5\,O_2 + H_2SO_4 \rightarrow Fe_2(SO_4)_3 + H_2O \tag{11.5}$$

$$FeS_2 + 7Fe_2(SO_4)_3 + 8H_2O \rightarrow 15FeSO_4 + 8H_2SO_4. \qquad (11.6)$$

Ferric sulfate plays an important role in pyrite oxidation, and the production of ferric sulfate from ferrous sulfate is greatly accelerated by activity of bacteria of the genus *Thiobacillus*, and under acidic conditions, oxidation of pyrite by ferric sulfate is rapid.

Ferric sulfate can hydrolyze with formation of sulfuric acid according to the following reactions:

$$Fe_2(SO_4)_3 + 6H_2O \rightarrow 2Fe(OH)_3 + 3H_2SO_4 \qquad (11.7)$$

$$Fe_2(SO_4)_3 + 2H_2O \rightarrow 2Fe(OH)SO_4 + H_2SO_4. \qquad (11.8)$$

Ferric sulfate also can react with iron pyrite to form elemental sulfur and the sulfur may be oxidized to sulfuric acid by microorganisms

$$Fe_2(SO_4)_3 + FeS_2 \rightarrow 3FeSO_4 + 2S \qquad (11.9)$$

$$S + 1.5\,O_2 + H_2O \rightarrow H_2SO_4. \qquad (11.10)$$

Ferric hydroxide can react with bases, such as potassium, in acid sulfate soils to form jarosite, a basic iron sulfate

$$3Fe(OH)_3 + 2SO_4^{2-} + K^+ + 3H^+ \rightarrow KFe_3(SO_4)_2(OH)_6 \cdot 2H_2O + H_2O. \qquad (11.11)$$

Jarosite is relatively stable, but in older acid-sulfate soils where acidity has been neutralized, jarosite tends to hydrolyze as follows:

$$KFe_3(SO_4)_2(OH)_6 \cdot 2H_2O + 3H_2O \rightarrow 3Fe(OH)_3 + K^+ + 2SO_4^{2-}$$
$$+ 3H^+ + 2H_2O. \qquad (11.12)$$

There are several purely chemical oxidations and microbially mediated oxidations of reduced sulfur compounds that cause acidity in sediment. The important point is that all of these reactions ultimately lead to the release of sulfuric acid. All of the reactions are summarized by the reaction of oxygen with sulfur (Eq. 11.10). An example of the calculation of the potential acidity of a potential acid sulfate soil is provided in Ex. 11.1.

Ex. 11.1 *The potential acidity of an aquatic soil (bulk density = 1.15 kg/m³) containing 5.5% sulfide-sulfur will be calculated.*

<u>Solution:</u> The approximate answer can be obtained using Eq. 11.10. The potential acid is reported as equivalent calcium carbonate. The stoichiometric relationships are:

$$S + 1.5\,O_2 + H_2O \rightarrow H_2SO_4$$

$$CaCO_3 + H_2SO_4 \rightarrow Ca^{2+} + CO_2 + SO_4 + H_2O.$$

It follows that

$$S \rightarrow H_2SO_4 \rightarrow CaCO_3$$

and 1 kg of soil with 5.5% S would contain 0.055 kg S. The stoichiometric ratio is:

$$\frac{0.055 \, kg}{S} = \frac{X}{CaCO_3}$$
$$\frac{}{32} \quad \frac{}{100}$$

$$X = 0.17 \, kg \, CaCO_3/kg \, soil.$$

One hectare of this soil to a depth of 15 cm would weigh $10,000 \, m^2 \times 0.15 \, m \times 1150 \, kg/m^3 = 1,725,000 \, kg$. The amount of potential acid would be equal to

$$1,725,000 \, kg/ha \times \quad 0.17 \, kg \, CaCO_3/kg \, soil$$
$$= 293,250 \, kg \, CaCO_3/ha \, or \, 293.2 \, t/ha.$$

Sulfuric acid dissolves aluminum, manganese, zinc, copper, and other metals from soil, and runoff from acid-sulfate soils and mine spoils is highly acidic and usually contains potentially toxic metallic ions. The potential for acid production in an acid-sulfate soil depends largely upon the amount and particle size of the iron pyrite, the presence or absence of exchangeable bases and carbonate within the pyrite-bearing material, the exchange of oxygen and solutes with the soil, and the abundance of *Thiobacillus*. Because the exchange of oxygen and solutes and the abundance of *Thiobacillus* are restricted with depth, the acid-sulfate condition usually is a surface soil phenomenon.

Waterlogging of acid-sulfate soils restricts the availability of oxygen, and sulfuric acid production ceases when soil becomes anaerobic. Sulfate is reduced to sulfide under anaerobic conditions by bacteria of the genus *Desulfovibrio*. In natural waters, the sediment-water interface usually is aerobic, and sulfuric acid production occurs, but at a much slower rate than when sediment is exposed to air.

Oxidation of pyrite in spoil piles from mine excavations results in sulfuric acid that may enter streams in runoff and reduce pH. Seepage from underground mines also may be acidic. Surface coal mining was a major source of acid drainage to streams in the past, because the overburden above coal deposits was stripped off and piled on the land surface. The last overburden removed from above the coal formation was contaminated with particles of coal that contained sulfide sulfur. This contaminated material tended to end up on top of the spoil pile promoting contact with atmospheric oxygen and oxidation of sulfide to sulfuric acid. The acid removed from mine spoils by rainwater and runoff flowed into streams causing acidification. Acidic water also may seep or be pumped from mines to cause acidification of receiving waters.

The modern procedure at surface mining operations is to separate the overburden that is contaminated with coal and stockpile it separately from the other overburden. Once the coal deposit has been mined, the contaminated overburden is placed back into the mined area and covered with the stockpiled nonacidic overburden. Liming of spoil material and development of grass cover on it also lessens the problem with acid drainage. Most countries have regulations regarding treatment of mine effluent, and in some cases, effluents or streams are treated with limestone to neutralize acidity.

Acidic, Atmospheric Deposition

Rainwater naturally has a pH of about 5.6 because it becomes saturated with carbon dioxide while falling through the atmosphere. The carbon dioxide concentration of the atmosphere is increasing, but this has little effect on the pH of rain. A pH <5.6 in rainfall usually results from atmospheric pollution with sulfur and nitrogen gases. The acidic rain problems have arisen because of combustion of fuels, and it is most severe in urbanized and industrial areas and in the direction of the prevailing wind from these areas. Rainfall in eastern Canada and the northeastern United States was reported to have an annual, average pH of 4.2–4.4 and rain from individual storms had lower pH (Haines 1981).

Hydrogen sulfide and other sulfides emitted with sulfur dioxide are oxidized to sulfur dioxide in the atmosphere. In the gaseous phase of the atmosphere, sulfur dioxide and nitrogen oxides (NO_x) form strong acids as illustrated in the sequence of reactions given below:

$$2H_2S + 3O_2 \rightarrow 2H_2O + 2SO_2$$

$$SO_2 + OH^- \rightarrow HOSO_2 \quad \text{(hydroxysulfonyl radical)}$$

$$HOSO_2 + O_2 \rightarrow HO_2^- + SO_3 \quad \text{(sulfur trioxide)}$$

$$SO_3 + H_2O \rightarrow H_2SO_4$$

$$NO_2 + OH^- \rightarrow HNO_3$$

$$\text{or } NO_x + xOH^- \rightarrow HNO_x.$$

In cloud droplets, sulfur dioxide reacts in a slightly different manner:

$$SO_2 \rightarrow H_2O = SO_2 \cdot H_2O$$

$$SO_2 \cdot H_2O \rightarrow H^+ + HSO_3^- \quad \text{(bisulfite ion)}$$

$$HSO_3^- \leftrightarrow H^+ + SO_3^{2-}$$

$$SO_3^{2-} + H_2O \leftrightarrow H_2SO_4.$$

Combustion of fossil fuels is the main source of acid-forming substances in the atmosphere. All sectors that burn fuel contribute such emissions roughly in proportion to energy use. The emissions not only include hydrogen sulfide, sulfur dioxide, and nitrogen oxides, they also contain particulate material that can continue to oxidize and produce acidity. Wet precipitation (rainfall, snow, dew, etc.) and dry fall out of particles can deliver acidity to the earth's surface. Burning of crop residues after harvest and other agricultural wastes often cause much smoke and haze at times in developing countries. These emissions, however, contribute only about 0.1% of sulfur emissions (Smith et al. 2011).

It is interesting to note that global sulfur dioxide emissions had been increasing rapidly, but they declined following an effort beginning in the 1970s in developed countries to remove sulfur from emissions. Since the early 2000s, global sulfur emissions are rapidly increasing again as the industrial sector and economy of China, India, and a few other countries have expanded in the absence of effective air pollution regulations.

The rising carbon dioxide concentration in the atmosphere is an important factor related to acidification of rainfall or surface freshwater bodies. Increased atmospheric carbon dioxide does, however, influence oceanic carbon dioxide concentration, pH, and calcium carbonate saturation level. As the atmospheric carbon dioxide increases, the solubility of carbon dioxide in the ocean increases. Greater carbon dioxide concentration reduces ocean pH slightly despite the buffering capacity of the moderate alkalinity of the ocean. Caldeira and Wickett (2003) reported that the average pH of the ocean has fallen by 0.1 pH units since the beginning of the industrial revolution (mid 1700s). The USEPA (https://www.epa.gov/climate-indicators/climate-change-indicators-ocean-acidity) present data for ocean pH near Bermuda and Hawaii between 1987 and 2015. The pH decreases were 8.12 to 8.06 and 8.12 to 8.07, respectively. The average decrease has been about 0.002 pH units per year, and the trajectory of the decline has been essentially constant since 1987. Some authors such as Feely et al. (2004) predict ocean pH will drop by another 0.3–0.4 unit by 2100.

Reduction in ocean pH favors the dissolution of aragonite and calcite by reducing the pH below that of calcium carbonate saturation. Aragonite and calcite comprise the shells of calcifying marine organisms, and thinner shells will negatively affect survival and growth of these organisms (Orr et al. 2005). The effect on ocean ecology could be enormous.

Measurement of Acidity

The simplest procedure for measuring acidity is to titrate a water sample from its initial pH to pH 8.3 with standard sodium hydroxide solution. A 0.010 N sodium hydroxide solution is often used, but other normalities may be used where necessary to obtain suitable titration volumes (buret readings). The amount of base needed to raise the pH to the methyl orange endpoint (or to pH 4.5 as measured with a pH meter) is recorded, and the titration is continued to pH 8.3 (phenolphthalein endpoint) in instances where carbon dioxide acidity is of interest.

The titration of acidity is illustrated in Ex. 11.2.

Ex. 11.2 *A 100-mL water sample of pH 3.1 is titrated with 0.022 N NaOH to pH 4.5, and then the titration is continued to pH 8.3. The volume of acid to titrate to pH 4.5 is 3.11 mL, and the total acid used in the titration is 3.89 mL. The total acidity and mineral acidity will be estimated.*

Solution: *The same approach used for alkalinity calculation (Ex. 9.5) can be used here because results will again be reported as equivalent $CaCO_3$. The total acidity as milliequivalent $CaCO_3$ equals the milliequivalents of base used in the entire titration:*

$$(3.89 \ mL)(0.022 \ N) = 0.0856 \ meq \ NaOH = meq \ acidity = meq \ CaCO_3.$$

The weight equivalent $CaCO_3$ in the sample is

$$(0.0856 \ meq \ CaCO_3)(50.04 \ mg \ CaCO_3/meq) = 4.28 \ mg \ CaCO_3.$$

To convert milligrams of $CaCO_3$ in the sample to milligrams per liter of $CaCO_3$ in the sample,

$$(4.28 \ mg \ CaCO_3)\left(\frac{1,000 \ mL/L}{100 \ mL}\right) = 42.8 \ mg/L \ CaCO_3.$$

The same procedure is repeated for the volume of base used for titration to pH 4.5 to obtain the mineral acidity. The result is 34.2 mg/L $CaCO_3$.

The sample had 42.8 mg/L total acidity and 34.2 mg/L mineral acidity. The difference of 8.6 mg/L represents acidity of carbon dioxide in the sample.

Acidic waters may contain measurable concentrations of metal ions that may react with hydroxide during titration as shown below:

$$Fe^{3+} + 3OH^- \rightarrow Fe(OH)_3. \tag{11.13}$$

This results in an overestimation of acidity. The discrepancy can be prevented by the hot peroxide treatment method (Eaton et al. 2005). In the hot peroxide method, the sample pH is reduced below 4.0 (if necessary) by adding 5.00-mL increments of 0.02 N sulfuric acid. Five drops of 30% hydrogen peroxide (H_2O_2) are added, and the sample is boiled for 5 min. After cooling to room temperature, the sample is titrated with standard sodium hydroxide to pH 3.7, 4.5, or 8.3 as desired. Hot sulfuric acid-hydrogen peroxide treatment removes the potentially interfering substances, and the titration volume must be corrected for the added acidity. This is done by subtracting the product (mL H_2SO_4 × N H_2SO_4) from the product (mL NaOH × N NaOH).

The equations for calculating acidity based on the pH selected for the endpoint are:

Without hot peroxide pretreatment,

$$\text{Acidity (mg CaCO}_3/\text{L)} = \frac{(N_b)(V_b)(50)(1,000)}{S} \qquad (11.14)$$

where N_b = normality NaOH (meq/mL), V_b = volume NaOH used (mL), 50 = mg $CaCO_3$/meq, 1000 = mL/L, and S = sample volume (mL).

With hot peroxide pretreatment

$$\text{Acidity (mg CaCO}_3/\text{L)} = \frac{[(N_b \times V_b) - (N_a \times V_a)](50)(1,000)}{S} \qquad (11.15)$$

where N_a = normality H_2SO_4 (meq/mL) and V_a = volume H_2SO_4 (mL).

Effects of Acidity

Water bodies with high concentrations of humic substances typically have low productivity as the result of low pH, low nutrient concentrations, and restricted light penetration. Such waters also will have relatively low biodiversity. There have been studies suggesting that decomposition products (presumably humic substances) of barley straw in natural waters act as algicides (Everall and Lees 1996). However, there is no evidence that low primary productivity and biodiversity in humically-stained waters is a direct toxic effect.

The most serious effects of acidity on species diversity and productivity usually are observed at pH < 5. The phenomenon of acid rain provides an excellent example. Lakes in some areas of Canada and the northeastern United States declined in pH by 1–2 units between 1950 and 1980 (Haines 1981; Cowling 1982). Aquatic organisms have been affected at all trophic levels; species abundance, productivity, and species diversity has been reduced. Fish have suffered acute mortality, reduced growth, reproductive failures, skeletal deformities, and increased accumulations of heavy metals (Haines 1981).

There is a well-known exception to the preceding paragraph. Rivers flowing from tropical rainforests usually have pH 4–5 and high concentrations of humic substances, but they usually have diverse and flourishing fish communities. These fish possibly are adapted to low pH, but there is evidence that low pH from humic and fulvic acids is less harmful to fish than is low pH from mineral acids (Holland et al. 2015). Organic matter decomposition does not progress rapidly at low pH. Greater acidity favors fungi over bacteria, and fungi are not as effective in decomposing organic matter as are bacteria. In standing waters in poorly-drained areas, shallow and marginal areas usually are infested by species of aquatic macrophytes and reed-swamp plants. These plants have a low nitrogen concentration, a high fiber content, and they decompose slowly resulting in a large residue of organic matter. Peat bogs and organic soils are common in regions with highly acidic soils and water.

Humic substances have a low mammalian toxicity, and no direct toxicity problems are incurred from elevated concentrations of these substances in waters for swimming, bathing, or domestic water supply (de Melo et al. 2016). Of course, people desire clear water for these purposes, and clear water usually is preferred in bodies of water used for landscape enhancement (Smith et al. 1995; Nassar and Li 2004). High concentrations of humic substances in municipal water supplies also are problematic, because these substances react with free chlorine residuals to produce trihalomethane, a suspected carcinogen (Rathbun 1996). Removal of humic substances from water supplies is difficult. The most common method is chemical flocculation, but it does not reduce concentration by more than 40–60%. Other methods include membrane filtration, adsorption of activated carbon, and ozone oxidation.

Mitigation of Acidity

Limestone, lime, and other alkaline materials have been used to neutralize acidity for improving crop growth in agriculture since at least the thirteenth century (Johnson 2010). The effect of increasing pH on water quality in bog lakes was demonstrated in an early study by Hasler et al. (1951). In Cather Lake in Wisconsin, liming improved conditions for fish production by increasing pH from 5.6 to 7.1–7.5, alkalinity from 3.0 to 15–19 mg/L, and Secchi disk visibility from 2.0 m to 4.3–5.7 m. Conditions for fish production were greatly improved. The use of liming materials also is a common practice in sportfish ponds and aquaculture ponds (Boyd 2017).

The most common agent for neutralizing acidity in water bodies is agricultural limestone made by pulverizing limestone. This liming material usually has an acid neutralizing value of 90–100% $CaCO_3$ equivalent. Two other common liming materials are burnt lime (calcium oxide or a mixture of calcium and magnesium oxides) and hydrated lime (calcium hydroxide or a mixture of calcium and magnesium hydroxides). Burnt lime is made by burning limestone in a furnace, and the resulting chemical reaction is illustrated using calcium carbonate as the limestone source

$$CaCO_3 \xrightarrow{\Delta} CaO + CO_2 \uparrow . \tag{11.16}$$

Hydrated lime is made by treating burnt lime with water

$$CaO + H_2O \rightarrow Ca(OH)_2. \qquad (11.17)$$

Burnt lime usually has an acid neutralizing value of 160–180% $CaCO_3$ equivalent, while for hydrated lime the value is about 130–140% $CaCO_3$ equivalent.

Sodium bicarbonate and sodium hydroxide also can be used to neutralize acid, but they tend to be more expensive than limestone products. The acid neutralizing values of sodium bicarbonate and sodium hydroxide are around 60% and 125% $CaCO_3$ equivalent, respectively.

The neutralization of acidity in a water body appears simple at first consideration. A water body containing 10,000 m^3 with an acidity of 100 mg/L of $CaCO_3$ would require 1000 kg of liming material (expressed as equivalent $CaCO_3$) to neutralize its acidity (10,000 $m^3 \times 100$ mg $CaCO_3$/L of acidity $\times 10^{-3}$ kg/g = 1000 kg $CaCO_3$). This amount of liming material would only neutralize the acidity in the water. Additional liming material would be required to raise the alkalinity of the water. For example, to raise the alkalinity to 50 mg/L would require another 500 kg of $CaCO_3$ equivalent.

Effective liming cannot be achieved as illustrated in the preceding paragraph. In an acidic water body, the sediment contains acidity, and this acidity must be neutralized or it will remove alkalinity from the water. Soil acidity is more difficult to measure than is the acidity of water. The lime requirement of sediment can be measured by procedures modified from agricultural soil testing methods (Han et al. 2014; Boyd 2017).

Acidic effluents are treated with liming materials to neutralize acidity and prevent damage to organisms in receiving water bodies. Acidic water is corrosive to water pipes and other plumbing fixtures. Thus, liming of acidic water before distribution protects the delivery system from corrosion.

Conclusions

The productivity and biodiversity of water bodies decrease as pH declines, and most waters with little or no alkalinity (pH <5.5) are considered dystrophic. The exception is waters from tropical jungles in which the acidity is primarily from humic substances. The lowest pHs are found in surface waters that contain sulfuric acid from oxidation of iron pyrite in soils, from acidic mine drainage, or from acidic rainfall. Acidity can be measured by titration with standard sodium hydroxide. Liming materials (limestone, burnt lime, and hydrated lime) often are applied to mitigate acidity.

References

Beckett R, Jue Z, Giddings JC (1987) Determination of molecular weight distributions of fulvic and humic acids using flow field-flow fractionation. Environ Sci Tech 21:289–295

Borda MJ (2006) Pyrite. In: Lai R (ed) Encyclopedia of soil science. Taylor & Francis, New York, pp 1385–1387

Boyd CE (1968) Freshwater plants: a potential source of protein. Econ Bot 22:359–368

Boyd CE (2017) Use of agricultural limestone and lime in aquaculture. CAB Rev 12(015). http://cabi.org/cabreviews

Caldeira K, Wickett ME (2003) Anthropogenic carbon and ocean pH. Nature 425:365

Chandra AP, Gerson AR (2011) Pyrite (FeS$_2$) oxidation: a sub-micron synchrotron investigation of the initial steps. Geochim Cosmochim Acta 75:6239–6254

Cowling EB (1982) Acid precipitation in historical perspective. Environ Sci Tech 16:110–123

de Melo BA, Motta FL, Santana MH (2016) Humic acids: structural properties and multiple functionalities for novel technological developments. Mater Sci Eng C 62:967–974

Eaton AD, Clesceri LS, Rice EW, Greenburg AE (eds) (2005) Standard methods for the examination of water and wastewater. American Public Health Association, Washington, DC

Everall NC, Lees DR (1996) The use of barley-straw to control general and blue-green algal growth in a Derbyshire reservoir. Water Res 30:269–276

Fanning DS (2006) Acid sulfate soils. In: Lai R (ed) Encyclopedia of soil science. Taylor & Francis, New York, pp 11–13

Feely RA, Sabine CL, Lee K, Berelson W, Kleypas J, Fabry VJ, Millero FJ (2004) Impact of anthropogenic CO$_2$ on the CaCO$_3$ system in the ocean. Science 305:363–366

Fleming JF, Alexander LT (1961) Sulfur acidity in South Carolina tidal marsh soils. Soil Sci Soc Am Proc 25:94–95

Geissman TA, Crout DHG (1969) Organic chemistry of secondary plant metabolism. Freeman, San Francisco

Haines TA (1981) Acid precipitation and its consequences for aquatic ecosystems: a review. Trans Am Fish Soc 110:669–707

Han Y, Boyd CE, Viriyatum R (2014) A bicarbonate method for lime requirement to neutralize exchangeable acidity of pond bottom soils. Aqua 434:282–287

Hasler AD, Brynildson OM, Helm WT (1951) Improving conditions for fish in brown-water bog lakes by alkalization. J Wildl Manag 15:347–352

Holland A, Duivenvoorden LJ, Kinnear SHW (2015) Effect of key water quality variables on macroinvertebrate and fish communities within naturally acidic wallum streams. Mar Freshw Res 66:50–59

Johnson DS (2010) Liming and agriculture in the Central Pennines. BAR Publishing, Oxford

Klavins M, Rodinov V, Druvietis I (2003) Aquatic chemistry and humic substances in bog lakes in Latvia. Boreal Environ Res 8:113–123

Lawrence J (1980) Semi-quantitative determination of fulvic acid, tannin and lignin in natural waters. Water Res 14:373–377

Nassar JL, Li M (2004) Landscape mirror: the attractiveness of reflecting water. Landsc Urban Plan 66:233–238

Nesbitt HW, Bancroft GM, Pratt AR, Scaini MJ (1998) Sulfur and iron surface states on fractured pyrite surfaces. Am Min 83:1067–1076

Orr JC, Fabry VJ, Aumont O, Bopp L, Doney SC, Feely RA, Gnanadesikan A, Gruber N, Ishida A, Joos F, Key RM, Lindsay K, Maier-Reimer E, Matear R, Monfray P, Mouchet A, Najjar RG, Plattner G, Rodgers KB, Sabine CL, Sarmiento JL, Schlitzer R, Slater RD, Totterdell IJ, Weirig M, Yamanaka Y, Yool A (2005) Anthropogenic ocean acidification over the twenty-first century and its impact on calcifying organisms. Nature 437:681–686

Rathbun RE (1996) Disinfection byproduct yields from the chlorination of natural waters. Arch Environ Contam Toxicol 31:420–425

Rickard D, Luther GW III (1997) Kinetics of pyrite formation by the H_2S oxidation of iron (II) monosulfide in aqueous solutions between 25 and 125°C: the mechanisms. Geochim Cosmochim Acta 61:135–147

Rickard D, Luther GW III (2006) Metal sulfide complexes and clusters. Rev Mineral Geochem 61:421–504

Schippers A (2004) Biogeochemistry of metal sulfide oxidation in mining environments, sediments, and soils. In: Amend JP, Edwards KJ, Lyons TW (eds) Sulfur biogeochemistry: past and present, Special Paper 379. Geological Society of America, Washington, DC, pp 49–62

Shinozuka T, Shibata M, Yamaguchi T (2004) Molecular weight characterization of humic substances by MALDI-TOF-MS. J Mass Spectrom Soc Jpn 52:29–32

Smith DG, Croker GF, McFarlane K (1995) Human perception of water appearance. 1. Clarity and color for bathing and aesthetics. N Z J Mar Fresh Res 29:29–43

Smith SJ, van Aardenne J, Kilmont Z, Andres RJ, Volke A, Arias SD (2011) Anthropogenic sulfur dioxide emissions: 1850-2005. Atmos Chem Phys 11:1101–1116

Soil Survey Staff (1994) Keys to soil taxonomy, 6th edn. United States Department of Agriculture, Soil Conservation Service, Washington, DC

Sorensen DL, Knieb WA, Porcella DB, Richardson BZ (1980) Determining the lime requirement for the blackbird mine spoil. J Environ Qual 9:162–166

Steelink C (1963) What is humic acid? Proc Cal Assoc Chem Teach 7:379–384

Steelink C (2002) Investigating humic acids in soils. Anal Chem 74:327–333

Microorganisms and Water Quality

12

Abstract

Phytoplankton and bacteria have a greater effect on water quality than do other aquatic microorganisms. Phytoplankton are the main primary producers while bacteria are responsible for the majority of organic matter decomposition and nutrient recycling. An overview of microbial growth, photosynthesis, and respiration is provided, and methods for measuring primary production and respiration in water bodies are discussed. The combined physiological activities of producer and decomposer organisms in water bodies cause pH and dissolved oxygen concentration to increase and carbon dioxide concentration to decrease in daytime, while the opposite occurs during nighttime. In unstratified water bodies, aerobic conditions usually exist in the water column and at the sediment-water interface. Nevertheless, sediment typically is anaerobic at depths greater than a few centimeters in oligotrophic water bodies or greater than a few millimeters in eutrophic water bodies. In anaerobic sediment (or water), the metabolic activity of chemotrophic bacteria is important in decomposing organic compounds resulting from fermentation. Although chemotrophic bacteria are beneficial in assuring more complete decomposition of organic matter, toxic metabolic wastes—particularly nitrite and hydrogen sulfide—produced by these microorganisms can enter the water column. Blue-green algae—often called cyanobacteria—tend to dominate phytoplankton communities in eutrophic waters. Blue-green algae can cause surface scums and shallow thermal stratification, be toxic to other algae and aquatic animals, or produce taste and odor problems in public water supplies.

Introduction

Aquatic ecosystems are inhabited by many types of microorganisms to include algae, bacteria, fungi, protozoa, rotifers, bryozoa, and arthropods. Algae are the primary producers, and bacteria and fungi are the decomposers. Microscopic animals feed on microscopic plants and animals and their remains; they serve to link between

© Springer Nature Switzerland AG 2020
C. E. Boyd, *Water Quality*, https://doi.org/10.1007/978-3-030-23335-8_12

primary producers and larger animals in the food web of aquatic ecosystems. All of these organisms are important ecologically, but microscopic algae suspended in the water column (phytoplankton) and bacteria suspended in the water column and living in the sediment have a greater impact on water quality than do other microorganisms. Phytoplankton produce large amounts of organic matter through photosynthesis and release prodigious quantities of oxygen into water during the process. Bacteria decompose organic matter to release (recycle) inorganic nutrients. Respiration by bacteria and respiration and photosynthesis by phytoplankton have a pronounced effect on pH and concentrations of carbon dioxide and dissolved oxygen in water. Certain bacteria and other microscopic organisms are pathogenic, and some species of algae can impart bad tastes and odors to water as well as to the flesh of fish and other aquatic food animals.

Although phytoplankton and bacteria are the focus of this chapter, larger aquatic plants that are rooted in the bottoms and edges of water bodies also will be discussed as they are sometimes the dominant plants in aquatic habitats. Moreover, all consumer organisms participate in the process of organic matter degradation and nutrient recycling through their use of organic matter for food.

Bacteria

A main distinguishing feature between bacteria and other single-celled microorganisms is that the nuclei of bacterial cells are not enclosed in a membrane—they are known as prokaryotic organisms as opposed to eukaryotic organisms in which the nucleus is bound by a membrane. Bacteria may be either unicellular or filamentous, and their cells usually are spherical or cylindrical (rods). They may live free in water, attached to surfaces, in sediment, and within other living organisms. A few types are capable of locomotion. Some bacteria are pathogenic to plants, animals, or humans.

The food source for most bacteria is dead organic matter, while a few kinds of bacteria are capable of synthesizing organic matter. These two groups of bacteria are known as heterotrophic and autotrophic bacteria, respectively. Obligate aerobic bacteria cannot live without oxygen, obligate anaerobic bacteria cannot exist in environments with oxygen, and facultative anaerobic bacteria can do well with or without oxygen. Some species that can function without molecular oxygen obtain oxygen from nitrate, sulfate, carbon dioxide, or other inorganic compounds.

The major ecological role of most bacteria is to decompose organic matter and recycle its essential inorganic components, e.g., carbon dioxide, water, ammonia, phosphate, sulfate, and other minerals. Some bacteria can have a major influence on certain species in a water body by causing disease. There also are some bacteria and viruses that can cause illness in humans who drink water in which these organisms occur.

Physiology

Nutrients have three functions in growth and metabolism. They are raw materials for elaboration of biochemical compounds of which organisms are made. Carbon and nitrogen contained in organic nutrients are used in making protein, carbohydrate, fat, and other components of microbial cells. Nutrients supply energy for growth and chemical reactions. Organic nutrients are oxidized in respiration, and the energy released is used to drive chemical reactions that synthesize biochemical compounds necessary for growth and maintenance. Nutrients also serve as electron and hydrogen acceptors in respiration. The terminal electron and hydrogen acceptor for aerobic respiration is molecular oxygen. An organic metabolite or inorganic substance replaces oxygen as electron or hydrogen acceptor in anaerobic respiration.

Growth

Water and sediment contain species of microorganisms capable of decomposing almost any organic substance. Some substances decompose faster than others, but few organic compounds completely resist decay by microorganisms. Microbial activity is slow where organic matter is scarce, but actively growing microorganisms, resting spores, and other propagules are present almost everywhere. An increase in organic matter provides substrate for microbial growth, and the number of microorganisms increases. As a general rule, the bacteria necessary to degrade organic matter are present in nature. It is usually not necessary to add bacteria. If decomposition is slow, it is because environmental conditions do not favor rapid microbial action or the organic substrate is resistant to decay.

Bacteria reproduce by binary fission. One cell divides into two cells, and the new cells continue to divide. The time between cell divisions is called the generation time or doubling time. The number of bacterial cells present after a given time (N_t) can be computed from the initial number of cells (N_o) and the number of generations (n)

$$N_t = N_o \times 2^n. \tag{12.1}$$

Microorganisms with a short generation time increase their numbers quickly (Ex. 12.1).

Ex. 12.1 *Organic matter is added to a water containing 10^3 bacterial cells per milliliter. These bacteria can double every 4 hours. The number of bacterial cells after 36 hours will be estimated.*
　　Solution:
The number of generations that will occur in 36 hours is

$$n = \frac{36\ hr}{4\ hr/generation} = 9.$$

The number of cells after 36 hours can be calculated with Eq. 12.1.

$N_t = 10^3 \times 2^9 = 512 \times 10^3 = 5.12 \times 10^5\, cells/mL\ (a\ 512\ fold\ increase\ in\ 36\ hr).$

When bacteria are inoculated into fresh organic substrate, it takes a short time for them to adjust to the new conditions. The period during which there is little or no increase in cell number is called the lag phase (Fig. 12.1). Rapid growth follows as microorganisms utilize the new substrate; this period is known as the logarithmic or log phase. After a period of rapid growth, a stationary phase is attained during which cell number remains relatively constant. As the substrate is used up and metabolic by-products accumulate, growth slows and a decline phase occurs in which the number of cells decreases.

The generation time for a bacterial population can be computed from data on cell numbers during the logarithmic phase. Equation 12.1 can be rewritten as

$$\log N_t = \log N_o + n \log 2 \tag{12.2}$$

which becomes

$$0.301\, n = \log N_t - \log N_o \tag{12.3}$$

and

$$n = \frac{\log N_t - \log N_o}{0.301}. \tag{12.4}$$

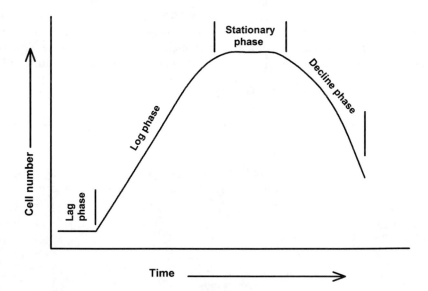

Fig. 12.1 Characteristic phases of growth in microbial cultures

The generation time (g) is computed from the time interval (t) over which a given number of generations (n) occurred

$$g = \frac{t}{n} \text{ or } n = \frac{t}{g}.$$ (12.5)

Substituting $\frac{t}{g}$ for n in Eq. 12.4 gives

$$g = \frac{0.301\,t}{logN_t - logN_o}.$$ (12.6)

The utility of Eq. 12.6 is illustrated in Ex. 12.2.

Ex. 12.2 *A bacterial culture increases from 10^4 cells/mL to 10^8 cells/mL in 12 hours. The generation time will be calculated using Eq. 12.6.*
Solution:

$$g = \frac{0.301(12)}{8 - 4} = 0.903\ hr = 54\ min.$$

The generation time of bacteria differs among species, and for a given species, generation time varies with temperature, substrate availability, and other environmental factors. When bacteria are growing rapidly, they also are quickly decomposing substrate. For example, in a water where microorganisms are growing logarithmically, the rate of evolution of carbon dioxide increases logarithmically.

Aerobic Respiration

Aerobic respiration is a major topic in books on plant and animal physiology, metabolic process, and general biochemistry. Nevertheless, a brief overview of the major features of aerobic respiration will be provided. The reader should note that aerobic respiration is much the same for all organisms, and it involves oxidation of organic compounds to carbon dioxide and water. The purpose of respiration is to release energy from organic compounds and store it in high-energy phosphate bonds of adenosine triphosphate (ATP)

$$\text{Adenosine diphosphate (ADP)} + PO_4 + \text{energy} \rightleftharpoons \text{ATP}.$$ (12.7)

Energy stored in ATP can be used to drive chemical reactions in cells. When energy is released from ATP, ADP is regenerated and used again.

In aerobic carbohydrate metabolism, a glucose molecule passes through glycolysis and the citric acid cycle (also called the Krebs cycle or the tricarboxylic acid cycle) before being completely oxidized to carbon dioxide. Glycolysis is a series of ordered reactions catalyzed by numerous enzymes—not indicated in the equations below—that transform one glucose molecule to two pyruvate molecules

$$C_6H_{12}O_6 \rightarrow 2C_3H_4O_3 + 4H^+. \tag{12.8}$$

Two moles of ATP are formed per mole of glucose in glycolysis, but no carbon dioxide is released. As a result, glycolysis does not require oxygen.

In aerobic respiration, pyruvate molecules from glycolysis react with coenzyme A (CoA) and nicotinamide adenine dinucleotide (NAD$^+$) forming acetyl CoA molecules

$$\text{Pyruvate} + \text{CoA} + 2\text{NAD}^+ \rightleftharpoons \text{Acetyl CoA} + 2\text{NADH} + 2\text{H}^+ + \text{CO}_2. \tag{12.9}$$

The formation of acetyl CoA from pyruvate is an oxidative decarboxylation in which two H$^+$ ions, 2 electrons, and one CO$_2$ molecule are removed from each molecule of pyruvate—a total of four hydrogen ions and two carbon dioxide molecules per molecule of the original glucose entering glycolysis. Acetyl CoA formation links glycolysis to the citric acid cycle. Although glycolysis does not use oxygen, acetyl CoA production and the citric acid cycle require oxygen.

The first step in the citric acid cycle is the reaction of acetyl CoA and oxaloacetic acid to form citric acid. In succeeding reactions—also catalyzed by enzymes—a series of transformations of organic acids occur with release of eight H$^+$ ions and two CO$_2$ molecules for each molecule of citric acid entering the cycle. This accounts for the oxidation of the six organic carbon atoms present in one glucose molecule. One molecule of oxaloacetic acid also is regenerated in the cycle (recycled) and it can react again with acetyl CoA.

Hydrogen ions produced through oxidation of glucose in glycolysis and the citric acid cycle reduce the enzymes NAD, NADP (nicotinamide adenine dinucleotide phosphate), and FAD (flavin adenine dinucleotide). A representative reaction where H$^+$ is released in the citric acid cycle is

$$
\begin{array}{ccccc}
C_6H_8O_7 & \rightarrow & C_6H_6O_7 & + & 2H^+ \\
\text{(Isocitric acid)} & & \text{(Oxalosuccinic acid)} & &
\end{array}
$$

and

$$\text{NADP}^+ + 2\text{H}^+ \rightleftharpoons \text{NADPH} + \text{H}^+. \tag{12.10}$$

Enzymes of the citric acid cycle are brought in contact with the electron transport system where they are reoxidized. Energy released in these oxidations is used for ATP synthesis, and the regenerated NAD, NADP, and FAD are used again as hydrogen acceptors. Reoxidation of enzymes is accomplished by a series of cytochrome enzymes that pass electrons from one cytochrome compound to another. In the electron transport system, ADP combines with inorganic phosphate to form ATP and H$^+$ ions combine with molecular oxygen to form water.

A total of 38 ATP molecules can be formed from 1 mole of glucose in aerobic respiration (glycolysis and citric acid cycle). The ATP molecules contain approximately one-third of the theoretical energy released during glucose oxidation. Energy not used to form ATP is lost as heat. In aerobic respiration, 1 mole of glucose consumes 6 moles of oxygen and releases 6 moles of CO_2 and water. This stoichiometry is illustrated in the summary equation for respiration

$$C_6H_{12}O_6 + 6O_2 \rightarrow 6CO_2 + 6H_2O + \text{energy.} \tag{12.11}$$

This equation usually is an adequate summary of the overall results of respiration in all types of aerobic organisms—microscopic or macroscopic and plant, animal, bacterial, or fungal.

Oxygen Stoichiometry in Aerobic Respiration

Organic matter contains a wide array of organic compounds, most of which are more complex than glucose. Bacteria excrete extracellular enzymes that break complex organic matter down into particles small enough to be absorbed. These fragments are broken down further inside the bacterial cell by enzymatic action until they are small enough to be used in glycolysis. These fragments pass through the respiratory reactions in the same way as glucose fragments.

The ratio of moles carbon dioxide produced to moles oxygen consumed is the respiratory quotient (RQ)

$$RQ = \frac{CO_2}{O_2}. \tag{12.12}$$

Reference to Eq. 12.11 reveals that in the oxidation of glucose, six oxygen molecules are consumed and six carbon dioxide molecules are released, i.e., $RQ = 1.0$. Other classes of organic matter do not contain the same ratio of carbon to oxygen to hydrogen as carbohydrate (Table 12.1), and RQ may be lesser or greater than 1.0. For example, consider the oxidation of a fat

$$C_{57}H_{104}O_6 + 80O_2 \rightarrow 57CO_2 + 52H_2O. \tag{12.13}$$

The respiratory quotient for the above reaction is $57CO_2/80O_2$ or 0.71. Organic compounds more highly oxidized (more oxygen relative to carbon) than carbohydrates will have a respiratory quotient greater than 1.0, and more reduced compounds such as fat and protein will have a respiratory quotient less than 1.0. It follows that the more reduced an organic substance, the more oxygen that is required to oxidize a unit quantity (Ex. 12.3).

Table 12.1 Carbon, hydrogen, and oxygen content of three major classes of organic compounds

	% C	% H	% O
Carbohydrate	40	6.7	53.3
Protein	53	7	22
Fatty acids	77.2	11.4	11.4

Ex. 12.3 *Calculate the amount of oxygen required to completely oxidize the organic carbon in 1 g each of carbohydrate, protein, and fat.*

Solution:

Assume that carbon concentrations are carbohydrate, 40%; protein, 53%; fat, 77.2% (Table 10.1). One gram of mass represents 0.4 g C, 0.53 g C, and 0.772 g C for carbohydrate, protein, and fat, respectively. The stoichiometry for converting organic carbon to CO_2 is

$$C + O_2 \rightarrow CO_2.$$

The appropriate ratio for computing oxygen consumption is

$$\frac{g \ C}{12} = \frac{O_2 \ used}{32}$$

$$O_2 \ used = \frac{32 \ (g \ C)}{12}$$

$$O_2 \ carbohydrate = \frac{(32)(0.4)}{12} = 1.07 \ g$$

$$O_2 \ protein = \frac{(32)(0.53)}{12} = 1.41 \ g$$

$$O_2 \ fat = \frac{(32)(0.772)}{12} = 2.06 \ g.$$

Based on the relationship between the degree of reduction of organic compounds and their oxygen requirement for decomposition, one can see that there is not a direct proportionality between the weight of organic matter and of oxygen required for its decomposition. There is, however, a direct proportionality between the percentage carbon in organic matter and the amount of oxygen necessary to completely decompose it. Of course, some complex organic matter is more resistant to decay than simple carbohydrates, and for this reason, the rate of decomposition of organic matter is not always related to carbon concentration. There also is a direct relationship between the amount of organic carbon in a compound and the amount of energy released by its oxidation. The specific heats of combustion of food groups are: carbohydrate, 4.1 kcal/g; protein, 5.65 kcal/g; fat, 9.4 kcal/g. Metabolizable energy is the important factor in nutrition, and the carbohydrates and protein are considered to have 4 kcal/g while fat has 9 kcal/g of metabolizable energy (Jumpertz et al. 2013).

Anaerobic Respiration

Under anaerobic conditions, certain kinds of bacteria, yeast, and fungi, continue to respire, but in the absence of molecular oxygen, terminal electron acceptors for anaerobic respiration are organic or inorganic compounds. Carbon dioxide may be produced in anaerobic respiration, but other end products include alcohol, formate, lactate, propionate, acetate, methane, other organic compounds, gaseous nitrogen, ferrous iron, manganous manganese, and sulfide.

Fermentation is a common type of anaerobic respiration. Organisms capable of fermentation hydrolyze complex organic compounds to simpler ones that can be used in a process identical to glycolysis to produce pyruvate. In fermentation, pyruvate cannot be oxidized to carbon dioxide and water with molecular oxygen serving as the terminal electron and hydrogen acceptor. Hydrogen ions removed from organic matter during pyruvate formation are transferred via nicotinamide adenine dinucleotide to an intermediary product of metabolism.

Ethanol production from glucose has traditionally been used in general biology textbooks as an example of fermentation. In this process, glucose is converted to pyruvate, and four ATP molecules result from each molecule of glucose. Hydrogen ions removed from pyruvate are transferred to acetaldehyde with the release of carbon dioxide. The summary equations are as follows:

$$\underset{\text{(Glucose)}}{C_6H_{12}O_6} \quad \rightarrow \quad \underset{\text{(Pyruvate)}}{2C_3H_4O_3} \quad + \quad 4H^+ \tag{12.14}$$

$$2C_3H_4O_3 \quad \rightarrow \quad \underset{\text{(Acetaldehyde)}}{2C_2H_4O} \quad + \quad 2CO_2 \tag{12.15}$$

$$2C_2H_4O + 4H^+ \quad \rightarrow \quad \underset{\text{(Ethanol)}}{2CH_3CH_2OH.} \tag{12.16}$$

In addition to producing carbon dioxide, fermentation also may yield hydrogen gas (H_2). The hydrogen gas apparently is formed when NADH is oxidized at low hydrogen pressure with liberation of hydrogen gas as follows:

$$NADH + H^+ \rightarrow NAD^+ + H_2. \tag{12.17}$$

Ethanol is only one of many organic compounds that can be produced by fermentation. For example, some microorganisms convert glucose to lactic acid

$$\underset{\text{(Glucose)}}{C_6H_{12}O_6} \quad \rightarrow \quad 2C_3H_4O_3 + 4H^+ \tag{12.18}$$

$$2C_3H_4O_3 + 4H^+ \quad \rightarrow \quad \underset{\text{(Lactic acid)}}{2C_3H_6O_3.} \tag{12.19}$$

In lactic acid production, carbon dioxide is not released as it is in ethanol production (Eqs. 12.15 and 12.16).

Fermentation does not oxidize organic matter completely. In ethanol production only one-third of the organic carbon in glucose is converted to carbon dioxide, and no carbon dioxide is produced in the fermentation of glucose to lactic acid. As a result, carbon dioxide and organic products accumulate in zones where fermentation is occurring. Fortunately, the end products of fermentation can be oxidized by microorganisms capable of using inorganic substances instead of molecular oxygen as electron acceptors. If decomposition stopped with fermentation, the environment would soon be very acidic and contain a lot of intermediate, organic decomposition products.

Bacteria that use nitrate as an electron acceptor can hydrolyze complex compounds and oxidize the hydrolytic products to carbon dioxide. Nitrate is reduced to nitrite, ammonia, nitrogen gas, or nitrous oxide. In the zone where nitrate-reducing bacteria occur, a part of the organic carbon is oxidized completely to carbon dioxide and a portion is converted to organic fermentation products.

The iron- and manganese-reducing bacteria utilize oxidized iron and manganese compounds as oxidants in the same manner as nitrate-reducing bacteria use nitrate. They absorb organic fermentation products and oxidize them to carbon dioxide. Ferrous iron (Fe^{2+}) and manganous manganese (Mn^{2+}) are released as by-products of respiration.

Sulfate-reducing bacteria and methane-producing bacteria cannot hydrolyze complex organic substances or decompose simple carbohydrates and amino acids originating from hydrolytic activity by other bacteria. They utilize short-chain fatty acids and simple alcohols produced by fermentation as organic carbon sources. Sulfate-reducing bacteria use sulfate as an oxygen source to oxidize fermentation products to carbon dioxide. Sulfide is released as a by-product.

Fermentation products also can be utilized by bacteria that produce methane. In the most common method of methane formation, a simple organic molecule is fermented and carbon dioxide is utilized as the electron (hydrogen) acceptor as shown below:

$$CH_3COOH + 2H_2O \rightarrow 2CO_2 + 8H^+ \tag{12.20}$$

$$8H^+ + CO_2 \rightarrow CH_4 + 2H_2O. \tag{12.21}$$

Subtraction of the two reactions gives

$$CH_3COOH \rightarrow CH_4 + CO_2. \tag{12.22}$$

Some bacteria also can use carbon dioxide as an oxidant to convert hydrogen gas produced in fermentation to methane

$$4H_2 + CO_2 \rightarrow CH_4 + 2H_2O. \tag{12.23}$$

Methane production is an important process because the hydrogen that accumulates from fermentation must be disposed of or it will inhibit the fermentation process.

The presence of alternative electron acceptors such as nitrate, ferric iron, sulfate, and other oxidized inorganic compounds in water or sediment favor anaerobic decomposition. Complete decomposition of organic matter in aquatic environments requires both aerobic and anaerobic organisms. The anaerobic organisms are especially important in sediment, because organic matter tends to settle to the bottom of water bodies, and anaerobic conditions typically occur a few centimeters or millimeters below the sediment-water interface. In eutrophic waters, the hypolimnion will become anaerobic during thermal stratification.

Complete decomposition of a given organic matter residue takes a long time. The easily-decomposed compounds are oxidized within a few days to months. As they decompose, bacteria die and also decompose. Slowly decomposing organic matter is synthesized into complex organic matter that decomposes even more slowly (a few years) and the remains of this material takes many years to decompose. Hundreds of years may be required to completely decompose an organic residue to its basic inorganic elements.

Environmental Effects on Bacterial Growth

Major factors affecting growth and respiration of bacteria are temperature, oxygen supply, moisture availability, pH, mineral nutrients, and composition and availability of organic substrates. Temperature effect on growth is illustrated in Fig. 12.2. Growth is limited by low temperature, there is a narrow optimal temperature range, temperature can be too high for good growth, and temperature can reach the thermal death point of the decomposers.

According to van Hoff's law, the rate of many chemical reactions doubles or triples with a 10 °C increase in temperature. Most physiological processes are chemical reactions and comply closely to van Hoff's law. The factor of increase in reaction rate that typically is about 2 or 3 for a 10 °C increase in temperature usually is called the Q_{10}. At suboptimal temperature, an increase of 10 °C normally doubles respiration and growth. Decomposition rate of organic matter also will double if the temperature is increased by 10 °C. Most bacteria in natural waters and sediment probably grow best at 30–35 °C, and decomposition rate slows at greater temperatures.

A continuous supply of oxygen is needed to support aerobic bacterial activity, but facultative and anaerobic bacteria continue to decompose organic matter under anaerobic conditions. Leachable and easily hydrolysable compounds in fresh organic matter decompose equally fast under either aerobic or anaerobic conditions, but anaerobic decomposition of structural remains (complex macromolecules) of living things is slower than aerobic decomposition (Kristensen et al. 1995).

Moisture is abundant in aquatic habitats, so aquatic decomposition is not hampered by lack of moisture. Bacteria thrive at pH 7 to 8, and decomposition is more rapid in neutral or slightly alkaline environments than in acidic ones. Bacteria must have inorganic nutrients such as sulfate, phosphate, calcium, potassium, etc., but these nutrients normally can be obtained from the substrate even if they are present in water at low concentration.

The nature of the organic substrate is particularly important. The protoplasmic components of organic matter can be decomposed more rapidly than the structural (cell wall) fraction. Fibrous remains of reeds or other large aquatic plants decompose slower than residues from dead phytoplankton. Organic residues with a large amount of nitrogen usually decompose faster than those with less nitrogen because bacteria

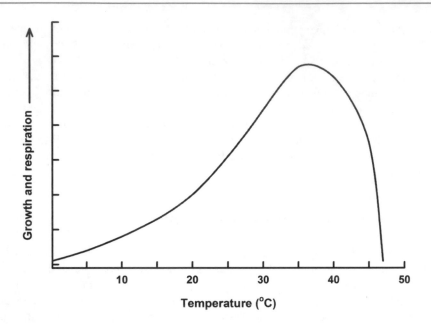

Fig. 12.2   Typical response of microbial growth and respiration to temperature

need a lot of nitrogen, and high nitrogen-content residues usually contain a minimum of structural components. Organic matter tends to accumulate in water bodies where there is low temperature and abundant growth of reedswamp plants that have high fibrous content and low nitrogen content. These water bodies typically have low pH and they fill in with slowly decomposing remains of plant residues creating bogs.

Estimating Bacterial Abundance

The common way of estimating bacterial abundance is to make serial dilutions of water samples, or samples of sediment mixed with water, pour the final dilution into a petri dish, add liquefied culture media, and mix. The dishes are incubated and the number of colonies that develop in the media is counted. It is assumed that each colony originates from a single cell or filament, and the bacterial count in colony-forming units (CFU) per milliliter is the number of colonies multiplied by the dilution factor. The counts usually are done in triplicate because of a high degree of variation.

Metabolic activity of certain types of bacteria can be used as an index of their presence or abundance. For example, the formation of gas in a culture tube with lactose medium indicates the presence of lactose fermentating bacteria. Conversion of ammonia to nitrate reveals the presence of nitrifying bacteria. The rate of consumption of oxygen or the release of carbon dioxide can indicate the rate of decomposition of organic matter, and the abundance of the decomposing bacteria is related to the rate of carbon dioxide release.

Sediment Respiration

Decomposition occurs in both water and sediment, and the main effect on water quality of aerobic respiration is to remove dissolved oxygen and release carbon dioxide, ammonia, and other mineral substances. The flocculent layer of sediment just above the sediment-water interface and the upper few centimeters or millimeters of sediment contain a large amount of fresh, readily decomposable organic matter. Microbial activity in aquatic ecosystems usually is greatest in the flocculent layer and upper sediment layer. Uptake of oxygen results from respiration in the flocculent layer, and from respiration and chemical oxidation of reduced metabolites in the sediment. Sediment respiration rates as great as 20–30 g O_2/m^2 per day have been measured, but rates usually are 5 g O_2/m^2 per day or less. The removal of dissolved oxygen for use in sediment respiration is illustrated in Ex. 12.4.

Ex. 12.4 *The loss of dissolved oxygen from the water will be estimated for sediment respiration of 2 g $O_2/m^2/day$ in a 3-m deep body of water.*
   Solution:
The water column above a 1 m^2 area of sediment will contain 3 m^3 of water. Remembering that 1 mg/L is the same as 1 g/m^3,

$$\frac{2 \ g \ O_2/m^2}{3 \ m^3 \ water/m^2} = 0.67 \ g/m^3 \ or \ 0.67 \ mg/L.$$

Oxygen in sediment pore water or in the water column also may be consumed by purely chemical reaction without biological influence as illustrated in the following equations:

$$2Fe^{2+} + 0.5O_2 + 2H_2O \rightarrow Fe_2O_3 + 4H^+ \tag{12.24}$$

$$Mn^{2+} + 0.5O_2 + H_2O \rightarrow MnO_2 + 2H^+ \tag{12.25}$$

$$H_2S + 2O_2 \rightarrow SO_4^{2-} + 2H^+. \tag{12.26}$$

Some bacteria also can mediate reactions in which reduced metabolites are oxidized, and they also consume oxygen in the process. It is difficult to separate oxygen consumption by respiration from oxygen consumption in chemical oxidation, and sediment oxygen uptake measurements usually indicate the combined uptake by both processes.

Oxidation of metabolites is important, because the deeper sediment is anaerobic, and the reduced metabolites of anaerobic respiration can diffuse upward or be mixed upward by bioturbation. Upon entering the aerobic zone of the sediment, the reduced metabolites may be oxidized chemically or through microbial activity. In the case where sediment is anaerobic to the sediment-water interface, the reduced metabolites enter the water column, and some of them are toxic. Chemical and biological oxidation of these metabolites is important in preventing toxic conditions from developing in the water column.

Organic matter decomposition controls the redox potential in water and sediment. As long as dissolved oxygen is plentiful, only aerobic decomposition occurs. However, when dissolved oxygen concentration falls to 1 or 2 mg/L, certain bacteria begin to use oxygen from nitrate. As nitrate is used up, the redox potential declines, and when nitrate is depleted, the bacteria begin to use oxidized forms of iron and manganese in respiration. Redox continues to fall until sulfate and finally carbon dioxide become oxygen sources. Because the utilization of oxidized inorganic compounds is sequential, the different bacterial reductions occur sequentially in time or occur in different layers of sediment. For example, when a body of water stratifies thermally, it no longer obtains dissolved oxygen from the illuminated epilimnion in which photosynthesis occurs. Organic matter settles into the hypolimnion, and dissolved oxygen and redox potential decline as aerobic bacteria decompose organic matter. Simply put, once dissolved oxygen is depleted, nitrate supports respiration; when nitrate is gone, iron and manganese compounds become oxidants in respiration; sulfate and finally carbon dioxide serve as oxidants as redox falls.

Sediment becomes depleted of oxygen quicker than water, because dissolved oxygen cannot move downward rapidly in the pore water. Oxygen availability declines with sediment depth, and this results in zonation of the different processes (Fig. 12.3). This zonation will develop in sediment in bodies of water that are not chemically stratified, but in most unstratified bodies of water, the thin surface layer of the sediment remains aerobic as shown in Fig. 12.3. If organic matter inputs are high, the entire sediment mass, including its surface, may become highly reduced in both stratified and unstratified water bodies. The vertical profile of redox potential in an unstratified lake and its sediment is illustrated in Fig. 12.3.

The results of anaerobic microbial respiration is the occurrence of reduced inorganic compounds to include nitrite, ferrous iron, manganous manganese, sulfide, and methane and many reduced organic compounds in waters of the hypolimnion of lakes, ponds, and other water bodies. When these reduced substances diffuse or are mixed into waters containing dissolved oxygen, they will be oxidized. If the input rate of reduced substances into aerobic water exceeds the oxidation rate, an equilibrium concentration of reduced substances may persist.

Some bacteria can oxidize ammonia to nitrate and thereby lower ammonia concentrations in water. This process will be discussed in Chap. 13.

Phytoplankton

The aquatic flora is very diverse and includes many thousand species. These include planktonic algae (phytoplankton) that are microscopic and suspended in the water, macrophytic algae that live on the bottom or form mats in the water, fungi which lack chlorophyll and are decomposers rather than producers like other plants, liverworts and mosses that lack flowers and water conducting tissues characteristic of more advanced plants, and vascular plants that have water-conducting tissues throughout their bodies. Phytoplankton, macrophytic algae, and vascular plants usually are

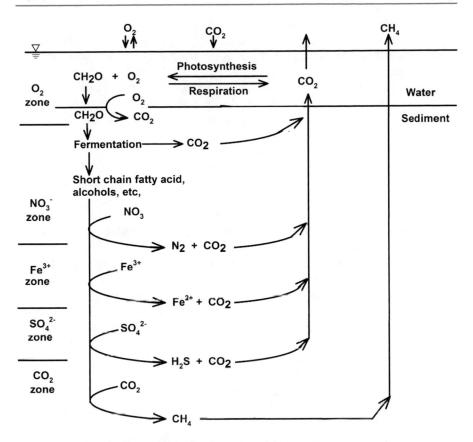

Fig. 12.3 Zonation of water and sediment based on electron acceptors in respiration

present in various proportions in most aquatic ecosystems, but phytoplankton are more biologically active and in most water bodies have a greater influence on water quality than do other plants.

Biology

The phytoplankton is distributed primarily among the phyla Pyrrhophyta, Euglenophyta, Cyanophyta, Chlorophyta, and Heterokontophyta. The Pyrrhophyta are mainly marine and include the dinoflagellates. The Euglenophyta—like the Pyrrhophyta—contain flagellated motile organisms the best known of which are species of the genus *Euglena*. The Chlorophyta are known as the green algae, and they are mostly freshwater organisms. Members of this phylum include macroalgae such as the marine genus, *Ulva*, and the freshwater genus, *Spirogyra*, as well as many phytoplankton genera. Some common planktonic genera are *Scenedesmus*, *Chlorella*, and *Closterium*. Heterokontophyta is a large phylum of both marine and freshwater algae that includes the yellow-green algae (Xanthophyceae), golden algae (Chrysophyceae), brown algae (Phaeophyceae), and diatoms (Bacillariophyceae). The Cyanophyta are the blue-green algae. However, this group of algae is prokaryotic like bacteria, and taxonomic authorities consider blue-green algae to be cyanobacteria. But, in this book these organisms will be referred to as blue-green algae. A more complete discussion of phytoplankton systematics is beyond the scope of this text, but there are many books and websites from which more information can be obtained.

Blue-green algae can have particularly large effects on water quality. They tend to be abundant in nutrient enriched waters, and many species of planktonic blue-green algae are considered undesirable. They form scums on the water surface, they are subject to sudden population crashes, a few species may be toxic to other aquatic organisms, and some species excrete odorous compounds into the water. These odorous compounds may cause a bad odor and taste in drinking water or impart bad flavor to fish and other aquatic food animals.

Phytoplankton requires sunlight for growth, so actively-growing phytoplankton are found only in illuminated water where there is more than about 1% of incident light. The density of phytoplankton is slightly greater than that of water, so they tend to sink from the photic zone (illuminated surface stratum), but all species have adaptations such as small size, irregular morphology often with projecting surfaces, gas vacuoles, or means of motility to reduce their sinking rate and permit them to remain suspended even in waters of low turbulence. Nearly all waters contain phytoplankton, and waters with abundant nutrients will contain enough phytoplankton to cause turbidity and discoloration. Waters discolored by phytoplankton are said to have phytoplankton blooms and appear green, blue-green, red, yellow, brown, black, gray, or various other colors.

Reproduction in phytoplankton usually is by binary fission as described above for bacteria. The life span of phytoplankton is short, individual cells probably survive for no more than 1 or 2 weeks. As dead cells settle, they quickly rupture and the protoplasmic contents spill into the water. Natural waters usually contain much organic detritus originating from dead phytoplankton.

Most species of phytoplankton disperse readily. The atmosphere contains spores and vegetative bodies of many species. If a flask of sterile nutrient solution is opened to the atmosphere, algal communities will soon develop. Wading birds often spread phytoplankton from one body of water to another. This results because phytoplankton propagules stick to their external surfaces, and viable algal spores and vegetative bodies pass through their digestive tracts. Species of phytoplankton that colonize a habitat depend upon the suitability of the habitat into which their propagules happened to enter.

Suppose that a tank of nutrient-enriched water containing no algae is exposed outdoors. Species of phytoplankton in the air will fall into the tank, and the species entering would be primarily the result of chance. However, the species that persist in the tank will depend upon environmental conditions. Moreover, once the phytoplankters begin to grow they will go through a succession until a climax phytoplankton community is attained. This community eventually will make the environment unsuitable for its growth and crash (see Fig. 12.1), or the season may change making the conditions more suitable for other species. Thus, phytoplankton communities are continually changing in species composition.

Photosynthesis

Photosynthesis is the process by which green plants capture energy from photons of sunlight and use it to reduce carbon dioxide to organic carbon in form of carbohydrate. The basic features of the reaction of photosynthesis are depicted (Fig. 12.4).

In the light reaction of photosynthesis, sunlight strikes chlorophyll and other light sensitive pigments in plant cells, and photons of light are captured by the pigments. This energy is used in a reaction known as photolysis to split water molecules into molecular oxygen, hydrogen ions or protons, and electrons. Two water molecules yield one oxygen molecule, four hydrogen ions, and four electrons. The hydrogen ions and electrons then react to convert nicotinamide adenine dinucleotide phosphate ion ($NADP^+$) to its un-ionized form (NADPH). Energy captured by the pigments also converts ADP to ATP—a process called photophosphorylation.

The second phase of photosynthesis is not dependent on light, so it is termed the dark reaction despite occurring in the light. In a series of enzyme catalyzed reactions, energy from ATP and hydrogen ions and electrons from photolysis of water reduce carbon dioxide to carbohydrate (CH_2O) and water. Of course, $NAPD^+$ and ADP are regenerated and used again in the light reaction.

The summary equation for photosynthesis can be written as

$$2H_2O + CO_2 \xrightarrow{\text{light}} CH_2O + H_2O + O_2. \tag{12.27}$$

However, it is tradition to give the product of photosynthesis as glucose ($C_6H_{12}O_6$). This is done by multiplying Eq. 12.27 by six

$$12H_2O + 6CO_2 \xrightarrow{\text{light}} C_6H_{12}O_6 + 6H_2O + 6O_2. \tag{12.28}$$

Of course, Eq. 12.28 can be presented in more familiar form by subtracting $6H_2O$ from both sides of the reaction equation to give

$$6H_2O + CO_2 \xrightarrow{\text{light}} C_6H_{12}O_6 + 6O_2. \tag{12.29}$$

Green plants can produce their own organic matter and are said to be autotrophic. The resources required for photosynthesis and plant production are carbon dioxide, water, sunlight, and inorganic nutrients. Molecular oxygen and hydrogen are available from the photolysis of water in photosynthesis. Other nutrients are primarily absorbed from the soil solution by terrestrial plants and from the water by aquatic plants. Essential mineral nutrients for the majority of plants are nitrogen, phosphorus, sulfur, potassium, calcium, magnesium, iron, manganese, zinc, copper, and molybdenum. Some plants require in addition one or more of the following: sodium, silicon, chloride, boron, and cobalt.

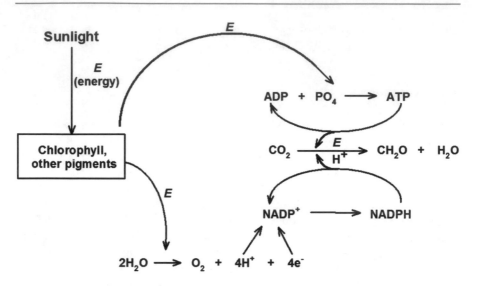

Fig. 12.4 Basic features of the photosynthesis process

Biochemical Assimilation

Plants use carbohydrate fixed in photosynthesis as an energy source and a raw material for assimilating other biochemical compounds to include starch, cellulose, hemicellulose, pectins, lignins, tannins, fats, waxes, oils, amino acids, proteins, and vitamins. These compounds are used by the plant to construct its body and to carry out physiological functions necessary for life. Plants must do biological work to maintain themselves, grow, and reproduce. The energy for doing this work comes from biological oxidation of the organic matter produced in photosynthesis.

Factors Controlling Phytoplankton Growth

Phytoplankton requires sunlight for growth, but intense sunlight near the water surface may inhibit certain species. Some waters contain enough turbidity from suspended mineral or organic particles to greatly restrict light penetration and lessen the growth of phytoplankton. Phytoplankton growth responds favorably to warmth. Highest growth rates usually occur during spring and summer, but growth continues at a slower rate during colder months. Measurable rates of photosynthesis may even occur under clear ice.

The nutrient in shortest supply relative to the requirements of the plant will limit growth. For example, if phosphorus is the nutrient present in the shortest supply relative to plant needs, production will be limited to the amount possible with the ambient phosphorus concentrations (Fig. 12.5). If phosphorus is added to the water, plant growth will increase. Growth will continue until the nutrient present in the

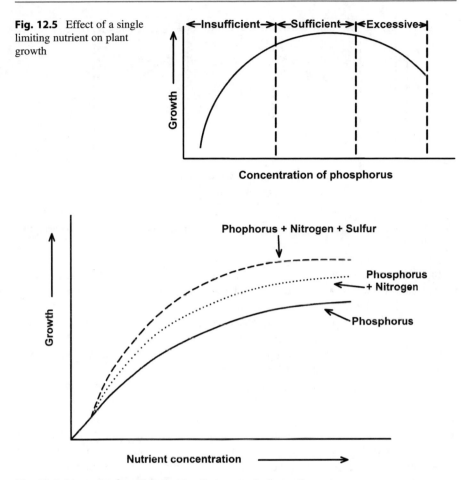

Fig. 12.5 Effect of a single limiting nutrient on plant growth

Fig. 12.6 Example of multiple limiting factors on plant growth

second shortest supply becomes limiting. Further growth can be achieved by adding more of the second limiting nutrient. This is a case of multiple limiting factors (Fig. 12.6).

There is an important caveat, too much of a nutrient may limit the growth of some species. Adding more phosphorus typically stimulates growth of some phytoplankton species while inhibiting the growth of other species. Eutrophic water bodies may have high primary productivity, but relatively few species of phytoplankton.

The concept of limiting factors and ranges of tolerance also applies to other environmental factors, such as light intensity (Fig. 12.7). There is a range of tolerance within which plants can grow, and within this range light may be inadequate, optimum, or excessive for growth. Beyond the range of tolerance, plants will die. The ideas expressed in this paragraph are known as Liebig's "Law of the Minimum" and Shelford's "Law of Tolerance, respectively."

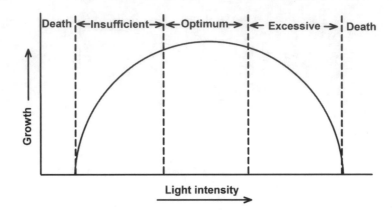

Fig. 12.7 Example of the response of a plant to light, illustrating the range of tolerance to a particular factor

Shortage of any one or a combination of the essential nutrients may limit phytoplankton growth in natural waters. However, in most bodies of water, phosphorus, and to a lesser extent, nitrogen, are the most important limiting nutrients. A study of 49 American lakes showed phosphorus to limit phytoplankton growth in 35 lakes, while nitrogen was limiting in eight lakes (Miller et al. 1974). Other factors were thought to limit growth in the remaining lakes. Turbidity may limit growth even when there is an ample nutrient supply.

Typical concentrations of essential plant nutrients in freshwater, seawater, and phytoplankton are shown in Table 12.2. Concentration factors obtained by dividing the concentration of each element in phytoplankton by the aqueous concentration indicate how much each element is accumulated by phytoplankton above the concentration of the element in water. There is less phosphorus and nitrogen in water relative to the concentrations of these two nutrients in phytoplankton than there is for other elements. A shortage of nitrogen and phosphorus tends to be the most common nutrient limitation of phytoplankton growth in freshwater, brackishwater, and seawater. In spite of the common belief that phosphorus is not as likely to be a limiting factor in marine environments as in freshwater ones, a review of the literature on this topic (Elser et al. 2007) suggested that nitrogen and phosphorus limitations of freshwater, marine, and terrestrial ecosystems are similar. Of course, there are waters where nutrients other than phosphorus and nitrogen limit phytoplankton growth. Iron and manganese in particular may limit marine phytoplankton productivity.

Redfield (1934) observed that the average molecular ratio of carbon:nitrogen:phosphorus in marine phytoplankton was about 106:16:1 (weight ratio ≈41:7:1); this ratio became known as the Redfield ratio. However, this ratio does not imply that the addition of nitrogen and phosphorus into ecosystems at a ratio of 7:1 would be the most effective ratio in promoting phytoplankton growth. As will be seen in Chaps. 13 and 14, there is considerably more recycling of nitrogen than of

Table 12.2 Concentrations of nutrients in seawater, freshwater, and phytoplankton

Element	Concentration (mg/L)		Phytoplankton	Concentration factors	
	Seawater	Freshwater	(mg/kg)[a]	Seawater	Freshwater
Phosphorus	0.07	0.03	230	3,286	7,667
Nitrogen	0.5	0.3	1,800	3,600	6,000
Iron	0.01	0.2	25	2,500	125
Manganese	0.002	0.03	4	2,000	133
Copper	0.003	0.03	2	667	100
Silicon	3	2	250[b]	83	125
Zinc	0.01	0.07	1.6	1.6	23
Carbon	28	20	12,000	429	600
Potassium	380	2	190	0.5	95
Calcium	400	20	220	0.55	11
Sulfur	900	5	160	0.18	32
Boron	4.6	0.02	0.1	0.02	5
Magnesium	1,350	4	90	0.07	22.5
Sodium	10,500	5	1,520	0.14	304

[a]Wet weight basis
[b]Concentration is much greater for diatoms

phosphorus in aquatic ecosystems. A narrower nitrogen:phosphorus ratio than 7:1 would need to be added to realize a 7:1 ratio in the water.

There has been much debate over the concentrations of nitrogen and phosphorus necessary to cause phytoplankton blooms in water bodies. The response of phytoplankton to nitrogen and phosphorus differs among aquatic ecosystems, and development of noticeable phytoplankton blooms may require concentrations of 0.01 to 0.1 mg/L soluble inorganic phosphorus and 0.1 to 0.75 mg/L of inorganic nitrogen.

Many limnologists feel that carbon does not limit phytoplankton growth in natural waters, but algal culture studies indicate that the possibility of carbon as a limiting factor in aquatic ecosystems cannot be entirely dismissed (King 1970; King and Novak 1974; Boyd 1972). Moreover, several workers have observed that phytoplankton production and fish production increase as total alkalinity increased in natural waters up to 100–150 mg/L. This observation does not necessarily indicate that waters with higher alkalinities have higher concentrations of available carbon, and therefore, greater phytoplankton productivity. The studies were conducted in natural waters that were not fertilized intentionally or polluted through human activity. Water with higher alkalinity tends to have a greater complement of most plant nutrients than does water of low alkalinity. Correlations between alkalinity and phytoplankton productivity may have been related to differences in nitrogen and phosphorus availability rather than different concentrations of carbon dioxide or alkalinity *per se* (Boyd and Tucker 2014).

Phytoplankton productivity also may be regulated by the growth of macrophytes because these plants compete with phytoplankton for nutrients. Macrophytes floating on pond surfaces and those with leaves at the surface shade the water column and greatly restrict phytoplankton growth. Certain macrophytes also excrete substances known as allelopaths that are toxic to phytoplankton. Regardless of the reasons, additions of nutrients to water bodies containing extensive macrophyte communities often do not cause phytoplankton blooms; rather, they stimulate further macrophyte growth.

Estimating Phytoplankton Abundance

The number of individuals (single cells, filaments, or colonies) per milliliter or liter of water may be determined directly by microscopic examination. Different species have characteristic sizes, and the volume of phytoplankton cells per unit of water provides a better estimate of biomass than does enumeration of phytoplankton abundance. However, it is exceedingly difficult to measure the volume of phytoplankton in a sample. A formula must be concocted for estimating the value of each species present and multiplied by the number of that species per unit volume of water.

Indirect methods of assessing phytoplankton abundance are popular. Chlorophyll *a* determination on particulate matter removed from a water sample by filtration provides an index of phytoplankton abundance. Because particulate organic matter often consists mostly of phytoplankton, this variable is indicative of phytoplankton abundance in many waters. Suitable techniques for separating phytoplankton from other particulate matter are not available, so results must be assessed with caution. The Secchi disk visibility decreases with increasing plankton abundance, and in many bodies of water, the Secchi disk visibility can be a useful measure of phytoplankton abundance. Of course, one must consider how much of the turbidity appears to be from other sources when assessing plankton abundance with a Secchi disk.

Effects of Phytoplankton on Water Quality

The most profound effects of phytoplankton activity on water quality are changes in pH and concentrations of dissolved oxygen and carbon dioxide. Photosynthesis usually dominates over respiration during daylight in a water body, while the opposite occurs at night. In other words, there is a net increase in dissolved oxygen and a net decrease in carbon dioxide during the day and *vice versa* at night. Changes in carbon dioxide concentration affect pH. The pH increases when carbon dioxide decreases. Daily changes in carbon dioxide, dissolved oxygen, and pH are illustrated in Fig. 12.8. The magnitude of daily fluctuations in concentrations of the three variables will tend to increase as phytoplankton abundance increases.

Light availability decreases with depth, so the rate of phytoplankton photosynthesis will tend to decrease with depth. In the afternoon, dissolved oxygen and pH will tend to be higher in surface water than in deeper water while the opposite will be true for carbon dioxide.

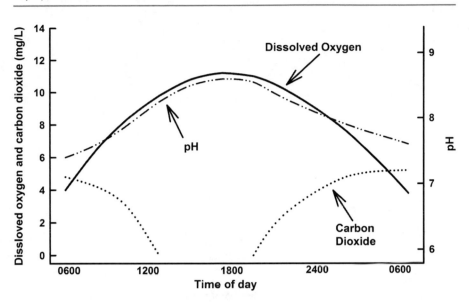

Fig. 12.8 Changes in pH and concentrations of dissolved oxygen and carbon dioxide over a 24-hour period in a eutrophic water body

Stratification of dissolved oxygen concentration in a thermally stratified lake is illustrated in Fig. 12.9. The lake can be considered eutrophic because oxygen depletion in the hypolimnion is a criterion for separating eutrophic (nutrient rich) and oligotrophic (nutrient poor) lakes.

Phytoplankton blooms sometimes die quickly. Sudden, massive mortality of phytoplankton—often called die-offs—can cause severe depression of dissolved oxygen concentrations. Die-offs are characterized by sudden death of all or a great portion of the phytoplankton followed by rapid decomposition of dead algae. Dissolved oxygen concentrations decline drastically, and they may fall low enough to cause fish kills. The reasons for phytoplankton die-offs have not been determined exactly, but they usually involve dense surface scums of blue-green algae. Die-offs often occur on calm, bright days when dissolved oxygen concentrations are high, carbon dioxide concentrations are low, and pH is high. This combination has been suggested to kill blue-green algae through a photo-oxidative process (Abeliovich and Shilo 1972; Abeliovich et al. 1974).

Events surrounding a complete die-off of a dense population of the blue-green algae *Anabaena variabilis* in a fish pond at Auburn, Alabama were documented (Boyd et al. 1975). The pond contained a uniform density of *A. variabilis* throughout the water column on windy days in March and April. In late April, a succession of clear, calm days resulted in a surface scum of phytoplankton on April 29. On the afternoon of April 29, the phytoplankton died, and the pond water was brown and turbid with decaying algae on April 30. No living *A. variabilis* filaments and few individuals of other algal species were observed in water samples taken between April 30 and May 5. Between May 5 and 8, a new phytoplankton community

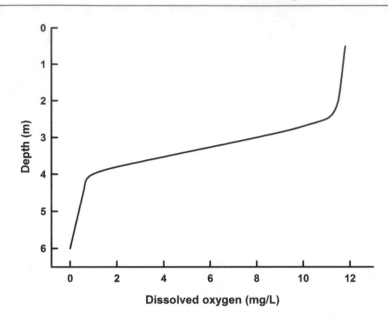

Fig. 12.9 Change in dissolved oxygen concentration with depth in a thermally-stratified, eutrophic lake

developed that consisted primarily of desmids. Dissolved oxygen concentrations quickly dropped to 0 mg/L following death of the *A. variabilis* population, and dissolved oxygen remained at or near this concentration for nearly a week until the new phytoplankton community developed. All phytoplankton die-offs are not as spectacular as the one described above, but they are rather common events in fish ponds.

Weather profoundly influences dissolved oxygen concentrations. On clear days, water bodies normally have dissolved oxygen concentrations near saturation at dusk. On cloudy days, photosynthesis is limited by insufficient light, and dissolved oxygen concentrations often are lower than normal at dusk. The probability of dissolved oxygen depletion is greater during nights following cloudy days than during nights following clear days (Fig. 12.10). Thermal destratification or overturns of water bodies may occur during unusually cool spells, heavy winds, and heavy rains. Fish kills may occur following overturns because sudden mixing of large volumes of oxygen-deficient hypolimnetic water with epilimnic water can result in rapid oxygen depletion. Weather-related problems with low dissolved oxygen are most common in water bodies with abundant phytoplankton.

Phytoplankton also influences water quality by removing nutrients from the water and by serving as a source of organic matter. Ammonia nitrogen concentrations and soluble reactive phosphorus concentrations in water bodies usually decline when phytoplankton are growing rapidly.

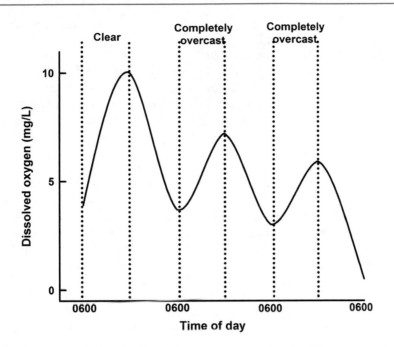

Fig. 12.10 Daily fluctuations in dissolved oxygen concentrations in fish ponds on clear and overcast days

Measuring Photosynthesis and Respiration

The oxygen light-dark bottle technique is a relatively simple procedure for measuring photosynthesis and respiration rates. In its simplest application, this procedure involves filling three bottles [usually 300-mL biochemical oxygen demand (BOD) bottles]—two transparent bottles and an opaque one—with water from the water body of interest. The dissolved oxygen concentration is measured immediately in one of the transparent bottles (initial bottle or IB). The other transparent bottle, called the light bottle (LB), and the opaque bottle, called the dark bottle (DB), are incubated in the water body. After a period of incubation, the bottles are removed for measurement of dissolved oxygen concentration.

The initial bottle provides an estimate of the concentration of dissolved oxygen in water of the light and dark bottles at the beginning of incubation. During incubation, both photosynthesis and respiration occurs in the light bottle, and part of the oxygen produced in photosynthesis is used in respiration. The gain in dissolved oxygen in the light bottle represents the amount of net photosynthesis. No light enters the dark bottle; photosynthesis does not occur, but respiration removes dissolved oxygen. The loss of dissolved oxygen in the dark bottle represents respiration of phytoplankton and other microorganisms. The total amount of oxygen produced in photosynthesis (gross photosynthesis) is the sum of dissolved oxygen produced in net photosynthesis and used in respiration.

If the objective is to determine the influence of depth on photosynthesis and respiration or to obtain average rates of photosynthesis and respiration in the water column, bottles must be filled with water from several depths and then incubated at those same depths. This arrangement is necessary because light penetration decreases with depth and photosynthesis rates are sensitive to light intensity. Both temperature and the abundance of planktonic organisms also differ with depth and influence photosynthesis and respiration.

Light intensity varies during the day and influences photosynthetic rate. A 24-hour incubation would be ideal, but sedimentation of organisms within bottles, depletion of dissolved oxygen in the dark bottle, or supersaturation of dissolved oxygen in the light bottle can alter respiration and photosynthesis rates. These "bottle effects" make it necessary to restrict incubation to a few hours. Incubations from dawn (provided the initial dissolved oxygen is high enough to sustain respiration in the dark bottle) until noon are more feasible than incubations from noon until dusk, because at noon waters are already high in dissolved oxygen concentration. If bottles are incubated at dawn and removed at noon, the incubation period will represent one-half of a photoperiod. Another alternative is to incubate the bottles during a specific time interval and measure solar radiation during the incubation interval and for the entire photoperiod. The fraction of the daily radiation occurring during the incubation period can be determined by dividing radiation during the incubation period by total radiation for the day. The total amount of photosynthesis is assumed to equal the amount during the incubation period times the ratio of total daily radiation to radiation received during incubation.

It is possible to calculate three variables, net photosynthesis (NP), gross photosynthesis (GP), and water column respiration (R) from light-dark bottle data. Of course, net photosynthesis will be underestimated because organisms other than phytoplankton contribute to respiration and detract from net photosynthesis. Equations for these calculations follow:

$$NP = LB - IB \tag{12.30}$$

$$R = IB - DB \tag{12.31}$$

$$GP = NP + R \tag{12.32}$$

$$\text{or } GP = LB - DB. \tag{12.33}$$

All variables are reported in milligrams per liter of dissolved oxygen. Where bottles are incubated at different depths, results can be averaged to obtain values for the water column. Results can be adjusted to a 24-hour basis by considering the length of the incubation period and the amount of radiation during the incubation period. The computations are illustrated in Ex. 12.5.

Ex. 12.5 *The following data are the averages for results of light-dark bottle incubations at 0.1, 0.5, 1.0, and 1.5 m in a 1.75-m deep column of water.*

IB = 4.02 mg/L
LB = 6.75 mg/L
DB = 2.98 mg/L
Incubation period = dawn to noon (6.5 hours)

NP, GP, and R will be calculated for a 24-hour day.
Solution:
For the incubation period,

$$NP = 6.75 - 4.02 = 2.73 \ mg/L$$

$$R = 4.02 - 2.98 = 1.04 \ mg/L$$

$$GP = 6.75 - 2.98 = 3.77 \ mg/L.$$

Assuming a clear day, the values may be doubled to provide totals for the photoperiod. Respiration will continue throughout the 11-hour period of darkness, so nighttime respiration will be only 11/13 of daytime values. Nighttime respiration must be subtracted from daytime NP in order to estimate 24-hour NP.

Photosynthesis and respiration rates may be expressed on an areal basis (usually square meter). This is done by multiplying the average values for NP, GP, and R in the entire water column by water depth. Values for NP, GP, and R also can be expressed in terms of carbon. In the general expressions for respiration and photosynthesis, one mole of oxygen can be equated to one mole of carbon dioxide or carbon. The ratio of the atomic weight of carbon (C) to the molecular weight of oxygen (O_2), 12/32 or 0.375, can be used to convert oxygen concentrations to carbon concentrations.

The Compensation Depth

Photosynthesis rate decreases with increasing depth because of diminishing light. In most water bodies there will be a depth at which oxygen produced by photosynthesis will be equal to oxygen used in respiration. At this depth—called the compensation depth or point—net photosynthesis will be zero. At depths less than the compensation depth, more dissolved oxygen will be produced by phytoplankton than used in respiration by microorganisms in the water column. At depths greater than the compensation depth, more dissolved oxygen will be used in respiration than produced by photosynthesis.

In a stratified body of water, the compensation depth usually corresponds to the thermocline, and no oxygen is produced in the hypolimnion. Because of

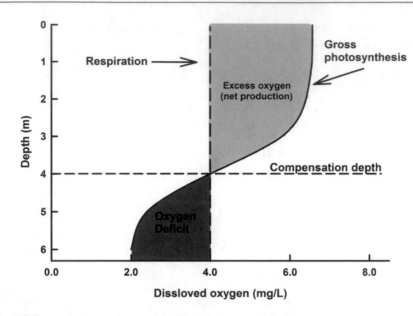

Fig. 12.11 Illustration of the compensation point in a water body

stratification, an oxygen debt develops in the hypolimnion. When destratification occurs, water from the hypolimnion is mixed with epilimnetic water, and the accumulated oxygen debt must be satisfied. This can cause a decrease in dissolved oxygen concentration at the time of overturns that is proportional to the size of the hypolimnetic oxygen debt.

In bodies of water that do not stratify thermally, there is daily mixing of bottom and surface waters. If the combined oxygen deficit at depths greater than the compensation depth exceeds the oxygen surplus in water shallower than the compensation depth, low dissolved oxygen concentration can result. In an unstratified water body, the difference between the oxygen surplus in water shallower than the compensation depth and the oxygen deficit of the deeper water (Fig. 12.11) represents the oxygen available for fish and other larger organisms in the water column whose respiration is not included in light-dark bottle measurements and sediment respiration.

Harmful Algae

Any type of phytoplankton can be harmful to aquatic animals if the combined standing crop becomes great enough to cause dissolved oxygen depletion. However, a few species of algae may be directly toxic to aquatic animals or even to humans and livestock. These algae include certain unicellular marine algae known as prymnesiophytes, and some species of blue-green algae, dinoflagellates, diatoms, and chloromonads. Toxic algae—especially those that are responsible for the red tide phenomenon in marine waters—can cause huge fish kills over wide areas.

Toxins passed along the food chain may represent a health threat to humans who consume certain aquatic products. The best example is shell fish poisoning in which bivalve molluscs such as oysters, mussels, clams, and scallops filter toxic algae—certain dinoflagellates, diatoms, and blue-green algae—from the water and accumulate the algal toxins in their tissues. There are four types of shellfish poisoning: amnesic (ASP), diarrheal (DSP), neurotoxic (NSP), and paralytic (PSP) that may result in humans from eating improperly or uncooked shellfish. The effects vary ranging from an unpleasant bought with diarrhea (DSP) or unusual sensations (NSP), to amnesia and possible permanent cognitive damage (ASP), or even death (PSP has a mortality rate of 10–12%). Some blue-green algae are known to cause allergic reactions, mostly skin rashes, in humans. Livestock and wildlife have been killed by drinking water from pools infested with toxic algae.

Algae, and blue-green algae in particular, are known to produce compounds that can cause bad taste and odor in drinking water supplies. The two most common offensive compounds are geosmin and 2-methylisobornel. These same compounds can be adsorbed by fish, shrimp, and other aquatic animals and impart a bad taste and odor to the flesh. Such products are deemed "off-flavor" and are of low acceptability in the market. Algal pigments may accumulate in the hepatopancreas of shrimp, and when the shrimp are cooked, the hepatopancreas ruptures and the algal pigments discolor the shrimp head. Such shrimp often are unacceptable in the "heads-on" market for shrimp.

The major means of combating off-flavor in drinking water and in aquaculture of food animals in ponds is use of the algicide copper sulfate which is especially toxic to blue-green algae responsible for the phenomenon (Boyd and Tucker 2014). This treatment also is commonly used in many municipal water supply reservoirs throughout the world. In extreme situations, drinking water is sometimes passed through activated carbon filters to remove tastes and odors.

Carbon and Oxygen Cycle

From the standpoint of ecology and water quality, respiration and photosynthesis are simply opposite processes as shown below:

$$6CO_2 + 6H_2O + energy \underset{\text{Respiration}}{\overset{\text{Photosynthesis}}{\rightleftarrows}} C_6H_{12}O_6 + 6O_2. \qquad (12.34)$$

Of course, the biochemical processes are distinctly different in the two reactions, and while the reactions are not truly opposite, the overall results are.

In photosynthesis, phytoplankton and other plants remove carbon dioxide from the environment, trap solar energy, and use the energy to reduce inorganic carbon in carbon dioxide to organic carbon in carbohydrate and release oxygen into the environment. In respiration, organisms oxidize organic carbon to inorganic carbon of carbon dioxide with the consumption of oxygen from the environment and the release of heat energy and carbon dioxide. Plants generally produce more organic matter and oxygen than they use—this is net photosynthesis available to animals, bacteria, and other heterotrophic organisms for food. The surplus oxygen is used by aerobic, heterotrophic organisms as an oxidant in respiration. Organic matter in ecosystems is the remains of organisms. It consists primarily of plant remains, because production of plant biomass greatly exceeds that of animal biomass in ecosystems.

The dynamics of carbon and oxygen are closely entwined in nature. When organic carbon is fixed, oxygen is released, and oxygen is consumed when organic carbon is mineralized. Life on earth depends upon the input of energy from the sun and upon the cyclic transformations of carbon and oxygen. The global carbon and oxygen cycles are depicted in simple form in Fig. 12.12. The food web which consists of all of the transfers of organic matter among producer, consumer, and decomposer organisms may be prepared by expanding the carbon cycle to include all the pathways by which food (organic matter) and energy move through ecosystems.

The carbon cycle has been greatly altered since the beginning of the industrial age about 1750 and the resulting increase in human population. Increased use of fossil fuels, accelerated deforestation, cement manufacturing and other anthropogenic activities have resulted in greater carbon dioxide evolution than in earlier history. The carbon dioxide concentration in the atmosphere has increased from about 280 ppm to about 400 ppm today (Fig. 12.13). This increase is considered by most scientists as the major reason for global warming, climate change, sea-level rise, and acidification of the ocean.

In the past century, average global surface temperature increased by about 0.78 °C and mean sea level rose by 17 cm because of melting of polar ice and thermal expansion of ocean water as a result of warming. Climate change is

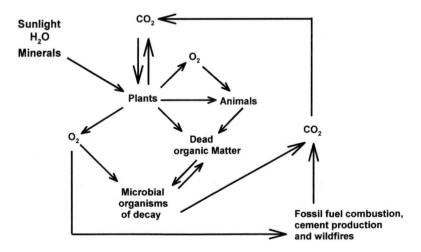

Fig. 12.12 Carbon and oxygen cycles

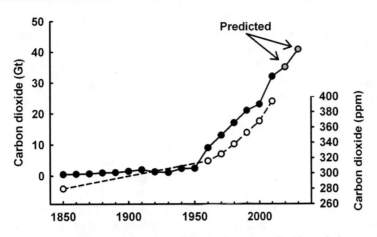

Fig. 12.13 Annual estimates of global, anthropogenic carbon dioxide emissions at 10-year intervals (dots) (1850–2030) and changes in atmospheric carbon dioxide concentrations measured at Mauna Loa, Hawaii from 1958 to 2012 (circles). (Boyd and McNevin 2014)

occurring resulting in more extreme weather. The extreme weather index maintained by the US National Oceanographic and Atmospheric Administration fluctuated but showed no clear pattern of increase from 1920 to 1970, but since, this index has exhibited an upward trend. Average ocean pH declined from 8.12 in 1988 to 8.09 in 2008, and it is resulting in thinner calcium carbonate shells on many marine invertebrates. The changes are expected to accelerate during the remainder of the twenty-first century (Boyd and McNevin 2015).

Human Pathogens and Chlorination

Water supplies for human use may be contaminated with pathogens. Some of the most serious waterborne infections such as typhoid fever, dysentery, and cholera are not common in developed countries, but waterborne diseases are still of major significance in the developing world. Organisms responsible for waterborne diseases include viruses, bacteria, protozoans, mites, and worms.

Waters that are contaminated with human pathogens usually are also contaminated by human fecal material. Coliform bacteria are used to identify waters that may have received human wastes, and water containing significant numbers of coliforms is a health hazard because of the possibility of waterborne diseases.

Coliform organisms are aerobic and facultative anaerobic, gram-negative, nonspore-forming, rod-shaped bacteria that ferment lactose with gas formation within 48 hours at 35 °C. Some of these organisms can be found in soil and on vegetation, but fecal coliforms *Escherichia coli* usually originate from feces of warm-blooded animals. The ratio of fecal coliforms to fecal streptococci is sometimes used to differentiate waters contaminated with human fecal coliforms from those contaminated with fecal coliforms of other warm-blooded animals. The fecal coliform:fecal streptococci ratio for humans is above 4.0, but for other warm-blooded animals, the ratio is 1.0 or less (Tchobanoglous and Schroeder 1985). Fecal coliforms should not be present in drinking water, but in some cases, public health authorities may permit up to 10 total coliform organisms/100 mL provided there are no fecal coliforms. When water contains coliforms, it should be disinfected before it is consumed by humans. The most common method of disinfecting municipal water supplies is chlorination.

The efforts to use chlorine as a disinfectant began about 1850 with the use of chlorine as a hand wash in hospitals in Vienna, Hungary and by John Snow to disinfect the Broad Street pump in London, England during a cholera outbreak. However, it was not widely used for disinfection of public water supplies until the early 1900s. An excellent review of chlorination is provided by George C. White's *Handbook of Chlorination and Alternative Disinfectants* (Black and Veatch Corporation 2010). The common, commercial chlorine disinfectants are chlorine gas (Cl_2), sodium hypochlorite (NaOCl), and calcium hypochlorite [$Ca(OCl)_2$]. But, chloramines—especially monochloramine (NH_2Cl)—are being used increasingly to disinfect drinking water.

Chlorine reacts with water to form hydrochloric acid (HCl) and hypochlorous acid (HOCl)

$$Cl_2 + H_2O \rightarrow HOCl + H^+ + Cl^-. \tag{12.35}$$

Hydrochloric acid dissociates completely, but hypochlorous acid partially dissociates according to the following equation

$$HOCL \rightleftharpoons H^+ + OCl^- \quad K = 10^{-7.53}. \tag{12.36}$$

Chlorination of water may yield four chlorine species: chloride that is not a disinfectant, and chlorine, hypochlorous acid, and hypochlorite ion that have disinfecting power and are called free chlorine residuals. The disinfecting powers of chlorine and hypochlorous acid are about 100 times greater than that of hypochlorite (Snoeyink and Jenkins 1980). The dominant free chlorine residual in water depends on pH rather than type of chlorine compound applied (Fig. 12.14). Chlorine occurs only at very low pH, HOCl is the dominant residual between pH 2 and 6; HOCl and OCl^- both occur in significant portions between pH 6 and 9, but HOCl declines relative to OCl^- as pH increases; OCl^- is the dominant residual above pH 7.53 (Fig. 12.14).

Disinfection at pH 7 normally requires about 1 mg/L of free chlorine residuals. The ratio HOCl:OCl^- decreases as pH increases, ranging from 32:1 at pH 6 to 0.03:1 at pH 9. Greater concentrations of free chlorine residuals are required as the pH rises, because disinfection power decreases as the proportion of HOCl among the free chlorine residual declines.

Free chlorine residuals participate in many reactions that diminish their disinfecting power. Free chlorine residuals oxidize organic matter, nitrite, ferrous iron, and sulfide, and disinfecting power is lost because free chlorine residuals are reduced to chloride. Free chlorine residuals also combine with organic nitrogen

Fig. 12.14 Effect of pH on distribution of free chlorine residuals in water

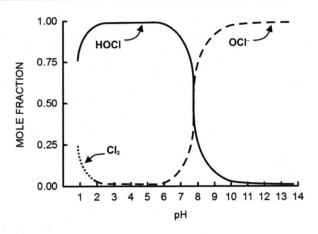

compounds, phenols, and humic acid to produce organochlorine compounds. At least one group of these compounds—the trihalomethanes—is suspected to be a carcinogen in humans (Jimenez et al. 1993).

A common extraneous reaction of chlorine is formation of chloramines

$$\text{Monochloramine} \quad NH_3 + HOCl \rightarrow NH_2Cl + H_2O \tag{12.37}$$

$$\text{Dichloroamine} \quad NH_2Cl + HOCl \rightarrow NHCl_2 + H_2O \tag{12.38}$$

$$\text{Trichloramine} \quad NHCl_2 + HOCl \rightarrow NCl_3 + H_2O. \tag{12.39}$$

Because these extraneous reactions diminish concentrations of free chloride residuals, they must be considered in establishing the chlorine dose for disinfection. Water samples typically are treated with the chlorinating agent until the dose necessary to provide the desired concentration of free chlorine residuals is established.

Chloramines are becoming popular for chlorination for two reasons. They are less likely to react with organic matter to form trihalomethanes and other byproducts, and they have a longer residual life in water distribution systems than traditional chlorination compounds.

Energy from sunlight drives the reaction in which hypochlorous acid is reduced to nontoxic chloride as shown below:

$$2HOCl \overset{\text{sunlight}}{\rightarrow} 2H^+ + 2Cl^- + O_2. \tag{12.40}$$

Conclusions

The biological aspects of water quality are extremely important, because photosynthesis and respiration by microorganisms are major factors controlling dissolved oxygen dynamics and redox potential in waters and sediment of aquatic ecosystems.

Dissolved oxygen concentration is probably the most important variable related to the overall well-being of aquatic ecosystems. In highly eutrophic ecosystems, excessive phytoplankton production results in high concentrations of organic matter, wide daily fluctuations in dissolved oxygen concentrations, and low redox potential at the sediment-water interface. Low dissolved oxygen concentration is stressful to aquatic species, and only those species most tolerant to low dissolved oxygen concentrations can prosper in eutrophic water bodies. Usually, species diversity and stability decline in eutrophic ecosystems in spite of the high level of primary productivity.

Techniques relying on the presence, absence, or relative abundance of indicator species have been developed for assessing the trophic status of ecosystems and to indicate relative degrees of pollution in aquatic ecosystems. However, these procedures are difficult and time consuming. A less tedious technique for assessing the trophic status of water bodies is measurement of primary productivity. Wetzel (1975) ranked lake trophic status according to the average net primary productivity as follows: ultra-oligotrophic, <50 mg C/m^2/day; oligotrophic, 50–300 mg C/m^2/day; mesotrophic, 250–1000 mg C/m^2/day; eutrophic, >1000 mg C/m^2/day.

Simpler measures of trophic status also can be used. For example, the depletion of dissolved oxygen in the hypolimnion of a water body during stratification indicates eutrophic conditions. The clarity of water as determined by Secchi disk visibility in waters where plankton is the main source of turbidity can indicate trophic status. Oligotrophic bodies of water may have Secchi disk visibilities of 4 or 5 m or more while Secchi disk visibility is less than 1.0 m in highly eutrophic waters. Of course, Secchi disk visibility is not a good index of productivity where light penetration is restricted by suspended soil particles or humic substances. Chlorophyll *a* concentrations are not difficult to measure, and they provide a good index of phytoplankton abundance and trophic status of ponds, lakes, reservoirs, and coastal waters. The relationship between chlorophyll *a* concentration and the trophic status of lakes and reservoirs is provided (Table 12.3).

Microorganisms in water bodies also can have a great influence on human and animal health. Better sanitary conditions to prevent the contamination of public

Table 12.3 Relationship between chlorophyll *a* concentrations and conditions in reservoirs and lakes

Chlorophyll *a* (μg/L)		
Annual mean	Annual maximum	Conditions
<2	<5	Oligotrophic, aesthetically pleasing, very low phytoplankton levels
2–5	5–15	Mesotrophic, some algal turbidity, reduced aesthetic appeal, oxygen depletion not likely
5–15	15–40	Mesotrophic, obvious algal turbidity, reduced aesthetic appeal, oxygen depletion likely
>15	>40	Eutrophic, high levels of phytoplankton growth, significantly reduced aesthetic appeal, serious oxygen depletion in bottom waters, reduction in other uses

water supplies with human wastes and methods for disinfecting drinking water have been major milestones in the continuing effort to improve public health. Of course, many diseases of fish and wildlife also can be spread by water.

The presence of excessive algae discolors water and creates surface scums that detract from aesthetic value, and turbidity created by planktonic algae restricts underwater visibility and makes waters less desirable for swimming and other water sports. Bad tastes and odors are imparted to drinking waters by some phytoplankton species that occur in public water supplies. The same algal compounds that cause bad tastes and odors in drinking water can be absorbed by fish and other aquatic organisms. Absorption of algal compounds taints the flesh of aquatic food animals and makes them less desirable to the consumer or even harmful to the health of the consumer. Shellfish poisoning is caused by an algal toxins and can lead to various painful symptoms or even death in humans.

References

Abeliovich A, Shilo M (1972) Photo-oxidative death in blue-green algae. J Bacteriol 11:682–689

Abelivoich A, Kellenberg D, Shilo M (1974) Effects of photo-oxidative conditions on levels of superoxide dismutase in *Anacystis nidulans*. Photochem Photobiol 19:379–382

Black and Veatch Corporation (2010) White's Handbook of chlorination and alternative disinfectants, 5th edn. Wiley, Hoboken

Boyd CE (1972) Sources of CO_2 for nuisance blooms of algae. Weed Sci 20:492–497

Boyd CE, McNevin AA (2015) Aquaculture, resource use, and the environment. Wiley-Blackwell, Hoboken

Boyd CE, Tucker CS (2014) Handbook for aquaculture water quality. Craftmaster Printers, Auburn

Boyd CE, Prather EE, Parks RW (1975) Sudden mortality of a massive phytoplankton bloom. Weed Sci 23:61–67

Elser JJ, Bracken M, Cleland EE et al (2007) Global analysis of nitrogen and phosphorus limitation of primary producers in freshwater, marine and terrestrial ecosystems. Ecol Lett 10:1135–1142

Jimenez MCS, Dominguez AP, Silverio JMC (1993) Reaction kinetics of humic acid with sodium hypochlorite. Water Res 27:815–820

Jumpertz R, Venti CA, Le DS, Michaels J, Parrington S, Krakoff J, Votruba S (2013) Food label accuracy of common snack foods. Obesity 21:164–169

King DL (1970) The role of carbon in eutrophication. J Water Pollut Control Fed 42:2035–2051

King DL, Novak JT (1974) The kinetics of inorganic carbon-limited algal growth. J Water Pollut Control Fed 46:1812–1816

Kristensen E, Ahmed SI, Devol AH (1995) Aerobic and anaerobic decomposition of organic matter in marine sediment: which is fastest? Limnol Oceanogr 40:1430–1437

Miller WE, Maloney TE, Greene JC (1974) Algal productivity in 49 lake waters as determined by algal assays. Water Res 8:667–679

Redfield AC (1934) On the proportions of organic deviations in sea water and their relation to the composition of plankton. In: Daniel RJ (ed) James Johnstone memorial volume. University Press of Liverpool, Liverpool, pp 177–192

Snoeyink VL, Jenkins D (1980) Water chemistry. Wiley, New York

Tchobanoglous G, Schroeder ED (1985) Water quality: characteristics, modeling, modification. Adison-Wesley, Reading

Wetzel RG (1975) Limnology. WB Saunders, Philadelphia

Nitrogen

13

Abstract

The atmosphere is a vast storehouse of nitrogen—it consists of 78% by volume of this gas. Atmospheric nitrogen is converted by electrical activity to nitrate (NO_3^-) that reaches the earth's surface in rainfall. Atmospheric nitrogen also can be fixed as organic nitrogen by bacteria and blue-green algae, and it can be reduced to ammonia (NH_3) by industrial nitrogen fixation. Plants use ammonium (NH_4^+) or nitrate as nutrients for making protein that is passed through the food web. Elevated concentrations of ammonium and nitrate contribute to eutrophication of water bodies. Because it has several valence states, nitrogen undergoes oxidations and reductions most of which are biologically mediated. Nitrogen in organic matter is converted to ammonia (and ammonium) by decomposition. Organic matter with high nitrogen content typically decomposes quickly with release of appreciable ammonia nitrogen ($NH_3 + NH_4^+$). In aerobic zones, nitrifying bacteria oxidize ammonia nitrogen to nitrate, while in anaerobic zones, nitrate is reduced to nitrogen gas by denitrifying bacteria. Ammonia and ammonium exist in a temperature and pH dependent equilibrium—the proportion of NH_3 increases with greater temperature and pH. Elevated concentrations of un-ionized ammonia can be toxic to aquatic organisms. Nitrite sometimes reaches high concentrations even in aerobic water and is potentially toxic to aquatic animals.

Introduction

Nitrogen gas (N_2) comprises 78.08% by volume of the atmosphere. Protein, an essential constituent of living things, contains an average of 16% nitrogen. Protein consists of amino acids, which are made by plants using ammonium (NH_4^+) or nitrate (NO_3^-) as sources of nitrogen and by certain microorganisms capable of converting nitrogen gas to ammonia. Animals and many saprophytic microorganisms do not have the ability to synthesize certain amino acids from

© Springer Nature Switzerland AG 2020

C. E. Boyd, *Water Quality*, https://doi.org/10.1007/978-3-030-23335-8_13

their food and require these essential amino acids in their diets. Nitrogen is a constituent of chlorophyll, hemoglobin, cyanoglobin, enzymes, and many other biochemical compounds found in organisms. It also is a major component of digestive and metabolic wastes excreted by animals.

Organic fertilizers, chemical fertilizers, and feeds used in agriculture and aquaculture contain nitrogen in amounts ranging from <1% in some livestock manures to 45% in urea fertilizer and from 4 to 10% in feeds. Nitrogen in commercial fertilizers and for many industrial uses is made by industrial conversion of atmospheric nitrogen to ammonia.

Runoff from agricultural fields and animal feed lots, effluents from aquaculture production units, food processing plants, many industries, municipal waste treatment plants, and other sources result in organic and inorganic nitrogen pollution of water bodies. Organic nitrogen is decomposed to ammonia by microorganisms, and increased concentrations of ammonia and nitrate stimulate aquatic plant growth leading to a major water quality problem known as eutrophication.

Un-ionized ammonia (NH_3) and nitrite (NO_2^-) at elevated concentrations can be toxic to aquatic organisms. Nitrification, the conversion of ammonia and ammonium through oxidation by certain species of bacteria, reduces ammonia nitrogen concentrations, but the process removes dissolved oxygen from the water and produces acidity. Excessive nitrogen gas in water can cause gas bubble trauma in fish and some other aquatic animals.

Nitrogen exists in an unusually large number of valence states (Table 13.1). Organisms of decay convert organically-bound nitrogen to ammonia nitrogen during decomposition of organic matter. Ammonia can be oxidized to a variety of nitrogen species, which, in turn, also can be reduced. These oxidations and reductions are mostly microbial transformations, and they have a great influence on the forms and concentrations of nitrogen in water. Certain gaseous forms of nitrogen evolved by microorganisms decaying organic matter are air pollutants, and nitrous oxides released in the air from combustion of fuels contribute to the acid rain phenomenon.

Table 13.1 Valence states of forms of nitrogen

Compound or ion	Formula	Valence
Amino nitrogen	$R - NH_2$ or $R - NH - R$	−3
Ammonia and ammonium	NH_3 and NH_4^+	−3
Hydrazine	N_2H_4	−2
Hydroxylamine	H_2NOH	−1
Nitrogen	N_2	0
Nitrous oxide	N_2O	+1
Nitric oxide	NO	+2
Nitrite	NO_2^-	+3
Nitrogen dioxide	NO_2	+4
Nitrate	NO_3^-	+5

R = organic moiety

The purpose of this chapter is to discuss nitrogen dynamics in aquatic ecosystems and to consider the role of nitrogen in water quality.

The Nitrogen Cycle

The nitrogen cycle (Fig. 13.1) usually is presented as a global cycle, but most of the components of this cycle function in much smaller systems. For example, many of the steps depicted in Fig. 13.1 occur in fish ponds or even in a small home aquarium.

The ultimate, natural source of plant available nitrogen is atmospheric and biological nitrogen fixation. In the early 1900s chemists in Germany discovered how to fix atmospheric nitrogen by an industrial process that now provides most nitrogen fertilizers and nitrogen for industrial use. Protein made by plants passes through the food web providing amino nitrogen required by animals. Fecal material of animals and dead plants and animals become a pool of organic matter that is decomposed by bacteria and other organisms of decay. Ammonia released by decomposition of organic matter and excreted by animals is oxidized to nitrate by nitrifying bacteria, and nitrate is reduced to nitrogen gas and returned to the atmosphere by denitrifying bacteria to complete the cycle.

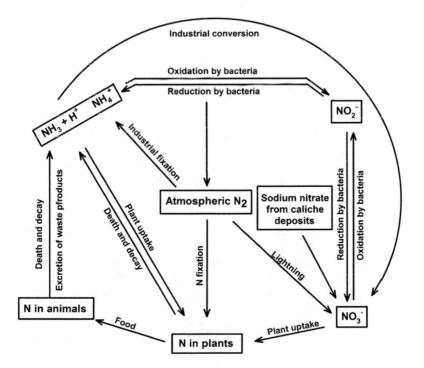

Fig. 13.1 The nitrogen cycle

Most of the earth's nitrogen is in the atmosphere, and in organic matter (including fossil fuels and living biomass). A small quantity is present in deposits of a sodium nitrate bearing mineral—called caliche—found in the Atacama Desert of Chile and a few other arid places. Most nitrogen transformations are biologically mediated, and gaseous forms of nitrogen can freely exchange between air and water.

Nitrogen Transformations

Atmospheric Fixation by Lightning

Nitrogen gas in the atmosphere must be transformed to ammonia nitrogen or nitrate to be useful to plants. Lightning can result in oxidation of nitrogen gas to nitric acid. This process occurs when heat from lightning breaks the triple bond between nitrogen atoms allowing them to react with molecular oxygen in the atmosphere forming nitric oxide

$$N_2 + O_2 \rightarrow 2NO. \tag{13.1}$$

Nitric oxide is then oxidized to nitrite

$$2NO + O_2 \rightarrow 2NO_2^-. \tag{13.2}$$

Though other reactions may occur, the most common way of expressing the conversion of nitrite to nitric acid in the atmosphere is

$$4NO_2^- + 2H_2O + O_2 \rightarrow 4HNO_3. \tag{13.3}$$

Nitric acid is washed from the atmosphere by rainfall. Fixation by lightning is an important source of plant available nitrogen in high rainfall regions—especially in those where there is frequent lightning during storms. Nitrogen oxides (NO_x) entering the atmosphere from combustion of fossil fuels and wildfires also is oxidized to nitric acid and comes down in rainfall.

Nitrate concentration in rainwater from 18 stations across the conterminous United States averaged 2.31 mg/L (as NO_3^-) with a range of 0.48–5.34 mg/L (Carroll 1962). The atmosphere also contains ammonia from various terrestrial sources, and rainwater from the 18 stations mentioned above had an average ammonia concentration of 0.43 mg/L (as NH_4^+) with a range of 0.05–2.11 mg/L. The average annual rainfall for the stations was 875 mm. Thus, the input of nitrogen in form of nitrate and ammonium in rainfall was about 7.5 kg N/ha/yr.

Biological Fixation

Nitrogen gas is soluble in water; its concentration at equilibrium for different temperatures and salinities are found in Table 7.7. Nitrogen gas is not highly reactive, but certain species of blue-green algae and bacteria are able to absorb molecular nitrogen from water, transform it to ammonia, and combine ammonia with intermediate carbohydrate compounds to make amino acids. The summary for the reaction producing ammonia nitrogen is

$$N_2 + 8H^+ + 8e^- + 16ATP \rightleftharpoons 2NH_4^- + 16\,ADP + 16\,PO_4^{2-}. \qquad (13.4)$$

The reaction is catalyzed by two enzymes both of which require iron and one of which contains molybdenum as cofactors. Ammonia produced by biological fixation is used to synthesize amino acids that are combined into protein. The process requires metabolic energy, but the process *per se* does not require molecular oxygen, and can occur in either aerobic or anaerobic environments.

Nitrogen fixation rates in aquatic ecosystems usually are in the range of 1–10 kg/ha per year, but larger rates have been observed. The greatest rates of nitrogen fixation in aquatic environments have been observed in wetlands where certain trees and shrubs have nitrogen-fixing bacteria associated with their roots. Large amounts of nitrogen also may be fixed by blue-green algae in rice paddies and aquaculture ponds.

Blue-green algae capable of fixing nitrogen have heterocysts; these large, thick-walled, spherical cells found in filaments of *Nostoc, Anabaena, Gloeocapsa*, and a few other genera are the site of nitrogen fixation. The ability of some blue-green algae to fix nitrogen gives them a nutrient availability advantage over other algae when nitrate and ammonium concentrations are low; however, nitrogen fixation by phytoplanktic blue-green algae tends to decline as nitrate and ammonia nitrogen concentrations increase. This results because it is more energy efficient for these microorganisms to use nutrients already present than to reduce N_2. There also tends to be a decrease in nitrogen fixation in water bodies as the ratio of total N:total P in water increases. At a total N:total P ratio of 13 or more, nitrogen fixation stops (Findlay et al. 1994). This suggests that adding phosphorus to waters will increase the rate of nitrogen fixation.

Industrial Fixation

Industrial nitrogen fixation is accomplished by the Haber-Bosch process. The process involves bringing hydrogen gas made from natural gas or petroleum and atmospheric nitrogen together at high pressure (15–25 M Pa) and temperature (300–550 °C) in the presence of a catalyst (K_2O, CaO, SiO_2, or Al_2O_3). One molecule of N_2 reacts with three molecules of H_2 to form two molecules of NH_3. Ammonia can be used directly or oxidized to nitrate by an industrial process. Most nitrogen for making fertilizers and for use in other industrial applications is from the

Haber-Bosch process. The importance of this process to mankind was expressed beautifully by the Serbian-American engineer, physicist, and inventor Nikola Tesla who said *"the earth is bountiful, and where her bounty fails, nitrogen drawn from the air will refertilize her womb."*

Plant Uptake of Nitrate and Ammonium

Ammonia, ammonium, nitrite, and nitrate nitrogen are said to be combined forms of nitrogen, and plants can absorb combined nitrogen from the water. Most species apparently prefer to use ammonium, because it is energetically less demanding to do so. Nitrate nitrogen must be reduced to ammonia nitrogen by the nitrate reductase pathway before it can be used for amino acid synthesis

$$NO_3 \rightarrow NO_2 \rightarrow NH_3. \tag{13.5}$$

Biological nitrate reduction requires the nitrate reductase enzyme, molybdenum as a cofactor, and metabolic energy from ATP. The ammonia from Eq. 13.5 is then reacted with intermediates of carbohydrate metabolism to produce amino acids:

$$Carbohydrates + NH_3 \rightarrow amino\ acids. \tag{13.6}$$

Most phytoplankton species contain from 5 to 10% of their dry weights as protein nitrogen. In phytoplankton without the capacity to fix nitrogen, protein is made entirely from combined nitrogen absorbed from the water. The result can be a considerable reduction in combined nitrogen concentration as illustrated in Ex. 13.1.

Ex. 13.1 *Net productivity of phytoplankton in a 2-m deep body of water is 2 g carbon/m^2/day. Assuming the phytoplankton contain 50% carbon and 8% nitrogen (dry weight), daily nitrogen removal from the water will be estimated.*
   Solution:

$$2\ g\ C/m^2/d \div 0.50\ g\ C/g\ dry\ wt = 4\ g\ dry\ wt/m^2/d$$

$$4\ g\ dry\ wt/m^2/d \times 0.08\ g\ N/g\ dry\ wt = 0.32\ g\ N/m^2/d$$

This amount is equivalent to 3.2 kg N/ha/d.
Phytoplankton removed nitrogen from the water at a rate of 0.16 mg N/L/d

$$0.32\ g\ N/m^2/d \div 2\ m^3/m^2 = 0.16\ g\ N/m^3\ or\ 0.16\ mg/L.$$

Phytoplankton uptake is a major factor affecting combined nitrogen concentrations in water bodies. Aquatic macrophyte communities also may contain large amounts of nitrogen (Ex. 13.2). Stands of macrophytes can remove nitrogen from the water and hold it in their biomass throughout the growing season, and thereby reduce the availability of nitrogen to phytoplankton.

Ex. 13.2 A lake of 100 ha in area and 3 m in average depth has 10 ha of rooted aquatic plants along its shallow edges. Assuming the dry matter standing crop of macrophytes is 800 g/m² containing 2% nitrogen, the amount of nitrogen in macrophyte biomass will be estimated.
Solution:

$$800 \; g/m^2 \times 10,000 \; m^2/ha \times 10 \; ha \times 0.02 \; g \; N/g \; dry \; wt = 1,600,000 \; g \; N.$$

Assuming the plants absorbed the nitrogen from lake water, the concentration of nitrogen in macrophyte biomass equals a concentration of 0.53 mg/L of combined nitrogen

$$\frac{1,600,000 \; g \; N}{100 \; ha \times 3 \; m \times 10,000 \; m^2/ha} = 0.53 \; g \; N/m^3 \; or \; 0.53 \; mg/L.$$

Phytoplankton and other plants may be consumed by animals or they may die and become organic matter that is decomposed by microorganisms. When animals eat plants, at least 10 units of plant dry weight usually are necessary to produce 1 unit of dry animal biomass and the ratio is often greater and especially for fibrous plants of low nitrogen content. The nitrogen in plant material consumed by animals and not converted to nitrogen in biomass is excreted in feces or other metabolic wastes (Ex. 13.3).

Ex. 13.3 Suppose the conversion of phytoplankton to tissue of animals consuming it is at a 15:1 ratio, the plants contain 6% nitrogen, and the animals are 11% nitrogen. The waste nitrogen excreted during the production of 1000 g of animal biomass will be calculated.
Solution:
The dry matter conversion ratio is 15:1, and it will require 15,000 g of plants to yield 1000 g of animal biomass. The nitrogen budget is

$$Plant \; nitrogen \quad 15,000 \; g \times 0.06 \; g \; N/g = 900 \; g \; N$$

$$Animal \; nitrogen \quad 1,000 \; g \times 0.11 \; g \; N/g = 110 \; g \; N$$

$$Waste \; nitrogen \quad (900 - 110) \; g \; N = 790 \; g \; N.$$

The waste will be excreted as organic nitrogen in feces and most aquatic animals excrete ammonia as a metabolic waste. Of course, when animals die, their bodies become organic matter to be decomposed by bacteria with excretion of ammonia.

Mineralization of Organic Nitrogen

Much of the knowledge of microbial decomposition of organic matter comes from the literature about the fate of organic residues applied to agricultural fields and pastures. The decomposition of different kinds of organic matter occurs at different rates: carbohydrates and protein decompose mostly within 1 year, the non-humic substances in 1–5 years, and the humic fraction decomposed very slowly over centuries. When bacteria and fungi decompose organic matter, part of the nitrogen in organic matter is converted to organic nitrogen in microbial biomass, some remains in undecomposed organic matter, and the remainder is released (mineralized) to the environment mainly in the form of ammonia.

Hoorman and Islam (2010) provided information for the annual decomposition of organic residues added to cropland. By converting the information to carbon concentration, it can be shown that the addition of around 35 g C in organic residue could result in about 19 g carbon in CO_2, 3 g carbon in bacterial biomass, 2 g carbon in undecomposed non-humic matter, and 11 g carbon in undecomposed humic matter. The ratio of carbon in living microbial biomass to carbon metabolized to CO_2 would be 3/16 or 0.16. This ratio is called the microbial growth efficiency (MGE), and Six et al. (2006) reviewed the literature on MGE and found values ranging from 0.14 to 0.77 for soils and from 0.01 to 0.70 for aquatic ecosystems. The average for aquatic ecosystems was 0.32. In aquaculture, experience with carbohydrate additions to remove ammonia by incorporating it in microbial protein in biofloc culture systems suggest that the MGE for bacteria utilizing raw sugar or molasses is around 0.5 (Hargreaves 2013). However, for more complex residues, the MGE is likely around 0.2–0.3, and much of the residue does not decompose in a single year.

The rate of decomposition of an organic residue depends upon its composition and especially its carbon:nitrogen (C/N) ratio. Bacteria usually are about 10% nitrogen and 50% carbon (5:1 C/N ratio). Fungi also contain around 50% carbon, but only about 5% nitrogen. Fungi typically have greater MGE than do bacteria. Microorganisms must have adequate nitrogen for biomass synthesis in order to decompose organic matter quickly. If there is not enough nitrogen, microorganisms must die and their nitrogen recycled to allow decomposition to progress. Residues with a small (narrow) C/N ratio typically decompose faster with a great mineralization of nitrogen in the environment than will residues with a greater (wider) C/N ratio. A residue with a wide C/N ratio will tend to lead to withdrawal of combined nitrogen from the environment (nitrogen immobilization) for use by bacteria and the accumulation of undecomposed organic matter in the environment.

By assuming a given residue is entirely decomposable in a few weeks or months leaving only living bacteria, carbon dioxide, and basic elements (a rare situation in nature), the importance of the C/N ratio of a residue will be illustrated in Ex. 13.4.

Ex. 13.4 *The amount of nitrogen mineralized by bacteria with a MGE of 0.25 in completely decomposing 1000 g (dry wt) of substrate containing 42% organic carbon and 4% nitrogen will be determined.*

Solution:
Substrate C: 1,000 g × 0.42 = 420 g C
Substrate N: 1,000 g × 0.04 = 40 g N
Bacterial biomass produced:

420 g C × 0.25 g bacterial C/g substrate C = 105 g bacterial C

105 g bacterial C ÷ 0.5 g C/g dry bacteria = 210 g bacteria

Bacterial nitrogen: 210 g bacteria × 0.1 g N/g bacteria = 21 g bacterial N.
Nitrogen mineralized: 40 g substrate N − 21 g bacterial N = 19 g N released to environment.

Using the same methods illustrated in Ex. 13.4, 21.1 g of nitrogen would be mineralized by fungi with a MGE of 45% and containing 5% nitrogen.

Organic matter in natural ecosystems normally contains 40 to 50% carbon on a dry matter basis. Nitrogen content is more variable ranging from less than 0.4 to 10% or more. As the nitrogen percentage in organic matter increases, there tends to be an increase in the rate of decomposition and in the proportion of organic nitrogen mineralized. This results because a high nitrogen content is associated with organic matter that contains less decay-resistant structural compounds and more easily degradable proteinaceous material. A high nitrogen content also assures that there is more than enough nitrogen to sustain the microbial biomass that develops during decay.

In a substrate with less nitrogen than needed to effect its complete and rapid decomposition by microorganisms, one or both of two phenomena occur during decomposition: (1) decomposition is slow and microorganisms must die so that their nitrogen can be mineralized and used again to decompose substrate; (2) ammonia nitrogen and nitrate in the environment can be removed (immobilized) and used by microorganisms to decompose the nitrogen-deficient residue. Where nitrogen is available for immobilization, decomposition of low-nitrogen content residues often is greatly accelerated.

The influence of the C:N ratio on nitrogen balance during decomposition can be illustrated by solving the nitrogen budget (see Ex. 13.4 for procedure) for two readily decomposable residues, one with 42% carbon and 2% N and a second with 42% carbon and 0.4% N, decomposed by bacteria with a MGE of 0.2. Decomposition of 1000 g of the first residue would mineralize 3.2 g nitrogen—the second, 1000-g residue with 0.4% nitrogen contains only 4 g nitrogen. This is 14.8 g less nitrogen than needed by the bacteria decomposing the residue.

The relationship of nitrogen concentration in organic matter to decomposition rate is illustrated in Fig. 13.2. The oxygen consumption rate of decomposition of aquatic plant residues increased as the nitrogen concentration of these residues increased (C/N ratios declined). The rate of oxygen consumption also dropped quickly in residues and after 5 days were essentially equal in residues of high and

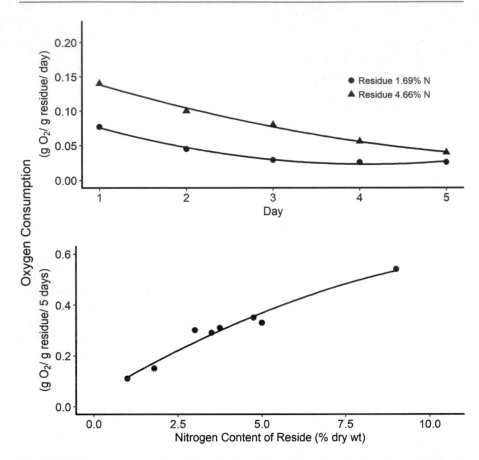

Fig. 13.2 Oxygen consumption during decomposition of aquatic plant residues of different nitrogen concentrations (Data from Boyd 1974)

low nitrogen concentration (Fig. 13.2). Nitrogen supplementation of the flask with a residue of low nitrogen concentration resulted in a greater respiration rate (Fig. 13.3).

Most organic matter consists of a variety of different compounds, and some of these substances will be much slower to decay than others. Nevertheless, organic matter with a narrow C:N ratio usually will typically decompose faster and release more nitrogen to the environment than organic substances with a wide C:N ratio.

Nitrification

In nitrification, chemoautotrophic bacteria oxidize ammonia nitrogen to nitrate. The reaction illustrated below occurs in two steps:

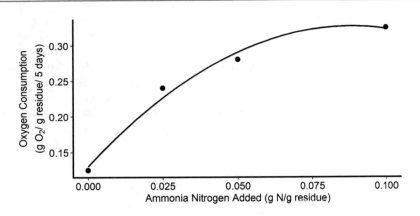

Fig. 13.3 Effect of adding nitrogen (in ammonium) on 5-day oxygen consumption by an aquatic plant residue containing 1.48% nitrogen (Data from Boyd 1974)

$$NH_4^+ + 1\tfrac{1}{2}\,O_2 \rightarrow NO_2^- + 2H^+ + H_2O \tag{13.7}$$

$$NO_2^- + 0.5\,O_2 \rightarrow NO_3^-. \tag{13.8}$$

Bacteria of the genus *Nitrosomonas* conduct the first oxidation, and the second oxidation is carried out by bacteria of the genus *Nitrobacter*. These two genera of bacteria usually occur together in the environment. Nitrite produced in the first reaction is oxidized to nitrate by the second reaction, and nitrite seldom accumulates. The two reactions can be added to give the overall nitrification equation

$$NH_4^+ + 2O_2 \rightarrow NO_3^- + 2H^+ + H_2O. \tag{13.9}$$

Energy is released when ammonium is oxidized to nitrate, and *Nitrosomonas* and *Nitrobacter* have mechanisms for capturing in ATP a part of the energy released and using the captured energy to reduce carbon dioxide to organic carbon. The overall result is that organic matter is synthesized, but the process differs from photosynthesis in that light is not required. The ratio of organic carbon synthesized to energy released through ammonia oxidation is quite low, and nitrification is not a significant source of organic carbon in ecosystems. It is, however, a major pathway for removing ammonia nitrogen from water. Nitrification is most rapid at temperatures of 25 to 35 °C and at pH values between 7 and 8, but the process can occur at slow rates down to pH 3 or 4 and at lower temperatures. Plenty of dissolved oxygen is a critical factor in nitrification.

Some bacteria can carry out anaerobic oxidation of ammonia. According to van der Graaf et al. (1995), the process occurs by the reaction

$$5NH_4^+ + 3NO_3^- \rightarrow 4N_2 + 9H_2O + 2H^+. \tag{13.10}$$

Nitrate is the oxygen source for oxidation of ammonium to nitrogen gas. This reaction could just as aptly be called denitrification because nitrate also is reduced to nitrogen gas.

Nitrification by both aerobic and anaerobic processes produces hydrogen ion, so the nitrification process contributes to acidity. Aerobic nitrification also requires an abundant supply of molecular oxygen and contributes greatly to oxygen demand in waters receiving high inputs of ammonia nitrogen. The oxygen demand and acidity of aerobic nitrification is illustrated in the next example.

Ex. 13.5 *The amount of dissolved oxygen consumed and the amount of acidity produced (in terms of alkalinity loss) in the biological oxidation of 1 mg/L ammonia nitrogen (NH$_4$-N) in aerobic nitrification will be estimated.*

<u>Solution:</u>

From Eq. 13.9, one mole of nitrogen consumes two moles of molecular oxygen and produces two moles of hydrogen ions. Thus,

(i) $$\dfrac{1\ mg/L}{N\ 14} = \dfrac{x}{2O_2\ 64}$$

$$x = 4.57\ mg/L\ of\ dissolved\ oxygen\ consumed\ per\ 1\ mg/L\ of\ NH_4^+\text{-}N.$$

(ii) *Two moles of hydrogen ions react with one mole of calcium carbonate (alkalinity is expressed as CaCO$_3$)*

$$CaCO_3 + 2H^+ = Ca^{2+} + CO_2 + H_2O$$

and

$$\dfrac{1\ mg/L}{N\ 14} = 2H^+ = \dfrac{x}{CaCO_3\ 100}$$

$$x = 7.14\ mg/L\ of\ total\ alkalinity\ neutralized\ per\ 1\ mg/L\ of\ NH_4^+\text{-}N.$$

In Ex. 13.5 above, 1 mg/L NH$_4$-N means that there is 1 mg/L of nitrogen in NH$_4^+$ from (NH$_4^+$-N). To convert 1 mg/L NH$_4^+$-N to milligrams per liter of NH$_4^+$, multiply 1 mg/L NH$_4^+$-N by NH$_4^+$/N or 18/14 to get 1.29 mg/L. The NH$_4^+$ concentration can be converted to NH$_4^+$-N concentration by dividing it by 18/14.

Rates of nitrification differ greatly among waters and over time in the same body of water. They tend to be much greater in summer than in winter, and they increase when ammonia nitrogen concentration increases. An abundant supply of dissolved oxygen is necessary for rapid nitrification. Nitrification can occur both in the water column and in aerobic sediment. Gross (1999) found that nitrification produced an average of 240 mg NO_3-N/m^2/day and 260 mg NO_3-N/m^2/day in the water column and sediment, respectively, of shallow, eutrophic ponds. However, Hargreaves (1995) reported that nitrification in the aerobic layers of sediment could be several times greater than in the water column.

Denitrification

Under anaerobic conditions, some bacteria use oxygen from nitrate as a substrate for molecular oxygen to oxidize organic matter (respire). One equation for denitrification is

$$6NO_3^- + 5CH_3OH \rightarrow 5CO_2 + 3N_2 + 7H_2O + 6OH^-. \tag{13.11}$$

Methanol is the carbon source for denitrification in Eq. 13.11, and methanol is the carbon source frequently used in the denitrification process at sewage treatment plants. In nature many types of organic matter can be used by denitrifying bacteria. The process is called denitrification because nitrate nitrogen is converted to nitrogen gas and lost from the ecosystem by diffusion into the atmosphere. Denitrification yields one mole of hydroxyl ion for each mole of nitrate reduced, and thereby increases the alkalinity of water.

Nitrification and denitrification can be thought of as coupled, because nitrate produced in nitrification can be denitrified. Only one-half of the alkalinity loss caused by nitrification is restored when nitrification and denitrification are perfectly coupled. This is because each mole of nitrogen oxidized in nitrification releases two moles of hydrogen ions, while each mole of nitrogen denitrified releases only one mole of hydroxyl ion. The potential acidity of nitrification is 7.14 mg/L equivalent $CaCO_3$ while the potential alkalinity of denitrification is 3.57 mg/L equivalent $CaCO_3$.

Denitrification does not always lead to formation of N_2. It can end with nitrite

$$NO_3^- + 2H^+ \rightarrow NO_2^- + H_2O. \tag{13.12}$$

Ammonia can be an end product when nitrate is reduced to nitrite followed by conversion of nitrite to hyponitrite ($H_2N_2O_2$) and hyponitrite to hydroxylamine (H_2NOH)

$$2NO_2^- + 5H^+ \rightarrow H_2N_2O_2 + H_2O + OH^- \tag{13.13}$$

$$H_2N_2O_2 + 4H^+ \rightarrow 2H_2NOH \tag{13.14}$$

$$H_2NOH + 2H^+ \rightarrow NH_3 + H_2O. \tag{13.15}$$

Another pathway results in formation of nitrous oxide as follows:

$$H_2N_2O_2 \rightarrow N_2O + H_2O. \tag{13.16}$$

However, N_2 formation is the most common pathway in denitrification.

Denitrification rates can be very high in aquatic ecosystems where nitrate is abundant and there is a readily available source of organic matter. The denitrification rate during the period June to October in small, eutrophic ponds at Auburn, Alabama, in which fish were fed a high-protein content ration ranged from 0 to 60 mg N/m²/day (average = 38 mg N/m²/day). On an annual basis, the average rate is equivalent to 139 kg N/ha (Gross 1999). Nitrogen inputs to most aquatic ecosystems are not great enough to support such large amounts of denitrification as seen in aquaculture ponds with daily additions of high-protein-content feeds.

Nitrite may sometimes accumulate in water to concentrations of 1 mg/L or more. When this phenomenon occurs, there may be toxicity to fish and other aquatic animals. Nitrite accumulation can result when the first step in nitrification (Eq. 13.7) proceeds faster than the second step. It also can result from nitrite production in anaerobic sediment as shown in Eq. 13.16 with subsequent diffusion of nitrite into the water column.

Ammonia-Ammonium Equilibrium

Ammonia nitrogen exists in water as ammonia (NH_3) and ammonium ion (NH_4^+). The un-ionized form is a gas, and it is potentially toxic to fish and other aquatic life at relatively low concentration. Ammonium is not appreciably toxic. Ammonia hydrolyses to ammonium in aqueous solution, and the two forms exist in a pH and temperature dependent equilibrium

$$NH_3 + H_2O = NH_4^+ + OH^- \qquad K_b = 10^{-4.75}. \tag{13.17}$$

The hydrolysis may alternatively be expressed as

$$NH_4^+ = NH_3 + H^+. \tag{13.18}$$

Analytical procedures for ammonia nitrogen (including both ion sensing and gas sensing electrodes) measure total ammonia nitrogen (TAN) and the NH_3 concentration must be calculated by aid of pH and water temperature (Zhou and Boyd 2016). Ammonia acts as a base in Eq. 13.17, but ammonium is an acid in Eq. 13.18. Because $K_aK_b = K_w = 10^{-14}$ (at 25 °C), $K_a = 10^{-14}/K_b$, and K_a for Eq. 13.18 is $10^{-9.25}$. The K_a and K_b values given above are for 25 °C, but values of K_a and K_b for other temperatures obtained from Bates and Pinching (1949) are presented (Table 13.2) along with K_w values for different temperatures. The ratio of NH_3-N to NH_4-N can be calculated most conveniently using Eq. 13.18 as follows:

Table 13.2 Equilibrium constants for ammonium dissociation (K_a), the hydrolysis of ammonia (K_b), and dissociation of water (K_w)

Temperature (°C)	K_a	K_b	K_w
0	$10^{-10.08}$	$19^{-4.86}$	$10^{-14.94}$
5	$10^{-9.90}$	$10^{-4.83}$	$10^{-14.73}$
10	$10^{-9.73}$	$10^{-4.80}$	$10^{-14.53}$
15	$10^{-9.56}$	$10^{-4.78}$	$10^{-14.35}$
20	$10^{-9.40}$	$10^{-4.77}$	$10^{-14.17}$
25	$10^{-9.25}$	$10^{-4.75}$	$10^{-14.00}$
30	$10^{-9.09}$	$10^{-4.74}$	$10^{-13.83}$
35	$10^{-8.95}$	$10^{-4.73}$	$10^{-13.68}$
40	$10^{-8.80}$	$10^{-4.73}$	$10^{-13.53}$

Modified from (Bates and Pinching 1949)

$$\frac{(NH_3\text{-}N)(H^+)}{(NH_4^+\text{-}N)} = K_a$$

$$\frac{(NH_3\text{-}N)}{(NH_4^+\text{-}N)} = \frac{K_a}{(H^+)}.$$

By assigning a value of 1.0 for $(NH_4^+\text{-}N)$ concentration, the proportion of the total ammonia nitrogen present in $NH_3\text{-}N$ form may be estimated as illustrated in Ex. 13.6 by aid of the equation

$$\frac{(NH_3\text{-}N)}{(NH_4^+\text{-}N)} = \left(\frac{10^{-9.25}}{(H^+)}\right) \div \left(1 + \frac{(10^{-9.25})}{(H^+)}\right). \qquad (13.19)$$

Ex. 13.6 *The proportion of the TAN concentration in NH_3-N form will be calculated for pH 8.0 and 9.0 at temperatures of 20 °C and 30 °C.*
  Solution:
  At 20 °C and pH 8.0 (K_a from Table 13.2)

$$\frac{(NH_3\text{-}N)}{(NH_4^+\text{-}N)} = \left(\frac{10^{-9.40}}{10^{-8.0}}\right) \div \left(1 + \frac{10^{-9.40}}{10^{-8.0}}\right) = 0.0398 \div 1.0398 = 0.0383.$$

At 20 °C and pH 9.0

$$\frac{(NH_3\text{-}N)}{(NH_4^+\text{-}N)} = \left(\frac{10^{-9.40}}{10^{-9.0}}\right) \div \left(1 + \frac{10^{-9.40}}{10^{-9.0}}\right) = 0.398 \div 1.398 = 0.2847.$$

At 30 °C and pH 8.0 (K_a from Table 13.2)

$$\frac{(NH_3\text{-}N)}{(NH_4^+\text{-}N)} = \left(\frac{10^{-9.09}}{10^{-8.0}}\right) \div \left(1 + \frac{10^{-9.09}}{10^{-8.0}}\right) = 0.0813 \div 1.0813 = 0.075.$$

At 30 °C and pH 9.0

$$\frac{(NH_3\text{-}N)}{(NH_4^+\text{-}N)} = \left(\frac{10^{-9.09}}{10^{-9.0}}\right) \div \left(1 + \frac{10^{-9.09}}{10^{-9.0}}\right) = 0.812 \div 1.812 = 0.449.$$

The proportion of NH_3-N increases with both increasing temperature and pH, but the increase is greater with pH (Ex. 13.6). At 20 °C the proportion of NH_3-N was 0.0382 at pH 8.0, but at the same pH and 30 °C, the proportion increased to 0.075—roughly a two-fold increase. When pH increased from 8.0 to 9.0 at 20 °C, the proportion of NH_3 rose from 0.0382 to 0.2847—an increase of more than seven-fold.

The K_a values for selected temperatures have been used to estimate the proportions of the total ammonia nitrogen concentration consisting of NH_3-N at different pH values (Table 13.3). The concentration of potentially toxic NH_3-N can be estimated easily for a particular pH and water temperature by multiplying TAN concentration by the appropriate factor from Table 13.3 as illustrated in Ex. 13.7. However, it also is possible to avail of convenient ammonia calculators that are posted on the internet. One such calculator may be found at http://www.hbuehrer.ch/Rechner/Ammonia.html. If the critical NH_3-N concentration—the concentration that is potentially toxic to fish is known, a table similar to Table 13.3 may be constructed

Table 13.3 Decimal fractions (proportions) of total ammonia existing as un-ionized ammonia in freshwater at various pH values and temperatures

pH	Temperature (°C)								
	16	18	20	22	24	26	28	30	32
7.0	0.003	0.003	0.004	0.004	0.005	0.006	0.007	0.008	0.009
7.2	0.004	0.005	0.006	0.007	0.008	0.009	0.011	0.012	0.015
7.4	0.007	0.008	0.009	0.011	0.013	0.015	0.017	0.020	0.023
7.6	0.011	0.013	0.015	0.017	0.020	0.023	0.027	0.031	0.036
7.8	0.018	0.021	0.024	0.028	0.032	0.036	0.042	0.048	0.057
8.0	0.028	0.033	0.038	0.043	0.049	0.057	0.065	0.075	0.087
8.2	0.044	0.051	0.059	0.067	0.076	0.087	0.100	0.114	0.132
8.4	0.069	0.079	0.090	0.103	0.117	0.132	0.149	0.169	0.194
8.6	0.105	0.120	0.136	0.154	0.172	0.194	0.218	0.244	0.276
8.8	0.157	0.178	0.200	0.223	0.248	0.276	0.306	0.339	0.377
9.0	0.228	0.255	0.284	0.313	0.344	0.377	0.412	0.448	0.490
9.2	0.319	0.352	0.386	0.420	0.454	0.489	0.526	0.563	0.603
9.4	0.426	0.463	0.500	0.534	0.568	0.603	0.637	0.671	0.707
9.6	0.541	0.577	0.613	0.645	0.676	0.706	0.736	0.763	0.792
9.8	0.651	0.684	0.715	0.742	0.768	0.792	0.815	0.836	0.858
10.0	0.747	0.774	0.799	0.820	0.840	0.858	0.875	0.890	0.905
10.2	0.824	0.844	0.863	0.878	0.892	0.905	0.917	0.928	0.938

in which the table entries represent the maximum acceptable total ammonia nitrogen concentrations. For instance, if the critical concentration is 0.10 mg/L NH_3-N, at pH 8 and 25 °C, the total ammonia nitrogen concentration should not exceed 1.92 mg/L (0.1 mg/L ÷ 0.052).

Factors for estimating the NH_3 concentration from the TAN concentration (Table 13.3) are for freshwater. Increasing ionic strength (greater salinity) lessens the proportion of NH_3 to TAN, and Spotte and Adams (1983) devised a series of tables of NH_4^+-N/NH_3-N ratios for estimating NH_3 concentration from the TAN concentration at different pH values, water temperatures, and salinities. The differences among salinities are rather small. For example, at pH 8.2 and 25 °C, the factor to be multiplied by TAN concentration to obtain NH_3-N concentrations are: 5 ppt, 0.078; 10 ppt, 0.077; 15 ppt, 0.075; 20 ppt, 0.072; 25 ppt, 0.070; 30 ppt, 0.068; 35 ppt, 0.065. The factor for freshwater extrapolated from Table 13.3 is 0.082. There also are online ammonia calculators that include salinity as a factor, e.g., (https://pentairaes.com/amonia-calculator).

Ex. 13.7 *A water sample contains 1 mg/L TAN. The concentration of NH_3-N will be estimated at 30 °C for pH 7, 8, 9, and 10.*

  Solution:
  From Table 13.3, the proportions of un-ionized ammonia at 30 °C are 0.008, 0.075, 0.448, and 0.890 in order of increasing pH. Thus,
  pH 7:
1 mg/L × 0.008 = 0.008 mg NH_3-N/L
  pH 8:
1 mg/L × 0.075 = 0.075 mg NH_3-N/L
  pH 9:
1 mg/L × 0.448 = 0.448 mg NH_3-N/L
  pH 10:
1 mg/L × 0.890 = 0.890 mg NH_3-N/L.

Ammonia Diffusion

The potential for diffusion of ammonia from the surface of a water body into the air obviously will be greatest when pH is high. Weiler (1979) reported ammonia diffusion losses up to 10 kg N/ha per day when wind velocity, pH, and total ammonia nitrogen concentrations were high. In most waters, the ammonia nitrogen concentration will not be great enough to support such large ammonia diffusion rates. Gross et al. (1999) reported diffusion losses of 9 to 71 mg N/m^2/day from small ponds with 0.05 to 5.0 mg/L of total ammonia nitrogen, afternoon pH values of 8.3 to 9.0, and water temperature of 21 to 29 °C. The effect of ammonia diffusion to the atmosphere on the total ammonia concentration in water is illustrated (Ex. 13.8).

Ex. 13.8 *The reduction in ammonia nitrogen concentration that could result by diffusion of ammonia at a rate 10 mg NH_3-N/m^2/day will be calculated for a 1-ha water body that has an average depth of 2 m.*
  Solution:

$$\frac{10\ mg\ NH_3\text{-}N/m^2/d}{2m^3/m^2} = 5\ mg\ NH_3\text{-}N/m^3/d = 0.005\ mg\ NH_3 - N/L/d.$$

The decrease in ammonia nitrogen concentration is small for 1 day, but over a year, the total loss of nitrogen could be substantial. If the loss rate used above is expanded for 1 year, the nitrogen loss would be

$$10\ mg\ NH_3\text{-}N/m^2/d \times 365\ d/yr = 3.65\ g/m^2.$$

This is 36.5 kg N/ha per year.

Nitrogen Inputs and Outputs

Aquatic ecosystems have various inputs and outputs of nitrogen. The main inputs are rainfall, inflowing water containing nitrogen from natural sources and pollution, nitrogen added intentionally (as in aquaculture), and nitrogen fixation. The outputs are outflowing water, harvest of aquatic products, intentional withdrawal of water, seepage, diffusion of ammonia into the atmosphere, and denitrification. The relative importance of each gain and loss varies greatly from one body of water to another.
  Nitrogen is stored in aquatic ecosystems primarily as organic matter in bottom sediments, but smaller amounts occur in the water as nitrogen gas, nitrate, nitrite, ammonia, and nitrogen in dissolved and particulate organic matter. Dissolved nitrogen gas often is the major form of nitrogen in water bodies (Table 7.7), but it is largely inert and seldom of water quality significance. When nitrogen in organic matter is mineralized, it is subject to the various processes and transformations described above, and it can be lost from the ecosystem. However, there is a tendency for recycling of nitrogen in aquatic ecosystems, and in most, an equilibrium among inputs, outputs, and stored nitrogen is reached. Pollution can quickly disrupt this equilibrium and cause higher nitrogen concentrations in the water and sediment.

Concentrations of Nitrogen in Water and Sediment

Dissolved nitrogen gas is not as biologically active as dissolved oxygen, and its concentrations usually remain near saturation. Ammonia nitrogen and nitrate nitrogen concentrations each normally are less than 0.25 mg/L in unpolluted waters. They may be above 1 mg/L in polluted waters, and in highly polluted waters concentrations of 5 to 10 mg/L are not uncommon. Nitrite nitrogen is seldom present at concentrations above 0.05 mg/L in oxygenated water, but it may reach concentrations of several milligrams per liter in polluted waters with low dissolved oxygen concentration.

The concentration of ammonia and nitrate nitrogen necessary to cause phytoplankton blooms in natural waters is difficult to ascertain, because many other factors also influence phytoplankton productivity. Concentrations of 0.1–0.75 mg/L of combined nitrogen in freshwater and even lower concentrations in brackish and marine waters have caused phytoplankton blooms. Limitation of combined nitrogen concentrations and loads is an important tool in eutrophication control.

The usual procedure is to determine total organic nitrogen rather than concentrations of specific organic nitrogen compounds. Measurements of total organic nitrogen include dissolved and particulate organic nitrogen, and they usually are below 1 or 2 mg/L in relatively unpolluted natural waters. Effluents and polluted waters may have much higher concentrations of total organic nitrogen. Domestic sewage has an average total nitrogen concentration around 40 mg/L (Tchobanoglous and Schroeder 1985).

Concentrations of nitrogen also are highly variable in sediment. As a general rule, organic matter is about 5% nitrogen, and sediments in aquatic ecosystems usually contain 1 to 10% organic matter which corresponds to around 0.05 to 0.5% nitrogen.

Ammonia Toxicity

Ammonia is the major nitrogenous excretory product of aquatic animals, and like other excretory products, is toxic if it cannot be excreted. A high concentration of ammonia in the water makes it more difficult for organisms to excrete ammonia. As a result the concentration of ammonia in the blood of fish and other aquatic animals increases with greater environmental ammonia concentration.

Although the mechanism of ammonia toxicity has not been elucidated entirely, a number of physiological and histological effects of high ammonia concentration in the blood and tissues of fish have been identified as follows: elevation of blood pH; disruption of enzyme systems and membrane stability; increased water uptake; increased oxygen consumption; gill damage; histological lesions in various internal organs.

The concentration of ammonia in most water bodies fluctuates daily because the proportion of ammonia to total ammonia nitrogen varies with pH and temperature (Table 13.3). Most data on ammonia toxicity were based on exposure to constant ammonia concentration, and less is known about the effects of fluctuating ammonia concentrations than about exposure to a constant concentration on aquatic animals. In water bodies, pH and temperature tend to increase in daytime to a peak in the afternoon and decline at night to their lowest near dawn. Hargreaves and Kucuk (2001) exposed three common species of fish used in aquaculture to daily fluctuating concentrations of NH_3 typical in magnitude to those occurring in ponds. The effects of ammonia were observed at greater concentrations than was expected from the results of toxicity tests conducted at constant temperature, pH and NH_3 concentration.

The potential for toxicity at a given ammonia nitrogen concentration depends on several factors in addition to pH and temperature. Toxicity increases with decreasing dissolved oxygen concentration, but this effect often is negated by high carbon dioxide concentration which decreases ammonia toxicity. There is evidence that ammonia toxicity decreases with increasing concentrations of salinity and calcium. Fish also tend to increase their tolerance for ammonia when acclimated to gradually increasing ammonia concentrations over several weeks or months.

Most ammonia toxicity data come from LC50 tests on fish—the LC50 is the concentration required to kill 50% of the animals exposed. The 96-hour LC50 concentrations for un-ionized ammonia nitrogen to various species of fish range from about 0.3 to 3.0 mg/L (Ruffier et al. 1981; Hargreaves and Kucuk 2001). Coldwater species usually are more susceptible to ammonia than warmwater species. Un-ionized ammonia should not cause lethal or sublethal effects at concentrations below 0.005 to 0.01 mg/L for coldwater species or below 0.01 to 0.05 mg/L for warmwater species. Corresponding concentrations of total ammonia nitrogen depend upon pH and temperature. The concentrations of ammonia nitrogen necessary to give 0.05 mg/L un-ionized ammonia nitrogen at selected pH values and 26 °C are as follows: pH 7, 8.33 mg/L; pH 8, 0.88 mg/L; pH 9, 0.13 mg/L. Ammonia toxicity obviously is of greater concern in waters where the pH is well above neutral than in neutral or acidic waters. The main effect of ammonia on aquatic life probably is stress rather than mortality. Several studies have shown that ammonia concentrations well below lethal concentrations lead to poor appetite, slow growth, and greater susceptibility to disease.

Nitrite Toxicity

Nitrite is absorbed from the water by fish and other organisms. In the blood of fish, nitrite reacts with hemoglobin to form methemoglobin. Methemoglobin does not combine with oxygen, and high nitrite concentrations in water can result in a functional anemia known as methemoglobinemia. Blood that contains significant methemoglobin is brown, so the common name for nitrite toxicity in fish is "brown blood disease." Nitrite also can bind hemocyanin, the oxygen binding pigment in crustacean blood, to lower the ability of their blood to transport oxygen.

Nitrite is transported across the gill by lamellar chloride cells. These same cells apparently transport chloride and they cannot distinguish between nitrite and chloride. The rate of nitrite absorption declines, at least in freshwater fish, as the concentration of chloride increases relative to nitrite. In freshwater, a chloride to nitrite ratio of 6 to 10:1 will prevent methemoglobinemia at nitrite concentrations up to at least 5 or 10 mg/L. In brackishwater, high concentrations of calcium and chloride tend to minimize problems of nitrite toxicity in fish (Crawford and Allen 1977).

Because the toxic effects of nitrite are worsened by low dissolved oxygen concentration and also depend upon chloride, calcium, and salinity concentrations, it is virtually impossible to make recommendations on lethal, sublethal, and safe concentrations. The 96-hour LC50 values (as nitrite) are in the range of 0.66 to 200 mg/L for freshwater fish and crustaceans and 40 to 4000 mg/L for brackishwater and marine species. Safe concentrations would likely be about 0.05 times the 96-hour LC50 concentration for a species.

Nitrogen and Gas Bubble Trauma

Because nitrogen is the most abundant atmospheric gas, it is the major gas in water supersaturated with air. As discussed in Chap. 7, nitrogen usually is the primary gas contributing to gas bubble trauma in aquatic animals exposed to gas supersaturated water. Common reasons for air supersaturation are sudden warming of water and water entraining air when it falls over high dams into streams below. Air leaks on the suction sides of pumps also can cause air supersaturation of discharge water.

Atmospheric Pollution by Nitrogen Compounds

Nitrous oxide that can be produced by denitrification in anaerobic environments is not generally considered a water quality issue, because it diffuses from water bodies into the atmosphere. However, nitrous oxide is one of the greenhouse gases that can increase the heat retaining capacity of the atmosphere to cause global warming. The global warming potential of nitrous oxide is about 300 times greater than that of carbon dioxide (the standard for greenhouse global warming potential).

Nitrogen dioxide (NO_2) and related nitrogen oxides—often referred to collectively as NO_x—are produced by combustion of fossil fuels and by wildfires. In the atmosphere, nitrogen oxides are oxidized to nitric acid and contribute to the acid rain phenomenon.

Conclusions

Plant growth often increases in response to increased nitrogen concentrations, because a shortage of available nitrogen is a limiting factor in many aquatic ecosystems. Nitrogen is used as a fertilizer in agriculture and aquaculture, but addition of nitrogen to most water bodies is considered to be nutrient pollution, because it contributes to dense phytoplankton blooms—causes eutrophication. Excessive concentrations of ammonia, nitrite, and dinitrogen gas (N_2) can be toxic to aquatic organisms. In addition, nitrous oxide is a greenhouse gas and nitrogen oxides contribute to acidity in rainfall.

References

Bates RG, Pinching GD (1949) Acid dissociation constant of ammonium ion at 0° to 50°C, and the base strength of ammonia. J Res Nat Bur Stan 42:419–420

Boyd CE (1974) The utilization of nitrogen from decomposition of organic matter by cultures of *Scenedesumes dimorphus*. Arch fur Hydro 73:361–368

Carroll D (1962) Rainwater as a chemical agent of geologic processes—a review. United States geological survey water-supply paper 1535 G, United States. Government Printing Office, Washington, DC

Crawford RE, Allen GH (1977) Seawater inhibition of nitrite toxicity to Chinook salmon. Trans Am Fish Soc 106:105–109

Findlay DL, Hecky RE, Hendzel LL, Stainton MP, Regehr GW (1994) Relationship between N_2-fixation and heterocyst abundance and its relevance to the nitrogen budget of lake 227. Can J Fish Aqua Sci 51:2,254–2,266

Gross A (1999) Nitrogen cycling in aquaculture ponds. Dissertation, Auburn University, Alabama

Gross A, Boyd CE, Wood CW (1999) Ammonia volatilization from freshwater fish ponds. J Env Qual 28:793–797

Hargreaves JA (1995) Nitrogen biochemistry of aquaculture pond sediments. Dissertation, Louisiana State University, Baton Rouge

Hargreaves JA (2013) Biofloc production systems for aquaculture. Publication 4503, Southern Regional Aquaculture Center, Stoneville

Hargreaves JA, Kucuk S (2001) Effects of diel un-ionized ammonia fluctuation on juvenile striped bass, channel catfish, and blue tilapia. Aqua 195:163–181

Hoorman JJ, Islam R (2010) Understanding soil microbes and nutrient recycling. Fact sheet SAG-16-10, The Ohio State University Extension, Columbus

Ruffier PJ, Boyle WC, Kleinschmidt J (1981) Short-term acute bioassays to evaluate ammonia toxicity and effluent standards. J Water Poll Con Fed 53:367–377

Six J, Frey SD, Thiet RK, Batten KM (2006) Bacterial and fungal contributions to carbon sequestration in agroecosystems. Soil Sci Soc Am J 70:555–569

Spotte S, Adams G (1983) Estimation of the allowable upper limit of ammonia in saline water. Mar Ecol Prog Ser 10:207–210

Tchobanoglous G, Schroeder ED (1985) Water quality. Addison-Wesley, Menlo Park

van der Graaf AA, Mulder A, de Bruijin P, Jetten MSM, Robertson LA, Kuenen JG (1995) Anaerobic oxidation of ammonium is a biologically mediated process. App Env Microb 61:1246–1251

Weiler RR (1979) Rate of loss of ammonia from water to the atmosphere. J Fish Res Bd Canada 36:685–689

Zhou L, Boyd CE (2016) Comparison of Nessler, phenate, salicylate and ion selective electrode procedures for determination of total ammonia nitrogen in aquaculture. Aqua 450:187–193

Phosphorus

14

Abstract

Phosphorus usually is the most important nutrient limiting phytoplankton productivity in both aquatic and terrestrial ecosystems. Phosphorus occurs naturally in most geological formations and soils in varying amounts and forms; the main source of agricultural and industrial phosphate is deposits of the mineral apatite—known as rock phosphate. Municipal and agricultural pollution is a major source of phosphorus to many water bodies. Most dissolved inorganic phosphorus in aquatic ecosystems is an ionization product of orthophosphoric acid (H_3PO_4). At the pH of most water bodies, $HPO_4{}^{2-}$ and $H_2PO_4{}^-$ are the forms of dissolved phosphate. Despite its biological significance, the dynamics of phosphorus in ecosystems are dominated by chemical processes. Phosphate is removed from water by reactions with aluminum, and to a lesser extent, with iron in sediment. In alkaline environments, phosphate is precipitated as calcium phosphate. Aluminum, iron and calcium phosphates are only slightly soluble, and sediments act as sinks for phosphorus. Concentrations of inorganic phosphorus in water bodies seldom exceed 0.1 mg/L, and total phosphorus concentration rarely is greater than 0.5 mg/L. In anaerobic zones, the solubility of iron phosphates increases; sediment pore water and hypolimnetic water of eutrophic lakes may have phosphate concentrations above 1 mg/L. Phosphorus is not toxic at elevated concentration, but along with nitrogen, it can lead to eutrophication.

Introduction

Phosphorus usually is in short supply in the environment relative to plant needs making it a key factor regulating primary productivity of natural and agricultural ecosystems. Phosphorus is the most important nutrient controlling plant growth in many water bodies, and phosphorus pollution of natural waters is considered a primary cause of eutrophication. Phosphorus is absorbed strongly by bottom sediment where it is bound in iron, aluminum, and calcium phosphate compounds and adsorbed onto iron and aluminum oxides and hydroxides. Solubilities of the mineral forms of phosphorus are regulated by pH, and there are comparatively few situations in nature in which phosphorus minerals are highly soluble. The phosphorus in organic matter is mineralized, but when it is, it usually will be adsorbed by sediment unless it is quickly absorbed by plants or bacteria. Because sediments tend to be a sink for phosphorus, high rates of plant growth in aquatic ecosystems require a continuous input of phosphorus.

Phosphorus is absorbed from environmental solutions by plants and supplied to animals and microorganisms of decay mainly through food webs. The storage form of phosphorus in plant tissues is phytic acid, a saturated cyclic acid with the formula $C_6H_{18}O_{24}P_6$—a six carbon ring with a phosphate radical ($H_2PO_4^-$) attached to each carbon. This compound is particularly abundant in bran and cereal grains. Phytic acid is not readily digestible by non-ruminant animals, but the enzyme phytase produced in the rumen by microorganisms allows ruminant animals to digest phytic acid. Most animals get their phosphorus from other phosphorus compounds in their diets rather than from phytic acid, but, microorganisms of decay can degrade phytic acid to release phosphate.

Phosphorus is a key nutrient with many functions in all living things. It is contained in deoxyribonucleic acid (DNA) that contains the genetic code (or genome) to instruct organisms how to grow, maintain themselves, and reproduce. Phosphorus also is a component of ribonucleic acid (RNA) that provides the information needed for protein synthesis, i.e., DNA is responsible for RNA production which in turn controls protein synthesis. Phosphorus is a component of adenosine diphosphate (ADP) and adenosine triphosphate (ATP) responsible for energy transfer, storage, and use at the cellular level. Phosphorus is a component of many other biochemical compounds such as phospholipids important in cellular membranes. Moreover, calcium phosphate is the major constituent of bone and teeth of vertebrates.

Over the past two centuries there has been much debate over phosphorus as "brain food." To this debate, Mark Twain, the famous American author and humorist once added, *"Agassiz does recommend authors to eat fish, because the phosphorus in it makes brains. But, I cannot help you to a decision about the amount you need to eat. Perhaps a couple of whales would be enough."* The modern view is that foods high in phosphorus concentration may help mental concentration.

Phosphorus has several valence states ranging from -3 to $+5$, but most phosphorus in nature has a valence state of $+5$. In contrast with nitrogen and sulfur, oxidation and reduction reactions mediated by chemotropic bacteria are not an important feature of its cycle. Microbial decomposition is an important factor by releasing phosphorus from organic matter, but despite its tremendous biological importance, the phosphorus cycle is largely chemical rather than biological.

This chapter will discuss the sources and reactions of phosphorus in aquatic ecosystems and consider the importance of phosphorus in water quality.

Phosphorus in the Environment

The global phosphorus cycle is not as well-defined as the global carbon, oxygen, and nitrogen cycles. The major sources of phosphorus are phosphorus-bearing minerals, e.g. iron, aluminum, and calcium phosphates that occur widely in soils at relatively low concentrations and in massive deposits of calcium phosphates of high phosphorus content at a few locations. These massive deposits of calcium phosphate consist of the mineral apatite commonly known as rock phosphate. Rock phosphate is mined and processed to make highly soluble calcium phosphate compounds for use as agricultural, industrial, and household phosphates. In relatively unpolluted natural waters, the primary source of phosphorus is runoff from watershed soils and dissolution of sediment phosphorus. Phosphorus concentrations in such waters reflect concentrations and solubilities of phosphorus minerals in soils and sediment. The atmosphere is not a significant source of phosphorus.

Phosphorus has many uses in agriculture, processing of food, manufacturing of beverages, other industries, and the home. Phosphate fertilizers are widely used in agriculture to promote plant growth, and it also is an ingredient of animal feed and a component of many pesticides. Phosphate fertilizers and pesticides also are used on lawns, gardens, and golf courses. Phosphoric acid is included in soft drinks to give them a sharper flavor and to inhibit the growth of microorganisms on sugar present in these beverages. Phosphorus is used for acidification, for buffering, as an emulsifier, and for flavor intensification in food. Surfactants and lubricants contain phosphorus. Phosphorus is a component of matches, and interestingly, phosphates are an ingredient of many fire retardants. Trisodium phosphate is used as a cleaning agent and water softener in both industry and the home. It should not be surprising that agricultural runoff, effluent from industrial operations, and municipal sewage contain elevated concentrations of phosphorus and can lead to higher phosphorus concentrations in water bodies into which they are discharged.

Increased phosphorus concentrations in natural waters resulting from human activities normally stimulate aquatic plant growth and especially phytoplankton growth. If phosphorus additions to natural water are too great, eutrophication occurs with excessive phytoplankton blooms or nuisance growths of aquatic macrophytes. Although phosphorus causes water pollution, it is applied to aquaculture ponds in fertilizers to increase natural productivity that is the base of the food web for fish production.

The dynamics of phosphorus in a water body are illustrated in Fig. 14.1. Dissolved and particulate phosphorus enters water bodies from their watersheds. Dissolved inorganic phosphorus is absorbed by plants and incorporated into their biomass. Plant phosphorus is passed to animals via the food web, and when plants and animals die, microbial activity mineralizes the phosphorus from their remains. If not absorbed by plants, dissolved inorganic phosphorus is strongly adsorbed by sediment. There is an equilibrium between inorganic phosphorus bound in sediment and phosphorus dissolved in water, but the equilibrium is shifted greatly towards sediment phosphorus. Rooted aquatic macrophytes can utilize sediment phosphorus that would otherwise not enter the water column, because their roots can extract

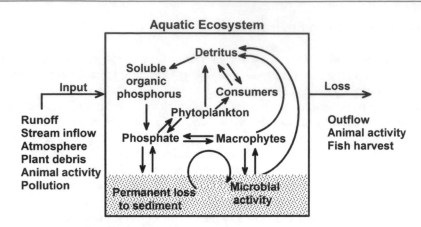

Fig. 14.1 A qualitative model of the phosphorus cycle in an aquatic ecosystem

phosphorus dissolved in sediment pore water. Phosphorus is lost from aquatic ecosystems in outflowing water, intentional withdrawal of water for human use, and harvest of aquatic products. The sediments of water bodies tend to be phosphorus sinks, and they tend to increase in phosphorus content over time. In unpolluted natural waters, there is an equilibrium among phosphorus inputs, outputs, and storage, and additions of phosphorus through water pollution disrupt this equilibrium.

Phosphorus Chemistry

Phosphorus interactions in water bodies (Fig. 14.1) involve biological processes, but the concentrations of phosphorus in water are controlled by chemical principles of dissolution, equilibrium, precipitation, and adsorption. These processes influence the mass balance of phosphorus which consists of inflow, outflow, storage, and concentration in the water. It is necessary to explain the basic features of phosphorus chemistry in water bodies.

Dissociation of Orthophosphoric Acid

Inorganic phosphorus in soils, sediment, and water usually occurs as an ionization product of orthophosphoric acid (H_3PO_4) that dissociates as follows:

$$H_3PO_4 = H^+ + H_2PO_4^- \qquad K_1 = 10^{-2.13} \qquad (14.1)$$

$$H_2PO_4^- = H^+ + HPO_4^{2-} \qquad K_2 = 10^{-7.21} \qquad (14.2)$$

$$HPO_4^{2-} = H^+ + PO_4^{3-} \qquad K_3 = 10^{-12.36}. \qquad (14.3)$$

When the molar hydrogen ion concentration equals the equilibrium constant for one of the steps in the dissociation, then the two phosphate ions involved in that step will have equal concentrations. For example, in Eq. 14.2, when pH $= 7.21$ [$(H^+) = 10^{-7.21}$], the ratio $K_2/(H^+)$ is $10^{-7.21}/10^{-7.21}$ or 1.0. This means that the ratio of $(H_2PO_4^{2-})/(HPO_4^-)$ also equals 1.0, and the concentrations of the two different phosphate ions in the expression are equal. The calculation of the proportions of the two forms of phosphorus in any one of the three dissociations of phosphoric acid may be done is illustrated in Ex. 14.1. This procedure can be applied across the entire pH range to provide the data needed for graphically depicting the proportions of H_3PO_4, $H_2PO_4^-$, HPO_4^{2-}, and PO_4^{3-} at different pH values (Fig. 14.2). Un-ionized H_3PO_4 occurs only in highly acidic solutions, while PO_4^{3-} predominates only in highly basic solutions. Within the pH range of 5–9 which includes most natural waters, dissolved phosphate exists as $H_2PO_4^-$ and HPO_4^{2-}. At pH values >7.21, there is more $H_2PO_4^-$, and HPO_4^{2-} dominates between pH 7.22 and pH 12.35. At pH >12.37, PO_4^{3-} exceeds HPO_4^{2-}.

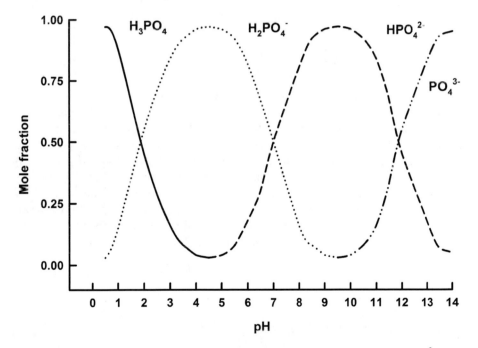

Fig. 14.2 Effects of pH on relative proportions (mole fractions) of H_3PO_4, $H_2PO_4^-$, HPO_4^{2-}, and PO_4^{3-} in an orthophosphate solution

Ex. 14.1 *The percentages of $H_2PO_4^-$ and HPO_4^{2-} at pH 6 will be estimated.*
   Solution:
   The mass action expression for Eq. 14.2 allows an expression for the ratio of HPO_4^{2-} to $H_2PO_4^-$,

$$\frac{(H^+)(HPO_4^{2-})}{(H_2PO_4^-)} = 10^{-7.21}$$

$$\frac{(HPO_4^{2-})}{(H_2PO_4^-)} = \frac{10^{-7.21}}{(H^+)}$$

and

$$\frac{(HPO_4^{2-})}{(H_2PO_4^-)} = \frac{10^{-7.21}}{10^{-6}} = 10^{-1.21} = 0.062.$$

There will be the equivalent of 1 part of $H_2PO_4^-$ and 0.062 part of HPO_4^{2-} at pH 6. The percentage of HPO_4^{2-} will be

$$\frac{0.062}{1 + 0.062} \times 100 = 5.83\%$$

and $100\% - 5.83\% = 94.17\%\ H_2PO_4^-$.

Polyphosphate compounds have a greater proportion of phosphorus than found in orthophosphate. For example, the orthophosphate ion PO_4^{3-} is 25% phosphorus, but polyphosphate ions such as PO_3^-, $P_2O_7^{4-}$, and $P_4O_{13}^{6-}$ contain 40%, 36%, and 38% phosphorus, respectively. The parent form of polyphosphate can be considered a compound such as sodium hexametaphosphate [$(NaPO_3)_6$] or polyphosphoric acid ($H_6P_4O_{13}$). Polyphosphates have many industrial uses and they may be used as fertilizer, and as a result be contained in effluents. Polyphosphates also occur naturally in certain bacteria, plants, and lower animals (Harold 1966). When introduced into water, polyphosphate soon hydrolyzes to orthophosphate as illustrated below for a single unit of $NaPO_3$ from $(NaPO_3)_6$:

$$NaPO_3 + H_2O \rightarrow Na^+ + H_2PO_4^-. \tag{14.4}$$

Phosphorus-Sediment Reactions

Inorganic phosphorus reacts with iron and aluminum in acidic sediments or waters to form slightly soluble compounds. The aluminum and iron phosphate compounds variscite ($AlPO_4 \cdot 2H_2O$) and strengite ($FePO_4 \cdot 2H_2O$) will be used as representative compounds. The dissolution of these compounds is pH-dependent

$$AlPO_4 \cdot 2H_2O + 2H^+ = Al^{3+} + H_2PO_4^- + 2H_2O \qquad K = 10^{-2.5} \qquad (14.5)$$

$$FePO_4 \cdot 2H_2O + 2H^+ = Fe^{3+} + H_2PO_4^- + 2H_2O \qquad K = 10^{-6.85}. \qquad (14.6)$$

Decreasing pH favors the solubility of iron and aluminum phosphates as illustrated in Ex. 14.2.

Ex. 14.2 *The solubility of phosphorus from variscite will be estimated for pH 5 and pH 6.*

   Solution:

(i) *From Eq. 14.5,*

$$\frac{\left(Al^{3+}\right)\left(H_2PO_4^-\right)}{\left(H^+\right)^2} = 10^{-2.5}$$

and

$$\left(H_2PO_4^-\right) = \frac{\left(H^+\right)^2 \left(10^{-2.5}\right)}{\left(Al^{3+}\right)}.$$

Letting $x = \left(Al^{3+}\right) = \left(H_2PO_4^-\right)$, *for pH 5* $\left(H^+ = 10^{-5} M\right)$

$$x = \frac{\left(10^{-5}\right)^2 \left(10^{-2.5}\right)}{x}$$

$$x^2 = 10^{-12.5}$$

$$x = 10^{-6.25}.$$

At pH 5,

$$\left(H_2PO_4^-\right) = 10^{-6.25} \ M \ or \ 5.62 \times 10^{-7} \ M.$$

There are 31 g of phosphorus per mole of $H_2PO_4^-$, *so*

$$5.62 \times 10^{-7} \ M \times 31 \ g \ P/mole = 1.74 \times 10^{-5} \ g \ P/L$$

or 0.017 mg/L of phosphorus.

(ii) *Repeating the above for pH = 6($H^+ = 10^{-6}$ M), we get 0.0017 mg P/L*

Calculation of the solubility of strengite for the same pH values used in Ex. 14.2, gives phosphorus concentrations of 0.00012 mg/L at pH 5 and 0.000012 mg/L at pH 6. Strengite is less soluble than variscite at the same pH by a factor of 150.

Acidic and neutral sediments usually contain appreciable iron and aluminum minerals, and aluminum tends to control phosphate solubility in aerobic water and sediments. Two representative iron and aluminum compounds found in sediment are gibbsite, $Al(OH)_3$ and iron (III) hydroxide, $Fe(OH)_3$. The dissolution of these two and other iron and aluminum oxides and hydroxides is pH dependent

$$Al(OH)_3 + 3H^+ = Al^{3+} + 3H_2O \qquad K = 10^9 \tag{14.7}$$

$$Fe(OH)_3 + 3H^+ = Fe^{3+} + 3H_2O \qquad K = 10^{3.54}. \tag{14.8}$$

The solubilities of the two minerals increase with decreasing pH as illustrated in Ex. 14.3.

Ex. 14.3 *The solubilities of gibbsite and iron (III) hydroxide will be estimated for pH 5 and 6.*

Solution:

From Eqs. 14.7 and 14.8,

$$pH = 5 \; (H^+ = 10^{-5}) \qquad \frac{(Al^{3+})}{(H^+)^3} = 10^9$$

and $(Al^{3+}) = (10^{-5})^3 (10^9) = 10^{-6}$ M.

Following the same approach and using $Fe(OH)_3$ at pH 5,

$$\frac{(Fe^{3+})}{(H^+)^3} = 10^{3.54}$$

$$(Fe^{3+}) = (10^{-5})^3 (10^{3.54}) = 10^{-11.46} \; M.$$

Repeating the calculation above at pH 6 ($H^+ = 10^{-6}$) we get

$$(Al^{3+}) = (10^{-6})^3 (10^9) = 10^{-9} \; M$$

$$(Fe^{3+}) = (10^{-6})^3 (10^{3.54}) = 10^{-14.46} \; M.$$

The Al^{3+} concentration is 2.7×10^{-2} mg/L and 2.7×10^{-5} mg/L at pH 5 and pH 6, respectively, while Fe^{3+} concentration is 1.94×10^{-7} mg/L and 1.94×10^{-10} mg/ L, respectively at these pH values. The iron concentration is more than four orders of magnitude lower than that of aluminum at each pH—essentially imperceptible.

From Ex. 14.3, it can be seen that the amounts of iron and aluminum in solution in sediment pore water and available to precipitate phosphorus as iron and aluminum phosphates will increase as pH decreases. But, at the same pH, aluminum compounds are much more soluble than iron compounds.

Despite iron and aluminum phosphate compounds being more soluble at lower pH (Ex. 14.2), iron and aluminum oxides and hydroxides tend to be much more abundant in sediment than are aluminum and iron phosphates. Thus, when phosphorus is added to acidic sediment there is adequate Al^{3+} and Fe^{3+} present to precipitate it. In reality, the availability of phosphorus from sediment tends to decrease with decreasing pH mainly because of the presence of more aluminum to precipitate it. This fact is illustrated in Ex. 14.4.

Ex. 14.4 *The solubility of phosphorus from a highly soluble source in a system at equilibrium with gibbsite will be calculated.*

Solution:

Gibbsite provides Al^{3+} to solution, and Al^{3+} can react with phosphorus to precipitate $AlPO_4 \cdot 2H_2O$ (variscite). The solubility of gibbsite in water of pH 5 is 10^{-6} M (see Ex. 14.3). Dissolution of variscite may be written as

$$AlPO_4 \cdot 2H_2O + 2H^+ = Al^{3+} + H_2PO_4^- + 2H_2O$$

for which $K = 10^{-2.5}$. Thus,

$$\frac{\left(Al^{3+}\right)\left(H_2PO_4^-\right)}{\left(H^+\right)^2} = 10^{-2.5}$$

$$\left(H_2PO_4^-\right) = \frac{\left(H^+\right)^2 \, 10^{-2.5}}{\left(Al^{3+}\right)}$$

$$= \frac{\left(10^{-5}\right)^2 \, 10^{-2.5}}{10^{-6}} = 10^{-6.5} \, M.$$

Expressed in terms of phosphorus, $10^{-6.5}$ M is 0.00000032 M × 30.98 g P/mole or 0.01 mg P/L. In this example, phosphate solubility is controlled by variscite, because variscite is less soluble than monocalcium phosphate.

Phosphate can be absorbed by iron and aluminum oxides as follows:

$$H_2PO_4^- + Al(OH)_3 = Al(OH)_2H_2PO_4 + OH^- \tag{14.9}$$

$$H_2PO_4^- + FeOOH = FeOH_2PO_4 + OH^-. \tag{14.10}$$

In soil and sediment much of the clay fraction is in the form of iron and aluminum hydroxides. Clays are colloidal and have a large surface area; they can bind large amounts of phosphorus.

Silicate clays also can fix phosphorus. Phosphorus is substituted for silicate in the clay structure. Clays also have some ability to adsorb anions, because they have a small number of positive charges on their surfaces. Absorption by silicate minerals and anion exchange is less important than is phosphorus removal by aluminum and iron in acidic soils and sediments.

Primary phosphate compounds in neutral and basic sediment are calcium phosphates. The most soluble calcium phosphate compound is monocalcium phosphate, $Ca(H_2PO_4)_2$. This is the form of phosphorus normally applied in fertilizer. In neutral or basic soils, $Ca(H_2PO_4)_2$ is transformed through dicalcium, octacalcium, and tricalcium phosphates to apatite. Apatite is not very soluble under neutral or alkaline conditions. A representative apatite, hydroxyapatite, dissolves as follows:

$$Ca_5(PO_4)_3OH + 7H^+ = 5Ca^{2+} + 3H_2PO_4^- + H_2O \qquad K = 10^{14.46}. \qquad (14.11)$$

A high concentration of Ca^{2+} and elevated pH favors formation of hydroxyapatite from dissolved phosphate in water or sediment pore water. Apatite is not appreciably soluble at pH above 7 even at low calcium concentration as shown in Ex. 14.5.

Ex. 14.5 *The solubility of phosphorus in water with 5 mg/L calcium ($10^{-3.90}$ M) at pH 7 and pH 8 will be calculated.*

Solution:

Assuming that the reaction controlling phosphorus concentration is Eq. 14.11,

$$\frac{\left(Ca^{2+}\right)^5 \left(H_2PO_4^-\right)^3}{\left(H^+\right)^7} = 10^{14.46}$$

at pH 7

$$\left(H_2PO_4^-\right)^3 = -\frac{\left(10^{-7}\right)^7 \left(10^{14.46}\right)}{\left(10^{-3.90}\right)^5} = 10^{-15.04} \ M$$

$$\left(H_2PO_4^-\right) = 10^{-5.01} \ M \ or \ 0.30 \ mg \ P/L.$$

Repeating the calculation for pH 8 gives

$$\left(H_2PO_4^-\right) = 10^{-7.35} \ M \ or \ 0.0014 \ mg \ P/L.$$

At a higher calcium concentration, the phosphorus concentration would be less, e.g. at pH 7 and 20 mg Ca^{2+}/L ($10^{-3.3}$ M), the phosphorus concentration would be only 0.03 mg/L—an order of magnitude lower than at 5 mg/L Ca^{2+}.

The maximum availability of phosphorus in aerobic soil or sediment typically occurs between pH 6 and 7 (Fig. 14.3). In this pH range, there is less Al^{3+} and Fe^{3+} to react with phosphorus and a smaller tendency of aluminum and iron oxides to adsorb phosphorus than at lower pH, and the activity of calcium is normally lower than at higher pH. Nevertheless, in the pH range of 6–7, most of the phosphorus added to

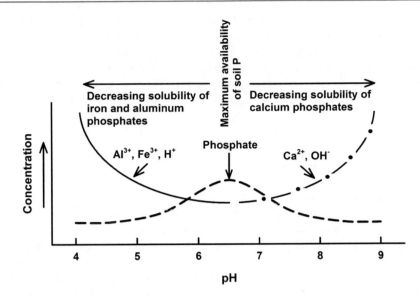

Fig. 14.3 Schematic showing effects of pH on the relative concentrations of dissolved phosphate in aerobic soil or sediment

aquatic ecosystems still is rendered insoluble through adsorption by colloids or precipitation as insoluble compounds.

Iron phosphates contained in sediment become more soluble when the redox potential falls low enough for ferric iron to be reduced to ferrous iron. Phosphorus concentrations in pore water of anaerobic sediment may be quite high (Masuda and Boyd 1994a). This pool of phosphorus is largely unavailable to the water column because iron phosphate reprecipitates when ferrous iron and phosphate diffuse into the aerobic layer normally existing at the sediment-water interface. The aerobic layer at the interface is lost during thermal stratification of eutrophic lakes and ponds. Diffusion of iron and phosphorus from anaerobic sediment can lead to high iron and phosphate concentrations in the hypolimnion. Concentrations of 10–20 mg/L iron and 1–2 mg/L soluble orthophosphate would not be unusual. When thermal destratification occurs, hypolimnetic waters mix with surface waters, and concentrations of phosphorus increase briefly in surface waters. However, because of the presence of dissolved oxygen, phosphorus concentrations quickly decline. Phosphorus either precipitates directly as iron phosphate (Eq. 14.6) or it is adsorbed onto the surface of the floc of iron (III) hydroxide precipitating from the oxygenated water.

Organic Phosphorus

The dry matter of plants commonly contains 0.05–0.5% phosphorus, while vertebrate animals such as fish may contain 2–3% phosphorus or more. Crustaceans typically contain about 1% phosphorus in their dry matter. Phosphorus contained in organic matter is mineralized by microbial activity in the same manner that nitrogen is mineralized. The same conditions favoring decomposition and nitrogen mineralization favor phosphorus mineralization. Just as with nitrogen mineralization, if there is too little phosphorus in organic matter to satisfy microbial requirements, phosphorus can be immobilized from the environment. The nitrogen:phosphorus ratio in living organisms and in decaying organic residues varies considerably ranging from around 5:1 to 20:1.

Phosphorus Analysis Issues

The phosphorus in water consists of various forms to include soluble inorganic phosphorus, soluble organic phosphorus, particulate organic phosphorus (in living plankton and in dead detritus), and particulate inorganic phosphorus (on suspended mineral particles). The soluble fraction can be separated from the particulate fraction by filtration through a membrane or glass fiber filter. However, common analytical methods do not distinguish perfectly between soluble inorganic and soluble organic phosphorus, and a portion of the soluble organic phosphorus will be included in measurements of soluble inorganic phosphorus. As a result, when phosphorus concentration is measured directly in filtrates of water, the resulting phosphorus fraction is called soluble reactive phosphorus. Digestion of a raw water sample in acidic persulfate releases all of the bound phosphorus, and analysis of the digestate gives total phosphorus. Most of the information on phosphorus concentrations in natural waters is for soluble reactive phosphorus and total phosphorus.

Phosphorus in sediment may be fractionated based on its extraction with various solutions. A common way of fractioning sediment phosphorus is a sequential extraction with 1 M ammonium chloride to remove loosely-bound phosphorus, 0.1 N sodium hydroxide to remove iron and aluminum-bound phosphorus, and 0.5 N hydrochloric acid to remove calcium-bound phosphorus (Hieltjes and Liklema 1982). Other extractants also may be used to remove phosphorus from sediment samples. Many methods of soil phosphorus analysis in soil testing laboratories are used for agriculture such as extraction in 0.03 M NH_4F and 0.025 M HCl, 0.61 M $CaCl_2$, 0.5 M $NaHCO_3$, and water (Kleinman et al. 2001), and various others (Masuda and Boyd 1994b). Soil can be digested in perchloric acid to release bound phosphorus for total phosphorus analysis.

Phosphorus Dynamics

Concentrations in Water

Phosphorus concentrations in surface waters generally are quite low. Total phosphorus seldom exceeds 0.5 mg/L except in highly eutrophic waters or in wastewaters. There generally is much more particulate phosphorus than soluble reactive phosphorus. For example, Masuda and Boyd (1994b) found that water in eutrophic aquaculture ponds contained 37% dissolved phosphorus and 63% particulate phosphorus. However, most of the dissolved phosphorus was non-reactive organic phosphorus, and only 7.7% of the total phosphorus was soluble reactive phosphorus. Typically, 10% or less of the total phosphorus will be soluble reactive phosphorus and readily available to plants. Most surface waters contain less than 0.05 mg/L soluble reactive phosphorus, and most unpolluted water bodies only contain 0.001–0.005 mg/L of this fraction.

Sediment contains much more phosphorus than found in the water column above it. Total phosphorus concentrations found in sediment range from less than 10 mg/kg to more than 3000 mg/kg. However, most of this phosphorus is tightly bound and not readily soluble in water. Phosphorus concentrations in sediment of a eutrophic fishpond (Masuda and Boyd 1994b) are illustrated in Table 14.1. Notice that 85.6% of the sediment phosphorus was not removable by normal extracting agents and had to be released by perchloric acid digestion.

Plant Uptake

Phytoplankton can absorb phosphorus from water very quickly. In water with a dense bloom of phytoplankton, phosphorus additions of 0.2–0.3 mg/L can be completely removed within a few hours (Boyd and Musig 1981). Macrophytes also can remove phosphorus from water very quickly, and rooted macrophytes can absorb phosphorus from anaerobic zones in the sediment (Bristow and Whitcombe 1971). Plant uptake is a major factor controlling concentrations of soluble reactive phosphorus in water, and much of the total phosphorus in water is contained in phytoplankton cells. Macrophyte communities can store large amounts of phosphorus in their biomass.

Table 14.1 Distribution of forms of soil and water phosphorus for a fish pond at Auburn, Alabama

Phosphorus pool	Phosphorus fraction	Amount (g/m^2)	(%)
Pond water[a]	Total phosphorus	0.252	0.19
	Soluble reactive phosphorus	0.019	0.01
	Soluble nonreactive phosphorus	0.026	0.02
	Particulate phosphorus	0.207	0.16
Soil[b,c]	Total phosphorus	132.35	99.81
	Loosely-bound phosphorus	1.28	0.96
	Calcium-bound phosphorus	0.26	0.20
	Iron- and aluminum-bound phosphorus	17.30	13.05
	Residual phosphorus[d]	113.51	85.60
Pond	Total phosphorus	132.60	100.00

[a]Average pond depth = 1.0 m
[b]Soil depth = 0.2 m
[c]Soil bulk density = 0.797 g/cm^3
[d]Phosphorus removed by perchloric acid digestion

Some plants can absorb more phosphorus than they need immediately, and they store it for use later. The absorption of phosphorus and other nutrients in excess of the amount required for growth has been demonstrated in many plant species including species of phytoplankton. This phenomenon termed luxury consumption is illustrated in Fig. 14.4.

The ability to absorb and store more nutrients than needed at the moment is of competitive advantage for plants. The phosphorus can be removed from the environment thereby depriving competing plants of it. Phosphorus in algal cells can be passed on to succeeding generations when cells divide and multiply. In larger plants, phosphorus can be translocated from storage in older tissue to rapidly growing meristematic cells.

Exchange between Water and Sediment

If sediment is placed in a flask of distilled water and agitated until equilibrium phosphorus concentration is attained, very little phosphorus usually will be present in the water. In a series of soil samples containing from 100 to 3400 mg/kg of total phosphorus, water extractable phosphorus concentrations ranged from non-detectable to 0.16 mg/L (Boyd and Munsiri 1996). The correlation between total phosphorus and water extractable phosphorus was weak (r = 0.581), but the correlation between dilute acid (0.075 N) extractable phosphorus and water soluble phosphorus was much stronger (r = 0.920).

It can be shown by successive extractions of a sediment with water that there is a continued release of phosphorus to the water for many extractions (Fig. 14.5). The

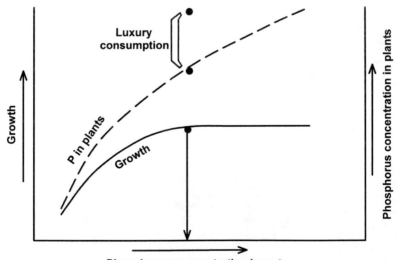

Fig. 14.4 Luxury consumption of phosphorus by phytoplankton

amount released, however, declines with the number of extractions. Because of the relationship shown in Fig. 14.5, the sediment is a reserve of phosphorus available when plant removal causes phosphorus concentrations in the water to fall below the equilibrium concentration. Nevertheless, the concentrations of phosphorus at equilibrium normally are quite low, and phosphate additions are necessary to stimulate abundant phytoplankton growth.

Macrophytes—especially rooted, submerged macrophytes—grow quite well in waters that are low in phosphorus because they can absorb phosphorus and other nutrients from sediment. Phosphorus in sediment is not readily available to phytoplankton because of the complex logistics of nutrient movement from sediment pore water to the illuminated zone where phytoplankton grow (Fig. 14.6).

When phosphorus enters water through intentional additions as in fishponds or through pollution, stimulation of phytoplankton growth creates turbidity and shades the deeper waters. Restriction in light caused by nutrient enrichment may eliminate many species of macrophytes from aquatic communities in eutrophic water bodies.

The relative concentrations of phosphorus in sediment, sediment pore water, at the sediment-water interface, and in surface water are illustrated in Fig. 14.7 for a small, eutrophic fishpond. There is roughly an order of magnitude difference in sediment phosphorus and pore water phosphorus concentrations, and another order of magnitude difference between phosphorus concentrations in pore water and at the water at the sediment-water interface. Pore water is anaerobic and phosphorus in

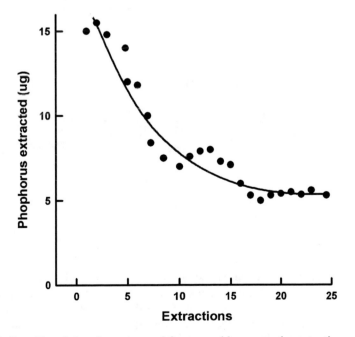

Fig. 14.5 Quantities of phosphorus removed from a mud by consecutive extractions with phosphorus free water

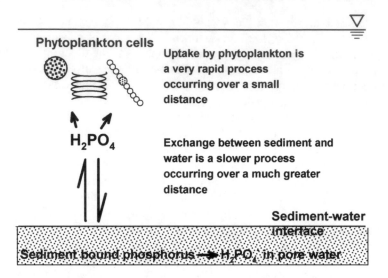

Fig. 14.6 Illustration of rapid uptake of phosphate by phytoplankton cells and slower exchange of phosphate between sediment and water

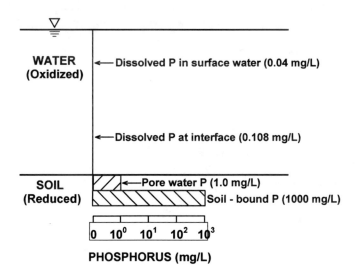

Fig. 14.7 Concentrations of phosphorus bound in soil, dissolved in pore water, and dissolved in overlaying pond water

pore water tends to precipitate at the aerobic interface and little enters the pond water. Even when anaerobic conditions exist at the sediment-water interface, phosphorus must diffuse from the pore water into the open water, and diffusion is a relatively slow process. Once phosphorus enters the open water, it can be mixed throughout the water body rather quickly by turbulence.

In pond aquaculture, phosphorus often is added to increase dissolved inorganic phosphorus concentrations. A portion of the added phosphorus is quickly absorbed by phytoplankton. The part that is not removed by plants will tend to accumulate in the sediment, and most of the phosphorus removed by plants also will eventually reach the sediment. Turbulence will allow soluble phosphorus to reach the sediment more quickly when compared to the movement of sediment-bound phosphorus into the water. The removal of fertilizer phosphorus from fish ponds after phosphate fertilizer application (Fig. 14.8) illustrates the rapidity of phosphorus removal from water.

The ability of sediment to hold phosphorus is usually quite large. In fish ponds on the Auburn University E. W. Shell Fisheries Center at Auburn, Alabama that received an average phosphorus input of 4.1 g P/m^2/year (41 kg/ha/year) for 22 years, bottom soils were only about half-saturated with phosphorus and still rapidly adsorbed phosphorus from the water (Masuda and Boyd 1994b). Nevertheless, sediment can become saturated with phosphorus or have a very low capacity to adsorb phosphorus, e.g., sandy sediment. In such bodies of water, additions of phosphorus are particularly effective in stimulating phytoplankton growth.

Sediment is not always necessary for phosphorus removal from the water. In waters with significant concentrations of calcium and pH of 7–9, phosphate will precipitate directly from the water as calcium phosphate.

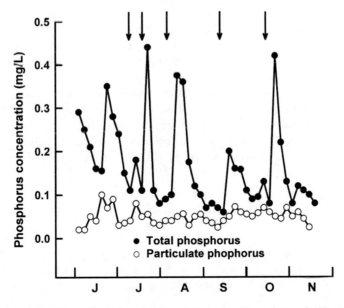

Fig. 14.8 Average concentrations of total and particulate phosphorus in two fertilized fish ponds. Vertical arrows indicate fertilizer application dates

Interaction with Nitrogen

Nitrogen and phosphorus are key nutrients regulating aquatic plant productivity. But, the amounts and ratios of these two nutrients vary among species. Redfield (1934) reported that marine phytoplankton contained on a weight basis about seven times more nitrogen than phosphorus. This value is often used as the average N:P ratio in plants, but the ratios for individual species vary from 5:1 to 20:1. In most ecosystems, an increase in phosphorus concentration will cause a greater response in plant growth than will an increase in nitrogen concentration. This results because phosphorus is quickly removed from the water and bound in the sediment, and there is limited recycling of sediment bound phosphorus the water column.

Nitrogen also is removed from water bodies by various processes, but as much as 10–20% of added nitrogen is present in organic matter deposited in sediment. Sediment organic matter is decomposed, and nitrogen is continually mineralized. The internal recycling of nitrogen in aquatic ecosystems is much greater than it is for phosphorus. As a result, in order to achieve a nitrogen to phosphorus ratio of 7:1 (the Redfield ratio) in the water, it usually requires that the addition of these two elements be at a lower N:P ratio.

Conclusions

Phosphorus is a key factor regulating the productivity of ecosystems. While it is widely used in agriculture, gardening, landscaping, and aquaculture to increase plant growth, its introduction into natural water bodies usually stimulates excessive plant growth and is considered pollution.

The main form of phosphorus in natural waters is orthophosphate. Dissolved phosphate is strongly sequestered by sediment through formation of iron, aluminum, and calcium phosphates. The equilibrium between sediment phosphorus and dissolved phosphorus is strongly shifted towards the sediment. Usually there must be a more or less continuous input of phosphate to maintain a high rate of plant productivity in water bodies.

References

Boyd CE, Munsiri P (1996) Phosphorus adsorption capacity and availability of added phosphorus in soils from aquaculture areas in Thailand. J World Aquacult Soc 27:160–167

Boyd CE, Musig Y (1981) Orthophosphate uptake by phytoplankton and sediment. Aquaculture 22:165–173

Bristow JM, Whitcombe M (1971) The role of roots in the nutrition of aquatic vascular plants. Am J Bot 58:8–13

Harold FM (1966) Inorganic polyphosphates in biology: structure, metabolism, and function. Bacteriol Rev 30:272–794

Hieltjes AHM, Liklema L (1982) Fractionation of inorganic phosphate in calcareous sediments. J Environ Qual 9:405–407

Kleinman PJA, Sharpley AN, Gartley K, Jarrell WM, Kuo S, Menon RG, Myers R, Reddy KR, Skogley EO (2001) Interlaboratory comparisons of soil phosphorus extracted by various soil test methods. Commun Soil Sci Plant Anal 32:2325–2345

Masuda K, Boyd CE (1994a) Chemistry of sediment pore water in aquaculture ponds built on clayey, Ultisols at Auburn, Alabama. J World Aquacult Soc 25:396–404

Masuda K, Boyd CE (1994b) Phosphorus fractions in soil and water of aquaculture ponds built on clayey, Ultisols at Auburn, Alabama. J World Aquacult Soc 25:379–395

Redfield AC (1934) On the proportions of organic deviations in sea water and their relation to the composition of plankton. In: Daniel RJ (ed) James johnstone memorial volume. University Press of Liverpool, Liverpool

Eutrophication

15

Abstract

The term eutrophication refers to a water body changing over time from nutrient-poor to nutrient-rich status. More nutrients cause increased aquatic plant growth that negatively influences water quality and especially dissolved oxygen availability. The result is a decrease in species diversity and an increase in the biomass of a few and usually less desirable species. The changes caused by eutrophication often are gradual and not noticed until there is a sudden shift (tipping point) to a more eutrophic status. The two nutrients responsible for eutrophication are nitrogen and phosphorus contained in nearly all effluents. Much effort is devoted to removing these nutrients by water treatment and watershed management practices. Sometimes, chemical coagulants may be added to water bodies to precipitate phosphorus, and chemicals such as copper sulfate applied as algicides. Although most popular discussions of eutrophication apply to lakes, streams also may suffer eutrophication.

Introduction

Kociolek and Stoermer (2009) state that oligotrophy means poor food or low level of nutrients as compared to eutrophy which suggests good food or high level of nutrients. Water bodies vary in availability of nutrients, and there exists a continuum of water bodies ranging from those that are very nutrient deficient to those with extremely abundant supplies of nutrients. This continuum extends from ultraoligotrophy through oligotrophy, mesotrophy (moderate food and nutrients), eutrophy, to ultraeutrophy. The earliest use of this continuum for categorizing nutrient states was probably about 1920, and it is still a fundamental concept of ecology and particularly in aquatic sciences. Oligotrophic water bodies with acidic waters stained by high concentrations of dissolved humic substances are a special case known as dystrophy.

The addition of nutrients to oligotrophic water bodies from natural or anthropogenic sources will make them more nutrient rich and less oligotrophic. While water bodies may naturally become more eutrophic over time, in the past 100 years, nutrient pollution from anthropogenic sources has greatly increased nutrient availability in many water bodies throughout the world. This has been a disturbing phenomenon, because it has encouraged plant growth (especially phytoplankton growth) in water bodies resulting in undesirable changes such as low dissolved oxygen concentration and reduction in biodiversity. The artificial enrichment of water bodies with nutrients has become known as eutrophication or more precisely cultural eutrophication.

The word eutrophication is defined in the Merriam-Webster Dictionary as "the process by which a water body becomes enriched in dissolved nutrients (such as phosphates) that stimulate the growth of plant life usually resulting in the depletion of dissolved oxygen." The term was said to have been first used with the above meaning in 1946 (https://www.merriam-webster.com/dictionary/eutrophication), but it has no doubt been used many millions of times since. The dictionary definition above refers only to phosphorus, but nitrogen is almost or equally as important in eutrophication as is phosphorus, and both are causes of eutrophication. The reference to "usually resulting in depletion of dissolved oxygen" also is not exactly accurate. Many eutrophic water bodies simply have lower dissolved oxygen concentration than less eutrophic ones, but if the process of eutrophication is allowed to progress, dissolved oxygen depletion will eventually become a common phenomenon.

Eutrophication has already been discussed briefly in several chapters and especially in the three preceding ones. Nevertheless, eutrophication is such a common concern in society and the focus of so many environmental protection efforts that further discussion is desirable.

The Eutrophication Process

Eutrophication of water bodies occurs naturally because nutrients and sediment accumulate as the result of erosion on catchments. Sediment decreases the volume of a water body, and as the volume becomes less, the store of nutrients in the sediment increases nutrient concentration in the water column. Sedimentation also results in shallow edges where rooted aquatic plants can thrive. A water body fills in and becomes more nutrient enriched as it ages. The edges become shallower allowing shrubs and trees to grow and over many years—usually centuries—the water body may fill in completely and becomes terrestrial habitat.

Cultural eutrophication greatly accelerates natural eutrophication or aging of water bodies. Human activities, particularly agriculture and deforestation on catchments expose the land surface allowing greater erosion and delivery of suspended solids to water bodies. These solids settle, and water bodies fill in much faster than through natural processes. Agricultural runoff contains nitrogen, phosphorus, and other nutrients applied to crops as fertilizer. Runoff from suburban lawns, parks, golf courses, and similar areas also contain nutrients from fertilizers and other sources. Municipalities and industries discharge effluents containing nutrients into water bodies, and septic tanks seep into ground water which may enter water bodies. The overall effect is that anthropogenic activities on watersheds cause water bodies to fill in faster with sediment and supply nutrients for abundant aquatic plant growth. Rooted aquatic plants fill the water column of shallow water areas, floating aquatic plants may completely cover smaller water bodies, and phytoplankton may become extremely abundant in open water areas. The overabundance of vegetation completely changes the physical and chemical environment, and there is a trend towards lower dissolved oxygen concentrations and even periods of oxygen depletion. These changes lead to a decrease in biological diversity with the exclusion of many ecologically-sensitive species. Biomass will be greater in eutrophic waters than in oligotrophic ones, but there will be fewer species and less biological stability in eutrophic waters. The phytoplankton community becomes dominated with blue-green algae, and fish communities decrease in species abundance with only a few coarse species dominating. Such water bodies often are said to be dead, but this is by no means true. Eutrophic water bodies are very much alive, but the life consists mainly of a few less desirable species and biodiversity and ecological stability decline. Eutrophic waters also are impaired for many human uses.

Nitrogen, Phosphorus, and Eutrophication

The trophic state of a water body can be established by several methods. The traditional opinion held that eutrophic lakes have hypolimnetic dissolved oxygen depletion during thermal stratification while oligotrophic ones do not. This is not always a reliable indicator of eutrophication in shallow water bodies that do not stratify thermally, and it does not allow classification as to degree of oligotrophy or eutrophy. The modern way of classifying trophic status of water bodies relies on measurements of total nitrogen, total phosphorus, and chlorophyll *a* concentrations, primary productivity, abundance of blue-green algae, and Secchi disk visibility. Indices of trophic status based on a combination of the preceding factors also are used.

There is no general agreement on the concentration ranges of the variables mentioned above that should be expected in water bodies of different trophic states. Water bodies vary in most chemical and physical features, and this variation may affect concentrations or levels of the indicator variables. Nevertheless, some typical values of the key variables in oligotrophic, mesotrophic, and eutrophic water bodies are presented (Table 15.1). Ultraoligotrophy and ultraeutrophy are not included in Table 15.1, but water bodies with extremely small or large values for the variables may be classified accordingly.

The author believes that there are three main factors affecting the concentrations of nutrients necessary to initiate phytoplankton blooms. Turbidity interferes with light penetration. In water bodies turbid with suspended clay particles or dissolved humic substances, high concentrations of nitrogen and phosphorus may not cause phytoplankton blooms because of light limitation of photosynthesis. The hydraulic

Table 15.1 Typical average concentrations of water quality variables in oligotrophic, mesotrophic, and eutrophic water bodies during the spring-fall period

Variable	Trophic state		
	Oligotrophic	Mesotrophic	Eutrophic
Total nitrogen (mg/L)	<0.5	0.5–1.5	>1.5
Total phosphorus (mg/L)	<0.025	0.025–0.075	>0.075
Chlorophyll a (μg/L)	<2	2–15	>15
Primary productivity (g/m^2/day)	<0.25	0.25–1.0	>1.0
Secchi disk visibility (m)	>8	2–8	<2

retention time influences the proportion of nutrients entering or originating in water bodies that is flushed out. A greater nitrogen and phosphorus input would be necessary to cause eutrophication in a water body with a hydraulic retention time of a few months than in a lake with a hydraulic retention time of several years assuming other factors are similar. Phosphorus will have a lower concentration at equilibrium in a water with high pH and high calcium concentration. This suggests that waters with moderate to high alkalinity and hardness (which also usually have ambient pH around 8) will require a greater input of phosphorus to cause algal blooms than will lakes of lower alkalinity, hardness, and pH.

As already mentioned, eutrophic water bodies have a high abundance of a few species, and many sensitive species disappear. In the early days of eutrophication research, diversity was almost always expressed as species diversity which is an index of how the individuals in a community are distributed among the species. As a result, phytoplankton diversity, zooplankton diversity, fish diversity, etc. was measured. Many equations have been made for calculating species diversity, and an index of phytoplankton species diversity (Margalef 1958) provides an example

$$\bar{H} = \frac{S-1}{\ln(N)} \tag{15.1}$$

where $\bar{H}$ = phytoplankton species diversity, S = the number of species, and N = the total number of individuals. This equation could obviously be applied to other types of communities, but it has been customary for investigators interested in different biological communities to formulate different equations in accordance with their judgement.

Ex. 15.1
Community A contains 25 species of phytoplankton and 1,000 individuals/mL while community B has 11 species and 14,000 individuals/mL. The diversity index for the two communities will be estimated by Eq. 15.1.
   Solution:

$$A. \quad \bar{H} = \frac{25-1}{\ln(1,000)} = \frac{24}{6.91} = 3.47$$

$$B. \quad \bar{H} = \frac{11 - 1}{\ln{(14,000)}} = \frac{10}{9.55} = 1.05.$$

A greater value of $\bar{H}$ implies a higher biodiversity, i.e., community A has a greater phytoplankton species diversity than does community B.

A community with more species relative to the total number of individuals has higher diversity than a community with fewer species relative to the number of individuals. As a rule, the greater the species diversity, the more stable an ecosystem is considered to be (Odum 1971). The logic for this supposition is that if a species disappears from a diverse community, there will be a greater likelihood that another species can fulfill the function of the species that disappeared than there would be in a less diverse community. Eutrophic aquatic ecosystems tend to be less stable than oligotrophic ones. This lack of stability may be reflected in sudden shifts in the abundance of species and also in sudden changes in dissolved oxygen concentrations and other water quality variables.

The modern concept of diversity has been expanded to include the diversity of ecosystems, communities, and habitats. Ecosystems have unique geological, edaphic, hydrologic, and climatic regimes that affect the types and abundance of species living in them. An aquatic ecosystem is much different from a terrestrial ecosystem, but a nutrient-poor aquatic ecosystem also differs greatly from a nutrient-rich one. Moreover, the species in an ecosystem all have genetic diversity, and the interaction of ecosystem diversity and genetic diversity results in gradual changes in the characteristics of organisms over time. The new concept of biodiversity is so broad that it is almost impossible to find a suitable, single index of biodiversity. As a result, it is still popular to enumerate species and count the number of individuals of each species in samples of the major types of communities as a way of assessing the biodiversity status.

Most unpolluted water bodies typically contain no more than 0.05 mg/L of total phosphorus and 0.75 mg/L of total nitrogen. An increase in the inputs of nitrogen and phosphorus may or may not cause an algal bloom to develop and the visibility in a lake to decrease. A water body is able to assimilate a certain quantity of nutrients with relative little change in primary productivity and water quality. However, if the inputs continue, ambient concentrations of phosphorus and combined nitrogen will increase and changes in primary productivity and water quality will become apparent.

The processes of a water body changing from oligotrophic to mesotrophic or from mesotrophic to eutrophic is gradual and the change is not easily detected visually or from water quality measurements and phytoplankton abundance. Nitrogen and phosphorus concentrations, phytoplankton abundance, species of plankton, water clarity, and dissolved oxygen concentrations naturally vary over spans of days or weeks and longer-term trends of change often are obscured by the short-term variation.

The phenomenon of an ecological system changing rapidly from one state to another state is often referred to as the tipping point in ecology and especially in discussions of eutrophication in water bodies. There have been hundreds of studies of the tipping point in ecological systems and of techniques for predicting when it will occur. These models usually include many factors, and they are not particularly accurate. A recent model for safe-guarding plant pollinators (Jiang et al. 2018) is based on comparatively fewer factors, and the concept upon which it is based likely has wider application. The topic of predicting tipping points is beyond the scope of this book, but the existence of tipping points in aquatic systems is important in assessing eutrophication. It makes changes in eutrophication status very difficult to detect from nitrogen and phosphorus concentrations and phytoplankton abundance. Levels of these variables may have been increasing gradually for decades with little observed effect, but at some time, the water body may suddenly begin to have dense phytoplankton blooms.

Lake Eutrophication

The eutrophication process has been studied in many lakes. The pattern is usually the development of dense phytoplankton blooms, often dominated by one or more species of blue-green algae, shallow thermal stratification, wide daily dissolved oxygen fluctuations in the epilimnion, and dissolved oxygen depletion in the hypolimnion. Other water quality changes are wider daily fluctuations in carbon dioxide concentration and pH, and of course, greater concentrations of combined nitrogen, total nitrogen, soluble reactive phosphorus, and total phosphorus.

The bigger concern relates to the changes in aquatic communities of eutropic waters. It has already been mentioned that a few species and sometimes a single species of blue-green algae will dominate the phytoplankton of eutrophic lakes. Genera such as *Anacystis*, *Aphanizomenon*, *Anabaena*, and *Oscillatoria* are commonly abundant in eutrophic situations. Species of these genera are particularly harmful, because they often form unsightly surface scums, produce obnoxious odors, cause taste and odors where lakes or reservoirs are used for water supply, and they may even be toxic to animals and humans. Algal scums also undergo massive die-offs that can have serious impacts on fish populations by causing dissolved oxygen depletion. Of course, even under normal situations dissolved oxygen depletion may occur in the late night and early morning in surface water of eutrophic reservoirs and lakes.

Underwater species of aquatic plants can only grow where light penetration is adequate to support photosynthesis. As phytoplankton abundance increases in lakes, submersed macrophyte species tend to disappear other than from very shallow water. This can be illustrated by the fact that fertilization to produce phytoplankton blooms is frequently recommended as a submerged macrophyte control method in fish ponds.

The food web in lakes and other aquatic habitats is complex and involves the movement of nutrients (organic matter) from the primary producers to the final consumers which usually are fish. There are many pathways by which organic matter produced by phytoplankton and aquatic plants can pass through the food web with a portion of it reaching large fish (Fig. 15.1). Phytoplankton can be eaten by zooplankton and zooplankton consumed by small fish which are then eaten by large fish. This is the food web pathway often used as an illustration. In lakes with filter-feeding fish, phytoplankton can go directly to large fish resulting in a relatively simple food web. The pathway usually is much more complicated: phytoplankton and zooplankton die, become organic detritus inhabited by bacteria, the detritus eaten by small fish, and the small fish are consumed by large fish. This pathway usually is more complex than shown in Fig. 15.1, because there are many links within the plankton, e.g., bacteria → protozoans and other flagellates and cilliates → rotifers → microcrustaceans (such as *Daphnia* and *Cladocera*).

There is much evidence to support the hypothesis that eutrophication encourages smaller zooplankton such as rotifers, cladorcerans, and copepods over larger microcrustaceans like *Daphnia* which are important fish food organisms (Tõnno et al. 2016). A review of the potential use of zooplankton as a bioindicator of eutrophic conditions (Ferdous and Muktadir 2009) concluded that species of rotifers, cladocerans, copepods, and ostracods were generally present in eutrophic lakes, but the species variation of each group decreased in eutrophic water. In addition, the smaller species tended to be favored. A eutrophic lake in India had the following percentages of zooplankton: rotifers, 52.38%; copepods, 26.5%; cladocerans, 16.45%; ostracods, 4.67% (Sunkad and Patil 2004). Large-bodied cladocerans are the most important component of the zooplankton for nutrition of many fish species or life stages.

Mollusc communities also are affected by eutrophication. Lake Dianchi in China had 31 species of mollusks in the 1940, 81 species during 1980–1999, and 16 species between 2000 and 2004 (Du et al. 2011). This reduction in mollusks species was attributed to an increase in eutrophication over the 60-year span.

Eutrophication is well known for reducing the diversity of fish species in lakes to favor species tolerant to low dissolved oxygen concentration. This shift is said to

Fig. 15.1 Generalized aquatic food web

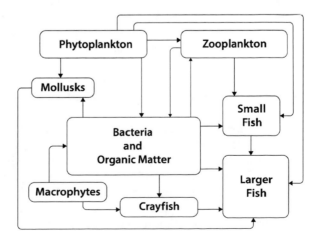

favor "coarse" species, because the species popular for sport fishing commonly become less abundant and sometimes disappear from the fish fauna. Kautz (1980) analyzed fish populations and trophic status in 22 lakes in Florida (USA). Total fish biomass was low in oligotrophic lakes, increased to a maximum in mesotrophic and eutrophic lakes, and was near maximum in hypereutrophic lakes. Species diversity was highest in mesotrophic lakes, but it decreased markedly in eutrophic and hypertrophic lakes. This reveals that a shrift from oligotrophy to mesotrophy may not reduce fish species numbers, but a shift to eutrophy will lessen the numbers of fish species.

Stream Eutrophication

Butcher (1947) states that a stream almost inevitably becomes more eutrophic as it flows downstream and receives nutrient inputs from its watershed. Phytoplankton blooms do not develop in rapidly flowing water, but submerged macrophytes or mat forming, filamentous macroalgae may grow well. Thièbaut and Muller (1998) reported that eutrophic streams in a mountainous region of France had a low diversity of submerged vascular plants because of overgrowth by filamentous algae. According to Dodds (2006), many natural streams become net heterotrophic systems (photosynthetic oxygen production < dissolved oxygen consumption in respiration). This situation may not cause extremely low dissolved oxygen concentration in reaches of eutrophic streams where reaeration replenishes the dissolved oxygen rapidly, but in stagnant reaches with low rates of reaeration, dissolved oxygen concentrations may become critically low. Large inputs of organic waste to streams also may cause dissolved oxygen depletion near effluent outfalls, but the stream reaerates as it passes downstream and dissolved oxygen concentration may recover and ammonia concentration decline as a result of nitrification. This topic is discussed in greater detail in Chap. 18.

The permissible organic waste input to streams often is estimated from stream aeration rate and the biological oxygen demand (BOD) of the wastewater input. This is possible because the rate that the stream reaerates can be used to estimate the daily BOD load to the stream that may be allowed without the stream dissolved oxygen concentration falling below a specified minimum level.

Climate Change and Eutrophication

The term global warming has largely been replaced by the concept of climate change, because some places may become cooler and others warmer even though the average global temperature is tending to increase. Rising temperature is the main consideration related to eutrophication. A study by Feuchtmayr et al. (2009) revealed that increasing water temperature resulted in greater dissolved phosphate concentrations, increased plant biomass, and reduced fish biomass. High nitrogen input and warming reduced the number of plant species. Warming also seemed to favor increased macrophyte growth more than it enhanced phytoplankton growth. The results suggested that large surface to volume ratios (shallow lakes and streams) would be more susceptible to adverse effects of global warming than would deeper water bodies with lower surface to volume ratios.

Kaushal et al. (2013) reported that the rising atmospheric carbon dioxide concentration was increasing limestone solubility and causing stream alkalinity to increase. They suggested that greater alkalinity in streams may increase the availability of inorganic carbon for photosynthesis in water bodies to accelerate eutrophication. They also suggested that high stream pH would make high ammonia nitrogen concentrations more problematic by increasing the proportion of potentially toxic NH_3 relative to NH_4^+.

Eutrophication and Water Use

Dense algal blooms cause water quality deterioration ultimately reducing biodiversity and leading to water bodies inhabited mainly by coarse fish species. Eutrophication has a serious impact on the usefulness of water bodies for sport fishing and for the capture of species of fish acceptable in the market. Dense algal blooms also make waters turbid and less desirable for swimming and other kinds of recreational water use. Taste and odor in municipal water supplies result in many consumer complaints. Water with algal blooms also are unsightly and less appealing to those who enjoy viewing lakes and streams.

The problem of algal toxicity has already been discussed in Chap. 12, and algal toxicity usually is not widespread other than in the case of shellfish poisoning in humans. Nevertheless, blue-green algae blooms can lead to medical problems ranging from rashes to death in some instances of shellfish poisoning. Livestock also have died from drinking water containing toxic algae.

Blue-green algae and other dense algal blooms increase the cost of water treatment. The removal of taste and odor problems sometimes cannot be resolved without passing water through activated carbon filters.

Eutrophication Control

Eutrophication control typically focuses on the removal of nitrogen and phosphorus from effluents. The major inputs of these two nutrients are municipal wastewater, runoff from agricultural and urban areas, certain industrial effluents, and in some regions, commercial aquaculture. Effluents contain nitrogen and phosphorus are either point-source that are discharged through a pipe or other conduit or non-point-source that are in the form of runoff which is either too voluminous or too diffuse to subject to standard, point-source wastewater treatment.

The phosphorus concentration in point-source effluents usually can be reduced by 80–90% by one of the following chemical treatments: lime [$Ca(OH)_2$] to precipitate hydroxyapatite [$Ca_{10}(PO_4)_6 \cdot 6H_2O$]; alum [$Al_2(SO_4)_3 \cdot 14H_2O$] to precipitate aluminum phosphate ($AlPO_4$); compounds such as ferric chloride [$Fe(Cl)_3$] to precipitate ferric phosphate ($FePO_4$). The precipitates of phosphorus and other settled solids are removed as sludge, and the wastewater is aerated. There also is a biological phosphorus removal procedure in which phosphorus is incorporated into microbial biomass and these organisms removed in a clarifier as sludge. By use of these methods, 90–95% of phosphorus can be removed.

Nitrification and denitrification are used to remove combined nitrogen from point source effluents. Much of the particulate organic nitrogen is removed by sedimentation in a clarifier along with other total solids, and the water is then aerated thoroughly for long enough to convert most of the ammonia nitrogen to nitrate through biological nitrification. The water is then put into an anaerobic digester and a carbon source such as methanol is added to allow denitrification. Around 85–90% of the total nitrogen in wastewater can be removed through sedimentation, nitrification, and denitrification.

Nitrogen and phosphorus concentrations are reduced in non-point-source effluent by use of management practices of many types depending upon effluent sources. In agriculture, the management practices are those that lessen nitrogen and phosphorus inputs in fertilizers and minimize their loss in runoff. Particular attention is given to maintaining vegetative cover and other erosion measures, installing sediment removal sills in ditches, and many others. Runoff from concentrated animal rearing facilities is directed in retention ponds and stored for land application.

Practices vary depending upon the activity or industry producing the non-point-source discharge of which agriculture is the major source. Of course, suburban lawns, golf courses, and other fertilized land are also significant sources of nitrogen and phosphorus in runoff.

In lakes and other static water bodies, phosphorus removal can be effected by chemical coagulation—usually aluminum sulfate (Huser et al. 2011). This treatment can be quite effective in lowering phosphorus concentrations for several years to lessen phytoplankton abundance. Biological control methods such as phosphorus (and nitrogen) removal in aquatic plants, filter-feeding fish, and mollusks have been promoted with varying degrees of success. Copper sulfate often is used to lessen the abundance of phytoplankton in water bodies for water supply and recreation use.

Schindler et al. (2008) did a long-term study that revealed phosphorus to be the primary limiting nutrient in some Canadian lakes—a fact already known for many years in pond aquaculture (Swingle 1947; Mortimer 1954; Hickling 1962; Hepher 1962). Because of the importance of phosphorus in freshwater, there is a tendency in eutrophication control to focus primarily on reducing phosphorus inputs. The study by Schindler et al. (2008) was done in Canada, and there have been reports of nitrogen-limited lakes and reservoirs in other places. Conley et al. (2009) warned that focusing on phosphorus control in freshwater will increase the nitrogen input to estuaries in which nitrogen is often a limiting factor for phytoplankton growth, and they cited some specific instances of this phenomenon. Conversely, if nitrogen is the focus of eutrophication control, low nitrogen concentrations in presence of elevated phosphorus concentrations will stimulate nitrogen fixation. In some lakes, especially shallow ones, phosphorus may be available from sediment in sufficient quantities to cause phytoplankton blooms. Both nitrogen inputs from anthropogenic activities and nitrogen fixation can trigger phytoplankton blooms. Phosphorus removal from water entering such lakes would not control eutrophication because of nitrogen fixation. The sensitivity of nitrogen fixation rate to the N/P to nitrogen fixation was considered in Chap. 13.

The upshot of the nitrogen versus phosphorus control issue can be summarized as follows: both can be limiting; all lakes are not the same with respect to nitrogen and phosphorus requirements; relying only on phosphorus control increases downstream nitrogen loads. While both nitrogen and phosphorus generally should be controlled, in specific situations, control of one over the other may be preferable. Most point-source effluent treatment removes both nitrogen and phosphorus relatively well. The major cause of eutrophication is non-point-source effluent, and agriculture is the main source of such effluent.

Conclusions

Eutrophication is probably the most troublesome aspect of water pollution. It causes increased phytoplankton growth and a reduction in dissolved oxygen availability leading to a reduction in species diversity in which species that tolerate low dissolved oxygen concentration are favored. Many of these species are undesirable for both efficient ecological function and water use by humans. Eutrophication control consists mainly of techniques for removing nitrogen and phosphorus from point-source effluents by treatment and application of management practices to reduce inputs of nitrogen and phosphorus to water bodies in runoff.

References

Butcher RW (1947) Studies in the ecology of rivers: VIII. The algae of organically enriched waters. J Ecol 35:186–191

Conley DJ, Paerl HW, Howarth RW, Boesch DF, Seitzinger SP, Havens KE, Lancelot C, Likens GE (2009) Controlling eutrophication: nitrogen and phosphorus. Science 323:1014–1015

Dodds WK (2006) Eutrophication and trophic state in rivers and streams. Limnol Oceanogr 51:671–680

Du LN, Li Y, Chen X, Yang JX (2011) Effect of eutrophication on molluscan community composition in Lake Dianchi (China, Yunnan). Limnol Ecol Manage Inland Waters 41:213–219

Ferdous Z, Muktadir AKM (2009) A review: potentiality of zooplankton as bioindicator. Am J Appl Sci 6:1815–1819

Feuchtmayr H, Moran R, Hatton K, Conner L, Heyes T, Moss B, Harvey I, Atkinson D (2009) Global warming and eutrophication: effects on water chemistry and autotrophic communities in experimental hypertrophic shallow lake mesocosms. J Appl Ecol 46:713–723

Hepher B (1962) Ten years of research in pond fertilization in Israel, II. The effect of fertilization on fish yields. Bamidgeh 14:29–48

Hickling CF (1962) Fish culture. Faber and Faber, London

Huser B, Brezonik P, Newman R (2011) Effects of alum treatment on water quality and sediment in the Minneapolis chain of lakes, Minnesota, USA. Lake Reservoir Manage 27:220–228

Jiang J, Huang ZG, Seager TP, Lin W, Grebogi C, Hastings A, Lai YC (2018) Predicting tipping points in mutualistic networks through dimension reduction. Proc Natl Acad Sci USA 115:39–47

Kaushal SS, Likens GE, Utz RM, Pace ML, Grese M, Yepsen M (2013) Increased river alkalinization in the Eastern U.S. Environ Sci Technol 47:10302–10311

Kautz RS (1980) Effects of eutrophication on the fish communities of Florida lakes. Proc Ann Conf SE Assoc Fish Wildlife Agencies 34:67–80

Kociolek JP, Stoermer EF (2009) Oligotrophy: the forgotten end of an ecological spectrum. Acta Bot Croatica 68:465–472

Margalef R (1958) Temporal succession and spatial heterogeneity in phytoplankton. In: Buzzati-Traverso AA (ed) Perspectives in marine biology. University of California Press, Berkeley, pp 323–349

Mortimer CH (1954) Fertilizers in fish ponds. Publication 5. Her Majesty's Stationery Office, London

Odum EP (1971) Fundamentals of ecology. 3rd edn. W. B. Saunders Company, Philadelphia

Schindler DW, Hecky RE, Findlay DL, Stainton MP, Parker BR, Paterson MJ, Beaty KG, Lyng M, Kasian SEM (2008) Eutrophication of lakes cannot be controlled by reducing nitrogen input: results of a 37-year whole-ecosystem experiment. Proc Natl Acad Sci USA 105:1254–1258

Sunkad BN, Patil HS (2004) Water quality assessment of Fort Lake of Belgaum (Karnataka) with special reference to zooplankton. J Environ Biol 25:99–102

Swingle HS (1947) Experiments on pond fertilization. Alabama Agricultural Experiment Station Bulletin 264. Auburn University, Auburn

Thièbaut G, Muller S (1998) The impact of eutrophication on aquatic diversity in weakly mineralized streams in the Northern Vosges Mountains (NE France). Biodivers Conserv 7:1051–1068

Tõnno I, Agasild H, Kõiv T, Freiberg R, Nõges P, Nõges T (2016) Algal diet of small-bodied crustacean zooplankton in a cyanobacteria-dominated eutrophic lake. PLoS One 11:e0154526. https://doi.org/10.1371/journal.pone.0154526

Sulfur

16

Abstract

Sulfur is a nutrient for plants and animals; hydrogen sulfide is an odorous, toxic material; and sulfur dioxide is an air pollutant responsible for acid rain. Plants primarily use sulfate as a sulfur source, and sulfur containing amino acids in plants are important to animal nutrition. Sulfur compounds undergo oxidations and reductions in the environment. The most famous sulfur oxidizing bacteria are of the genus *Thiobacillus* that oxidize elemental sulfur, sulfides, and other reduced sulfur compounds releasing sulfuric acid into the environment—acid sulfate soils and acidic mine drainage result from sulfur oxidation. Bacteria of the genus *Desulfovibrio* use certain sulfur compounds as electron and hydrogen acceptors in respiration allowing them to decompose organic matter in anaerobic environments. Ferrous iron and other metals may react with sulfide in anaerobic sediments to form metallic sulfides, e.g., iron sulfide (iron pyrite). If such sediment later is exposed to oxygen, sulfides will be oxidized resulting in sulfuric acid production. Sulfide in anaerobic zones sometimes diffuses or is mixed into overlaying aerobic water at a rate exceeding the oxidation rate of sulfide—toxicity to aquatic animals can result. Sulfide toxicity is favored by low pH, because un-ionized hydrogen sulfide (H_2S) is the toxic form. Elevated sulfate concentration and the presence of sulfide degrades the quality of drinking water.

Introduction

Sulfur is a major element of geochemistry, making up around 3% of the earth's total mass, but its abundance is only about 0.04% of the earth's crust. It occurs in the earth's crust as deposits of elemental sulfur, sulfate salts such as gypsum (calcium sulfate), epsomite (magnesium sulfate), varite (barium sulfate), and in metamorphic and igneous rock. Weathering delivers about 60 million tonnes of sulfur to the ocean each year. The sulfur concentration is about 1% in crude oil and 0.5–5.0% in coal and lignite. Sulfur occurs in soil mainly as sulfate salts, sulfate in the soil solution, and sulfur in soil organic matter. Sulfate is the major form of sulfur in the ocean and freshwater bodies. Freshwater contains a small percentage of the earth's sulfur, but the ocean is 0.09% sulfur by weight and sulfur in sulfate makes up 2.7% of the total dissolved solids in ocean water. Natural water contains only about 7% of the sulfur found in the earth's solid crust (Hem 1985).

Sulfur is widely used in industry as a reactant and component of products, and it is contained in fuels. The nineteenth century German chemist Justus von Liebig recognized the importance of sulfur when he declared: *"you may fairly judge the commercial prosperity of a country from the amount of sulfuric acid it consumes."* But, the modern day American environmental activist Denis Hayes is quoted as saying *"there was almost a universal acceptance of unhealthy conditions. Sulfur dioxide in smokestacks were the price, or smell, of prosperity."*

Sulfur dioxide results mainly from combustion of fuels and volcanic eruptions, and the atmosphere contains a small percentage (usually less than 10 ppb) of sulfur dioxide. For example, the annual, ambient sulfur dioxide concentrations in different regions of Canada ranged from 1 to 3 ppb in 2014 (https://www.ec.gc.ca/indicateurs-indicators/default.asp?lang=En&n=307CCE5B-1&pedisable-true). Hydrogen sulfide enters the atmosphere from microbial decay (especially in marshes), and it is oxidized to sulfur dioxide.

Sulfur is an essential nutrient for plants, animals, and bacteria, but unlike nitrogen and phosphorus, sulfur seldom limits primary productivity. Sulfur is a component of two essential amino acids (cysteine and methionine) and other organic compounds, the best-known of which are the vitamins biotin and thiamine, and iron-sulfur clusters called ferredoxins. Ferredoxins function in the transfer of electrons in various biochemical reactions.

Sulfur, like nitrogen, commonly occurs in the environment in several oxidation states (valences), and bacteria oxidize and reduce sulfur compounds. Oxidations and reductions of sulfur also may occur by chemical reactions not involving biological activity. Sulfate is the form of sulfur used most commonly by plants, but some plants can fix atmospheric sulfur by a method analogous to nitrogen fixation. As shown in Chap. 11, sulfur oxidations often produce sulfuric acid. Oxidation of sulfur dioxide emitted into the atmosphere by combustion of fossil fuels is the main cause of the acid rain phenomenon. Sulfide oxidation is responsible for acid mine drainage, and oxidation of sulfur or sulfide in soil can cause extreme acidity. Of course, certain anaerobic bacteria can reduce sulfate and other oxidized forms of sulfur to sulfides.

There is a global sulfur cycle, and most aspects of this natural cycle operate in aquatic ecosystems. Sulfate influences biological activity by acting as a nutrient, but sulfide can be toxic to aquatic organisms. The purpose of this chapter is to discuss sulfur chemistry and its influence on water quality.

The Sulfur Cycle

The main features of the sulfur cycle are depicted in Fig. 16.1. Natural sources of sulfur to the atmosphere include volcanic activity which releases hydrogen sulfide (H_2S) and sulfur dioxide (SO_2), sulfate originating from evaporation of sea spray, dust from arid land containing gypsum ($CaSO_4 \cdot 2H_2O$), hydrogen sulfide released by microbial decomposition of organic matter, and sulfur dioxide released by combustion of fossil fuels and by wildfires. Human activities release around 100 million metric tons of sulfur into the air annually (Smith et al. 2011). The amount of sulfur emissions entering the atmosphere from natural sources has not been established accurately, but the quantity is thought to be much smaller than anthropogenic sources. Hydrogen sulfide and sulfur dioxide are oxidized, resulting in sulfuric acid which is washed from the atmosphere by rainfall (Chap. 11). This prevents the long-term accumulation of reduced sulfur compounds in the atmosphere, but in areas where sulfur emissions are high, the pH of rain may be depressed enough to cause acidification of soils and waters resulting in serious ecological impacts.

Gypsum and elemental sulfur deposits are mined and used as sources of industrial and agricultural sulfur. Gypsum is a major component of drywall or gypsum board widely used in building construction. Sulfuric acid is made by industrial oxidation of SO_2 derived from oxidizing elemental sulfur. Sulfur dioxide also can be extracted from natural gas or crude oil. Sulfuric acid has a myriad of industrial uses, but about half of the global production of it is used in phosphate fertilizer manufacturing.

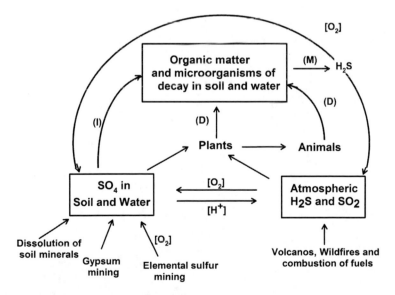

$[O_2]$ = oxidation ; $[H^+]$ = reduction ; (D) = death ; (I) = immobilization ; (M) = mineralization

Fig. 16.1 The sulfur cycle

Plants usually rely on sulfate as a sulfur source, but some plants can absorb sulfur dioxide, reduce it to sulfide, and use sulfide for making sulfur-containing amino acids. Sulfur in plant biomass passes through the food web to supply sulfur to animals. Animal excreta and remains of dead plants and animals contain sulfur that is used by bacteria. Organic sulfur in residues is mineralized by bacteria to sulfides, but in the presence of oxygen, sulfides are oxidized to sulfate. Sulfate can be reduced to sulfides by some bacteria.

Sulfur Transformations

Sulfur has five possible valences ranging from -2 to $+6$ (Table 16.1), and transformations among these valence states often are mediated by microorganisms.

Plant Uptake and Mineralization

Plants must reduce sulfate to sulfide in order to synthesize sulfur-containing amino acids. In this process, sulfate is first combined with ATP and converted to sulfite (SO_3^-). Sulfite is reduced to sulfide in a reaction catalyzed by the enzyme sulfite reductase. Sulfide is combined with organic moieties to form cysteine and methionine—the two sulfur-containing amino acids.

Plants usually contain 0.1–0.3% sulfur. The sulfur-containing amino acids of plant proteins pass through the food web to animals. Bacteria and other microorganisms must have a source of sulfur because their cells contain from 0.1 to 1.0% sulfur on a dry weight basis. If organic residues contain more sulfur than needed during microbial decomposition, sulfur will be mineralized to the

Table 16.1 Valence states of sulfur

Valence	Name	Formula
$\cdot 2$	Hydrogen sulfide	H_2S
	Ferrous sulfide	FeS
	Cysteine	$R\text{-}SH^a$
	Methionine	$R\text{-}S\text{-}R^a$
0	Elemental sulfur	S_8
+2	Sodium thiosulfate	$Na_2S_2O_3$
	Thiosulfuric acid	$H_2S_2O_3$
+4	Sulfur dioxide	SO_2
	Sulfurous acid	H_2SO_3
	Sodium bisulfate	$NaHSO_3$
+6	Sulfuric acid	H_2SO_4
	Calcium sulfate	$CaSO_4 \cdot 2H_2O$
	Potassium sulfate	K_2SO_4

[a]Note: R = organic moiety

environment (Ex. 16.1). Sulfur will be immobilized from the environment in instances where residues contain less sulfur than needed for microbial decomposition.

Ex. 16.1 *Suppose that a residue containing 0.3% sulfur and 45% carbon (dry weight basis) is completely decomposed by bacteria. It will be assumed that there is adequate nitrogen and other elements present to meet microbial needs. The bacteria have a carbon assimilation efficiency to 20% and contain 0.15% S. The sulfur mineralized from 1000 g of residue will be estimated.*
  Solution:
The carbon and sulfur contents of 1000 g residue are

$$1,000 \times 0.45 = 450 \ g \ carbon$$

$$1,000 \times 0.003 = 3 \ g \ sulfur.$$

The bacterial biomass will be

$$450 \ g \ C \times 0.2 = 90 \ g \ bacterial \ carbon$$

90 g bacteria C ÷ 0.5 = 180 g bacterial biomass.
The sulfur in bacterial biomass is

$$180 \ g \ bacteria \times 0.0015 = 0.27 \ g \ sulfur.$$

The residue contains 2.73 g more sulfur than needed for its decomposition, and this sulfur will be mineralized.

The C:N ratio is of more importance than the C:S ratio in controlling the rate of organic matter decomposition in nearly all situations.

Oxidations

Most sulfur oxidations can proceed by purely chemical processes, but biological intervention often speeds up sulfur oxidation. Bacteria that oxidize inorganic sulfur usually are obligate or facultative autotrophs. They use the energy obtained from oxidizing sulfur to convert carbon dioxide to organic carbon in their cells. This process is analogous to nitrification and results in organic matter synthesis. However, like in nitrification, carbon fixation by sulfur-oxidizing bacteria is not highly energy efficient. The amount of organic matter produced globally by sulfur oxidizing bacteria is miniscule in comparison to the quantity produced through photosynthesis.

  The best-known sulfur-oxidizing bacteria are species of *Thiobacillus,* a genus of gram-negative rod-shaped bacteria of the hetaproteabacteria. Some representative reactions carried out by these microorganisms are

$$Na_2S_2O_3 + 2O_2 + H_2O \rightarrow 2NaHSO_4 \tag{16.1}$$

$$5Na_2S_2O_3 + 4O_2 + H_2O \rightarrow 5Na_2SO_4 + H_2SO_4 + 4S \tag{16.2}$$

$$S + 1\frac{1}{2}O_2 + H_2O \rightarrow H_2SO_4. \tag{16.3}$$

In each of these reactions, energy is released and a small portion of the energy is captured (by converting ADP to ATP) and used to reduce carbon dioxide to carbohydrate.

Nitrate also can be an oxygen source for sulfur oxidation as shown below:

$$5S + 6KNO_3 + 2H_2O \rightarrow K_2SO_4 + 4KHSO_4 + 3N_2. \tag{16.4}$$

In the reaction depicted by Eq. 16.4, oxygen is removed from nitrate and used to oxidize sulfur to sulfate. The process may be called sulfur oxidation, but it also is denitrification, because nitrogen gas is released.

Green and purple sulfur bacteria also can oxidize sulfur. These bacteria are highly unusual in that they are anaerobic photoautotrophs. The green sulfur bacteria include *Chlorobium* and several other genera of the family Chlorobiaceae, while the purple sulfur bacteria are of the family Chromatiaceae that includes *Thiospirillum* and several other genera. The colored, green and purple sulfur bacteria use light and energy from oxidizing sulfide to reduce carbon dioxide to carbohydrate (CH_2O) according to the following reactions:

$$CO_2 + 2H_2S \underset{\text{light}}{\rightarrow} CH_2O + 2S + H_2O \tag{16.5}$$

$$2CO_2 + H_2S + 2H_2O \underset{\text{light}}{\rightarrow} 2CH_2O + H_2SO_4. \tag{16.6}$$

Both light and anaerobic conditions are necessary for colored, photosynthetic sulfur bacteria, so their presence is greatly restricted in nature. Colored sulfur bacteria are particularly troublesome by growing on the inside tops of sewer pipes near manhole covers where light is available. The result is corrosion of the upper inner surface of the pipe near the manhole cover by sulfuric acid produced by the bacteria.

Reductions

Sulfur reductions occur in anaerobic environments, and sulfur-reducing bacteria use oxygen from sulfate or other oxidized sulfur compounds as electron and hydrogen acceptors in respiration. The process is similar in this regard to denitrification. The predominant sulfur-reducing bacteria are species of gram-negative bacteria of the genus *Desulfovibrio,* which commonly occur in sediment or waterlogged soil that has a high concentration of organic matter. Some typical reactions are

$$SO_4^{2-} + 8H^+ \rightarrow S^{2-} + 4H_2O \qquad (16.7)$$

$$SO_3^{2-} + 6H^+ \rightarrow S^{2-} + 3H_2O \qquad (16.8)$$

$$S_2O_3^{2-} + 8H^+ \rightarrow 2SH^- + 3H_2O. \qquad (16.9)$$

The source of electrons, hydrogen, and energy for the reactions depicted in Eqs. 16.7, 16.8, and 16.9 are carbohydrates, organic acids, and alcohols. A representative complete reaction is

$$2CH_3CHOHCOONa + MgSO_4 \rightarrow H_2S + 2CH_3COONa + CO_2$$
$$+ MgCO_3 + H_2O. \qquad (16.10)$$

Hydrogen Sulfide

The preceding equations for sulfur reductions showed sulfate and other oxidized forms of sulfur being converted to S^{2-}, HS^-, or H_2S. These are all forms of sulfide-sulfur, and they also are dissociation products of the diprotic acid, hydrogen sulfide (H_2S). Hydrogen sulfide dissociates as follows:

$$H_2S = HS^- + H^+ \qquad K_1 = 10^{-7.01} \qquad (16.11)$$

$$HS^- = S^{2-} + H^+ \qquad K_2 = 10^{-13.89}. \qquad (16.12)$$

Depending on pH, the dominant sulfide species resulting from reduction of sulfur compounds by bacteria will be H_2S, HS^-, or S^{2-} (Fig. 16.2). Un-ionized hydrogen sulfide dominates in acidic environments, but in alkaline situations, S^{2-} will be the main form.

With respect to water quality, sulfides typically are produced in sediment or in the hypolimnion of stratified lakes. When sulfide enters the aerobic water column above the sediment or is mixed into aerobic water at thermal overturn, the sulfide is oxidized rather quickly. Only oxygen-depleted water will contain appreciable levels of sulfide.

Highly anaerobic environments contain both ferrous iron and sulfide, and ferrous sulfide precipitates. Sulfides also react forming metal sulfides. Reference to Table 4.3 shows that metal sulfides have small solubility product constants. Marine sediment is a particularly favorable environment for the formation of metal sulfides, because the high sulfate content of the pore water favors sulfide production.

The mass action expression for Eq. 16.11 will be used to estimate the proportion of sulfide in un-ionized form at a particular pH in the following example.

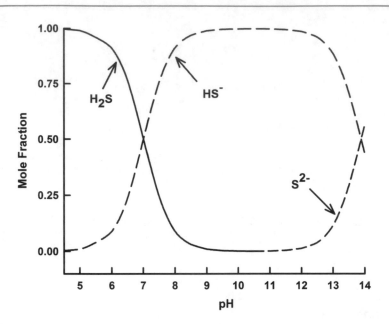

Fig. 16.2 Effects of pH on the relative proportions of sulfide species

Ex. 16.2 *The percentage un-ionized hydrogen sulfide will be estimated for pH 6.*
Solution:
The mass action expression of Eq. 16.11 is

$$\frac{(HS^-)(H^+)}{(H_2S)} = 10^{-7.01}$$

$$\frac{(HS^-)}{(H_2S)} = \frac{10^{-7.01}}{10^{-6}} = 10^{-1.01} = 0.098.$$

Thus, for 0.098 mole HS⁻, there will be 1 mole H₂S and

$$\frac{1}{1 + 0.098} = 0.911 \ (or\ 91.1\%H_2S).$$

The first ionization constants for hydrogen sulfide ($H_2S = HS^- + H^+$) at temperatures of 0–50 °C are provided (Table 16.2). The ionization constants were used to calculate the proportions of un-ionized hydrogen sulfide at different pH values and temperatures (Table 16.3). To illustrate the use of Table 16.3, suppose the total sulfide concentration is 0.10 mg/L at pH 8.0 and 30 °C. The factor from Table 16.3 for these conditions is 0.072; the amount of sulfur in hydrogen sulfide will be 0.1 mg/L × 0.072 = 0.0072 mg/L.

Table 16.2 First
ionization constants for H_2S
($H_2S = HS^- + H^+$) at
different temperatures

Temperature (°C)	K	Temperature (°C)	K
0	$10^{-7.66}$	30	$10^{-6.89}$
5	$10^{-7.51}$	35	$10^{-6.78}$
10	$10^{-7.38}$	40	$10^{-6.67}$
15	$10^{-7.25}$	45	$10^{-6.57}$
20	$10^{-7.13}$	50	$10^{-6.47}$
25	$10^{-7.01}$		

Table 16.3 Decimal fraction (proportions) of total sulfide-sulfur as un-ionized hydrogen sulfide-sulfur (H_2S-S) at different pH values and temperatures

pH	Temperature (°C)							
	5	10	15	20	25	30	35	40
5.0	0.997	0.996	0.947	0.993	0.990	0.987	0.983	0.979
5.5	0.981	0.987	0.983	0.977	0.970	0.961	0.950	0.937
6.0	0.970	0.960	0.947	0.931	0.911	0.886	0.858	0.831
6.5	0.911	0.883	0.849	0.810	0.764	0.711	0.656	0.597
7.0	0.764	0.706	0.641	0.575	0.505	0.437	0.376	0.318
7.5	0.506	0.431	0.360	0.299	0.244	0.197	0.160	0.129
8.0	0.245	0.193	0.152	0.119	0.093	0.072	0.057	0.045
8.5	0.093	0.071	0.053	0.041	0.031	0.024	0.019	0.015
9.0	0.031	0.023	0.018	0.013	0.010	0.008	0.006	0.003

Sulfur Concentrations

Sulfate concentrations in inland surface waters of humid regions are usually low and often only a few milligrams per liter. In arid regions, surface waters normally will have higher sulfate concentrations with values of 50–100 mg/L or more. Ocean water averages 2700 mg/L in sulfate (Table 5.8).

Sulfide rarely occurs in measurable concentration in aerobic surface waters. But, if the rate of sulfide release from anaerobic sediment is greater than the rate of sulfide oxidation within the water column, sulfide can accumulate in surface water. Sulfide concentration may be several milligrams per liter in hypolimnetic waters of eutrophic lakes and in certain well waters. When water containing sulfide is aerated, sulfide is rapidly oxidized to sulfate. Information on concentrations of other forms of sulfur in natural waters is scarce.

Freshwater sediments usually contain less than 0.1% total sulfur, and marine sediments usually contain between 0.05 and 0.3% total sulfur. In some sediments where iron pyrite and other metal sulfides have accumulated, total sulfur concentrations may range from 0.5 to 5%. Pore water in anaerobic sediment may contain 1–5 mg/L of sulfide-sulfur.

Effects of Sulfur

Sulfide oxidizes quickly in aerobic environments where fish are found. But, under some conditions, sulfide may be delivered into oxygenated zones at rates exceeding the rates at which it can be oxidized. This phenomenon can result in toxic concentrations of sulfide.

Sulfide in the un-ionized form (H_2S) is highly toxic to fish and other aquatic animals, but the two, ionized species (HS^- and S^{2-}) have little toxicity. The 96-hour LC50 of hydrogen sulfide to freshwater fish ranged from 4.2 to 34.8 µg/L (Gray et al. 2002), but marine species apparently are somewhat less sensitive to hydrogen sulfide—96-hour LC50s are 2–10 times greater than for freshwater fish (Bagarinao and Lantin-Olaguer 1999; Gopakumar and Kuttyamma 1996). Most authorities suggest that no more than 0.002 mg/L of sulfide should be present in natural waters and any detectable concentration is undesirable. Hydrogen sulfide toxicity to organisms results from several effects. Hydrogen sulfide inhibits the reoxidation of cytochrome a_3 by molecular oxygen, because it blocks the electron transport system and stops oxidative respiration. Blood lactate concentration also increases and anaerobic glycolysis is favored over aerobic respiration. The overall toxic effect is hypoxia, and low dissolved oxygen concentrations enhance sulfide toxicity (Boyd and Tucker 2014). As already mentioned, hydrogen sulfide is rarely detectable in environments containing dissolved oxygen, and it seldom is an important toxicity factor other than in polluted aquatic ecosystems.

The most common concern over sulfur concentration relates to drinking water quality. Sulfate in drinking water can impart a bitter taste. Depending upon their sensitivity, people notice the bitter taste at sulfate concentrations of 250–1000 mg/L. Water containing elevated sulfate concentration also can act as a laxative to people not used to drinking it. The highest quality drinking water will not contain more than 50 mg/L sulfate, but most authorities indicate that up to 250 mg/L is acceptable for drinking water supplies. In addition, high sulfate concentration in water interferes with chlorination.

Hydrogen sulfide can be a contaminant in well water. Although it is not toxic to humans at concentrations normally encountered, most people can detect a bad odor at sulfide concentrations of 0.1–0.5 mg/L. At sulfide concentrations up to 1 mg/L the odor often is described as musty, but higher concentrations cause the typical "rotten egg" odor of hydrogen sulfide. In addition to the odor, elevated hydrogen sulfide concentrations are highly corrosive to water pipes and fixtures. Methods for removing hydrogen sulfide from drinking water include activated carbon filtration, aeration, and oxidation with chlorine or potassium permanganate.

Atmospheric emissions of sulfur dioxide caused by combustion of fossil fuels are the major cause of acid rain. Since the 1970s, many countries have worked to reduce sulfur dioxide emissions by requiring sulfur dioxide removal devices in vehicles, on smokestacks, etc. Although considerable control of sulfur dioxide emissions was achieved and global emission declined for several years, sulfur dioxide emissions are now increasing because of increased fossil fuel use by countries with rapidly developing economies such as China and India (Boyd and McNevin 2015).

Conclusions

Sulfate concentration in most surface waters is adequate to support plant growth, and it is not a common limiting factor or a cause of eutrophication. Hydrogen sulfide is produced in anaerobic zones, and it may diffuse or mix into aerobic zones faster than it is oxidized leading to toxicity to fish and other aquatic life. Excessive sulfate concentration and the presence of sulfides may be a concern in drinking water quality. Excessive acidity and low pH in water bodies usually is a result of sulfuric acid from natural sources or pollution.

References

Bagarinao T, Lantin-Olaguer I (1999) The sulfide tolerance of milkfish and tilapia in relation to fish kills in farms and natural waters in the Philippines. Hydrobio 382:137–150

Boyd CE, McNevin AA (2015) Aquaculture, resource use, and the environment. Wiley Blackwell, Hoboken

Boyd CE, Tucker CS (2014) Handbook for aquaculture water quality. Craftmaster Printers, Auburn

Gopakumar G, Kuttyamma VJ (1996) Effect of hydrogen sulphide on two species of penaeid prawns Penaeus indicus (H. Milne Edwards) and Metapenaeus dobsoni (Miers). Env Con Tox 57:824–828

Gray JS, Wu RS, Or YY (2002) Effects of hypoxia and organic enrichment on the coastal marine environment. Mar Ecol Prog Ser 238:249–279

Hem JD (1985) Study and interpretation of the chemical characteristics of natural water. Water Supply Paper 2254, United States Geological Survey, United States Government Printing Office, Washington

Smith SJ, van Aardenne J, Kilmont Z, Andres RJ, Volke A, Arias SD (2011) Anthropogenic sulfur dioxide emissions: 1850–2005. Atmos Chem Phy 11:1101–1116

Micronutrients and Other Trace Elements

17

Abstract

The solubilities of most minerals from which trace metals in natural waters originate are favored by low pH. The concentration of the free ion of a dissolved trace element usually is much lower than is the total concentration of the trace element. This results from ion pair associations between the free trace ion and major ions, complex ion formation, hydrolysis of metal ions, and chelation of metal ions. Several trace elements—zinc, copper, iron, manganese, boron, fluorine, iodine, selenium, cadmium, cobalt, and molybdenum—are essential to plants, animals or both. A few other trace elements are suspected, but not unequivocally proven to be essential. There are some reports of low micronutrient concentrations limiting the productivity of water bodies; but primary productivity in most water bodies apparently is not limited by a shortage of micronutrients. Trace elements—including the ones that are nutrients—may be toxic at high concentration to aquatic organisms. Excessive concentrations of several trace metals in drinking water also can be harmful to human health. Instances of trace element toxicity in aquatic animals and humans usually have resulted from anthropogenic pollution. Nevertheless, excessive concentrations of trace metals in drinking water sometimes occur naturally—an example is the presence of chronically-toxic concentrations of arsenic in groundwater that serves as the water supply for several million people in a few provinces of Bangladesh and adjoining India.

Introduction

Ninety-four of the 118 elements of the periodic table occur in nature and can be present in water at some concentration. The usual concentration range of different elements varies greatly. In a complete analysis of freshwater, some elements will be at concentrations of 10–100 mg/L, while others will be present at less than 0.1 mg/L. Even wider ranges in concentration of elements are found in seawater. Dissolved

© Springer Nature Switzerland AG 2020
C. E. Boyd, *Water Quality*, https://doi.org/10.1007/978-3-030-23335-8_17

elements in natural waters often are divided into one of two categories. Elements with concentrations above 1 mg/L usually are called major elements and those at lower concentration are known as trace elements (Gaillardet et al. 2003). Sometimes, elements of the lowest concentrations may be referred to as ultra trace elements.

Iron, manganese, zinc, copper, and a few other trace elements essential to aquatic organisms are called micronutrients or trace nutrients. Phosphate, nitrate, and ammonia nitrogen discussed earlier also are usually present in water at concentrations <1 mg/L, but they are not considered micronutrients, because they are required in large amounts by organisms. Concentrations of iron and sometimes other micronutrients may be low enough to limit productivity in some freshwater bodies (Goldman 1972; Hyenstrand et al. 2000; Vrede and Tranvik 2006) and in the ocean (Nadis 1998). Excessive concentrations of both micronutrients and non-essential trace elements can be toxic to organisms including humans who drink water with high concentrations of certain trace elements. Higher than normal concentrations of trace elements often result from pollution, but elevated concentrations occasionally are natural phenomena.

The purpose of this chapter is to discuss both micronutrients and non-essential trace elements. The focus will be on sources, chemistry, concentrations, and effects of trace elements in water bodies.

Chemical Control of Trace Element Concentrations

The solubility of trace elements in water is a complex topic and depends upon particular combinations of factors (Hem 1985; McBride 1989; Deverel et al. 2012). A source of an element obviously is necessary, and the main sources are minerals in the watershed, in the bottom of the water body, and in some instances, pollution. A major controlling factor in mineral solubility is the solubility product (K_{sp}). The K_{sp} effects the degree of dissolution of a mineral, but equally as important, if concentrations of certain ions in water exceed the K_{sp} of some mineral, precipitation of that mineral will occur. The common ion principle is particularly important in causing precipitation.

Some K_{sp} values (at 25 °C) of interest in water quality are provided (Table 17.1). The solubility reactions are in the form $aAbB(s) = aA(aq) + bB(aq)$. Solubility products are useful for calculating the probable water solubility of compounds, and tabular values of K_{sp} apply to solubilities of compounds in distilled water at 25 °C. The K_{sp} of a compound may have been measured; but, more likely, it was calculated from the standard Gibbs free energy ($\Delta G°$) of the chemical species involved from their standard states by use of Eq. 4.11 and illustrated in the example below.

Table 17.1 Solubility products (K_{sp}) for selected trace element compounds

Compound	K_{sp} exponent	Compound	K_{sp} exponent	Compound	K_{sp} exponent	Compound	K_{sp} exponent
$Al(OH)_3$	−33.39	CoS	−43.40	FeS_2	−26.89	$PbCO_3$	−13.13
Ag_2CO_3	−11.07	$Cr(OH)_2$	−33.10	$Hg(OH)_2$	−15.67	PbS	−26.77
$AgCl$	−9.75	$Cu(OH)_2$	−19.34	$HgCO_3$	−16.44	$PbSO_4$	−7.99
Ag_2S	−50.22	$Cu(OH)_2CO_3$	−33.16	Hg_2Cl_2	−17.84	$Sn(OH)_2$	−26.26
$BaCO_3$	−8.56	CuO	−20.36	HgS	−52.70	SnS	−27.52
$BaSO_4$	−9.97	Cu_2S	−47.60	$Mn(OH)_2$	−12.78	$SrCO_3$	−9.27
$Be(OH)_2$	−21.16	CuS	−36.26	Mn_2O_3	−84.55	$SrSO_4$	−6.63
$BiOOH$	−9.40	$CuSO_4 \cdot 5H_2O$	−7.06	Mn_3O_4	−54.15	$Tl(OH)_3$	−43.87
Bi_2S_3	−97.00	$Fe(OH)_3$	−37.08	$MnCO_3$	−10.39	Tl_2S	−21.22
CaF_2	−10.60	Fe_2O_3	−87.95	MnO_2	−17.84	UO_2	−55.86
$Cd(OH)_2$	−14.25	Fe_3O_4	−108.18	$MnOOH$	−18.26	$V(OH)_2$	−15.40
$CdCO_3$	−12.10	$FeCO_3$	−10.89	MnS	−10.19	$V(OH)_3$	−34.40
CdS	−29.92	FeO	−14.45	$Ni(OH)_3$	−15.26	$Zn(OH)_2$	−15.78
$Co(OH)_2$	−14.23	$FeOOH$	−42.97	$NiCO_3$	−6.84	$ZnCO_3$	−10.00
$Co(OH)_3$	−43.80	$FeMoO_4$	−6.91	NiS	−22.03	ZnS	−23.04
$CoCO_3$	−12.85	FeS	−17.91	$Pb(OH)_2$	−19.83		

[a]Sources: www2.chm.ulaval.ca/gecha/chm1903/6_solubilite_solides/solubility_products.pdf and http://www.aqion.de/site/16

The K_{sp} values are given as exponents of 10, e.g., K_{sp} exponent in table = −33.39, then $K_{sp} = 10^{-33.39}$

Ex. 17.1 *Calculation of* K_{sp} *of* Cu_2S.
  Solution:

$$Cu_2S = 2Cu^+ + S^{2-}$$

$$\Delta G^{\circ} = 2\Delta G_f^{\circ}(Cu^+) + \Delta G_f^{\circ}(S^{2-}) - \Delta G_f^{\circ}(Cu_2S)$$

$$\Delta G^{\circ} = 2(49.99 \; kJ/mol) + 85.8 \; kJ/mol - (-86.19 \; kJ/mol)$$

$$\Delta G^{\circ} = 271.97 \; kJ/mol$$

$$\Delta G^{\circ} = -5.709 \log K$$

$$\log K = \frac{\Delta G^{\circ}}{-5.709} = \frac{271.97}{-5.709} = -47.64$$

$$K = 10^{-47.64}$$

$$K_{sp} = 2.28 \times 10^{-48} \; or \; 10^{-47.64}$$

The K_{sp} *from* Table 17.1 *also is* $10^{-47.64}$.

Values of K_{sp} reported in different tables may vary slightly. They can be obtained for temperatures other than 25 °C using their relationship to standard Gibbs free energy. The K_{sp} values in tables do not give correct concentrations for compounds whose solubilities are influenced by pH, carbon dioxide concentration, or redox potential. To illustrate the effect of pH on solubility, consider tenorite (CuO; $K_{sp} = 10^{-20}$) that reacts in water to form Cu^{2+} and $2OH^-$. The equilibrium concentration of Cu^{2+} using the K_{sp} of tenorite is only $10^{-8.5}$ M (0.2 µg/L). However, CuO reacts with H^+ to produce Cu^{2+} and water for which $K = 10^{7.35}$. The solubility of copper oxide reacting at lower pH will be calculated in the following example.

Ex. 17.2 *Copper solubility from CuO at pH 5 and 25 °C will be calculated.*
  Solution:

$$CuO + 2H^+ = Cu^{2+} + H_2O$$

$$\frac{(Cu^{2+})}{(H^+)^2} = 10^{7.35}$$

$$(Cu^{2+}) = (10^{-5})^2(10^{7.35}) = 10^{-2.65} \; M \; or \; 2.24 \; mM$$

$$(Cu^{2+}) = 2.24 \; mM \; Cu^{2+} \times 63.54 \; mg \; Cu^{2+}/m \; mol = 142 \; mg/L \; Cu^{2+}$$

The effects of temperature and pH can be demonstrated by repeating the calculation in Ex. 17.2 for different temperatures and pH values. In order to accomplish this, the equation relating standard Gibbs free energy to the equilibrium constant at any temperature (Eq. 4.10) was used to calculate K_{sp} for different temperatures. The solubility of tenorite decreased with both increasing pH and temperature (Table 17.2). The response to pH and temperature differs among compounds, but concentrations of metal ions at equilibrium with their oxides decreases with greater pH as illustrated for equilibrium concentrations of several common trace elements (Fig. 17.1).

Some metals like copper, zinc, tin, lead, aluminum, and beryllium are amphoteric; that is, they can react as an acid or as a base. Their solubility may decrease as pH increases, but at some pH, their concentration will begin to increase as they react as an acid. For example, zinc oxide is a base in the reaction $Zn(OH)_2(s) + 2H^+(aq) = Zn^{2+}(aq) + 2H_2O$, but it is acidic when it reacts with hydroxide $Zn(OH)_2(s) + 2OH^-(aq) = Zn(OH)_4^{4+}(aq)$. As a result, the basic reaction will decrease as pH increases; but, the acidic reaction of zinc hydroxide increases with greater pH. This results in the possibility of a higher Zn^{2+} concentration at greater pH than expected from the usual solubility expression of zinc hydroxide acting as a base.

Table 17.2 Values for K_{sp} of the reaction $CuO + 2H^+ = Cu^{2+} + H_2O$ and copper oxide solubility at different temperatures and pH values

Temperature (°C)	K_{sp}	Cu^{2+} concentration (µg/L)		
		pH 6	pH 7	pH 8
10	$10^{7.74}$	3490	34.9	0.349
15	$10^{7.61}$	2590	25.9	0.259
20	$10^{7.48}$	1920	19.2	0.192
25	$10^{7.35}$	1420	14.2	0.142
30	$10^{7.23}$	1080	10.8	0.108

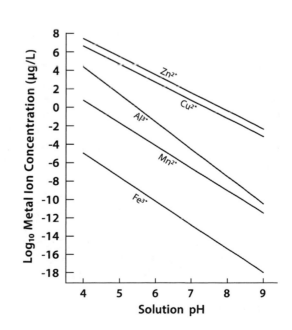

Fig. 17.1 Concentrations of metal ions at equilibrium in distilled water at 25 °C from CuO, ZnO, MnO, Fe_2O_3, and $Al(OH)_3$ at different pH values

The redox potential increases in presence of dissolved oxygen and it greatly influences the solubility of compounds of iron and several other elements. The solubility of ferrous iron from ferrous molybdate in pure water will be calculated from K_{sp} (Table 17.1) in Ex. 17.3.

Ex. 17.3 *The solubility of ferrous iron (Fe^{2+}) from ferrous molybdate ($FeMoO_4$) will be calculated.*

Solution:
The mass action form of the reaction (Table 17.1) is

$$\left(Fe^{2+}\right)\left(MoO_4^{2-}\right) = 10^{-6.91}$$

$$Letting\ X = \left(Fe^{2+}\right) = \left(MoO_4^{2-}\right)$$

$$X^2 = 10^{-6.91}; X = 10^{-3.455}\ or\ 3.51 \times 10^{-4}\ M\ Fe^{2+}\ and\ MoO_4^{2-}$$

In terms of Fe^{2+}

$3.51 \times 10^{-4}\ M \times 55.84\ g\ Fe^{2+}/gmw = 0.0195\ g/L\ Fe^{2+}\ or\ 20\ mg/L\ Fe^{2+}.$

The calculation of iron concentration in Ex. 17.3 is not the final result in oxygenated water, because Fe^{2+} oxidizes and precipitates as ferric hydroxide:

$$Fe^{2+} + O_2 + H_2O = Fe(OH)_3 \downarrow + H^+ \tag{17.1}$$

Ferric hydroxide [$Fe(OH)_3$] is quite insoluble at pH above 5.

$$Fe(OH)_3 + 3H^+ = Fe^{3+} + 3H_2O \qquad K = 10^{3.54} \tag{17.2}$$

At pH 7, the concentration of Fe^{3+} is only $10^{-17.46}$ M—undetectable analytically.

Low redox potential in sediment increases the solubility of ferric hydroxide because ferric iron (Fe^{3+}) is oxidized to soluble ferrous iron (Fe^{2+}). Anaerobic bacteria in sediment produce sulfide (S^{2-}) which reacts with ferrous iron and a few other metal ions to form insoluble metal sulfides, e.g., FeS and FeS$_2$. These metal sulfides precipitate in anaerobic sediment and do not enter the water column (Guo et al. 1997).

Despite the low solubility of Fe^{3+} and other trace metals, there will be detectable concentrations of trace metals in water bodies. This results from metal ions reacting with major ions, water, dissolved organic matter, ammonia, and cyanide to form soluble associations.

Carbon dioxide can increase the solubility of carbonates above that expected from K_{sp} values. The K_{sp} for $BaCO_3 = Ba + CO_3^{2-}$ is $10^{-8.56}$, and from $\Delta G°$, the reaction $BaCO_3 + CO_2 + H_2O = Ba^{2+} + 2HCO_3^-$ has a $K = 10^{-3.11}$. Calculations using the two equilibrium constants reveal that reaction of carbon dioxide increases the Ba^{2+} concentration at equilibrium by 2.3 mg/L.

Metal ions also can be absorbed onto soil minerals and organic matter or precipitated as a compound whose solubility has been exceeded. Even when a large amount of a trace element enters a water body, a large proportion may be quickly removed from solution by precipitation. This phenomenon occurs when copper sulfate is applied to ponds and reservoirs for algal control. The initial copper concentration following copper sulfate application usually is 100–250 µg/L as Cu^{2+}, but within 2–3 days, the total copper concentration in the water will be similar to the pre-treatment concentration and usually <20 µg/L. Most of the copper precipitates as CuO or $CuCO_3$ and a portion is adsorbed by the bottom soil (McNevin and Boyd 2004). Most of the dissolved copper will be in soluble union with other ions or dissolved organic matter and not as Cu^{2+}.

McBride (1989) concluded that trace ion interactions are so complex that no single theoretical approach or model can be used to account for the concentration of a trace metal. This still is a reasonable assessment, and the only way to know the actual concentration is by measuring it, and the measurement will not separate the free ion concentration from other soluble forms of the ion. Nevertheless, principles related to dissolution of trace elements help us understand why some trace elements typically occur at higher concentration than others, and why concentrations of a given trace element may be greater in one water body than another.

Trace ion concentrations in the ocean must be thought of somewhat differently from those in freshwater. Freshwater systems have a short hydraulic time as compared to the ocean, and measured concentrations of trace metals may change rapidly in freshwater bodies. The ocean has basically reached an equilibrium with respect to water chemistry—especially if we consider only a few decades. There may be natural changes in concentrations, and pollution may increase trace element concentrations in bays and estuaries, but in the open ocean, trace element concentrations tend to remain rather constant.

Several tables containing average concentrations of trace elements in the ocean are available, e.g., Goldberg (1963), Turekian (1968), Ryan (1992), and there are variations in concentrations among these tables. The author believes that the concentrations from Goldberg's table are the most reliable, and they were the main source of data for Table 17.3. Many trace elements have lower concentrations in the ocean than in freshwater.

The formation of the soluble associations between trace elements and other dissolved substances is a major factor determining trace element concentrations. This topic deserves further discussion.

Ion Pairs

Ion pairs of major ions usually are not of great importance in water quality, because only a small fraction of the total concentration of any major ion will be bound in ion pairs (Chap. 4). Trace ions are at low concentration, but major ions are at much higher concentration, and a large fraction of the trace element can be paired with one

Table 17.3 Average trace element concentrations in seawater (Goldberg 1963)

Element	Major forms[a]	Concentration (µg/L as element)	Element	Major forms[a]	Concentration (µg/L as element)
Al	$Al(OH)_4^-$, $Al(OH)_3$	10	Pb	$PbCO_3$	0.03
Sb	$Sb(OH)_6^-$	0.5	Mn	Mn^{2+}	2.0
As	$HAsO_4^{2-}$	3.0	Hg	$HgCl_4^{2-}$	0.03
Ba	Ba^{2+}	.30	Mo	MoO_4^{2-}	10
Be	$BeOH^-$, $Be(OH)_2$	0.0006	Ni	$NiCO_3$	20
Bi	BiO^+, $Bi(OH)_2^-$	0.02	Se	SeO_4^{2-}, SeO_3^{2-}	4.0
(B)	$B(OH)_3$, $B(OH)_4^-$	(4.45 mg/L)	Si	$Si(OH)_4$	(3.0 mg/L)
(Br)	(Br^-)	(65 mg/L)	Ag	$AgCl_2^-$	0.30
Cd	$CdCl_2^-$	0.11	Sr	Sr^{2+}	(8.0 mg/L)
Cr	CrO_4^{2-}	0.05	Sn	$Sn(OH)$	3.0
Co	Co^{2+}, $CoCO_3$	0.50	Tl[b]	Tl^+, $TlCl$	0.014
Cu	$CuCO_3$	3.0	U[c]	–	3.0
F	F^-, MgF^+, CaF^+	(1.4 mg/L)	V	HVO_4^{3-}, $H_2VO_4^-$	2.0
I	IO_3^-	60	Zn	Zn^{2+}	10
Fe	$Fe(OH)_3$	10			

[a]Ryan (1992)
[b]Flegal and Patterson (1985)
[c]Stralberg et al. (2003)

or more major ions. Zinc ion and sulfate ion attract each other to form the zinc sulfate ion pair $(ZnSO_4^0)$ as shown below:

$$Zn^{2+} + SO_4^{2-} \rightleftharpoons ZnSO_4^0 \qquad K_f = 10^{2.38}. \qquad (17.3)$$

The $ZnSO_4^0$ ion pair is soluble and in equilibrium with the free metal ion (Zn^{2+}) concentration in solution. A list of some common ion pair reactions of trace metals and the ion pair formation constants are given in Table 17.4. Most of the trace ions form several ion pairs, but only one or two of these ion pairs will have significant concentrations compared to the concentration of the free trace metal ion.

The calculation of ion pair concentrations is illustrated in the next example using the zinc as the example.

Ex. 17.4 *A water contains 10 µg/L of Zn^{2+} ($10^{-6.81}$ M), 6 mg/L CO_3^{2-} (10^{-4} M), 3.2 mg/L SO_4^{2-} (10^{-4} M), 3.54 mg/L Cl^- (10^{-4} M), and 1.9 mg/L F^- (10^{-4} M). The contribution of ion pairs to the dissolved zinc concentration will be calculated.*

Table 17.4 Equations and formation constants (K_f) for selected ion pairs of micronutrients and trace elements

Reaction	K_f	Reaction	K_f
$Ca^{2+} + F^- = CaF^+$	$10^{1.04}$	$Zn^{2+} + Cl^- = ZnCl^+$	$10^{0.43}$
$Fe^{3+} + F^- = FeF^{2+}$	$10^{5.17}$	$Zn^{2+} + 2Cl^- = ZnCl_2^0$	$10^{0.61}$
$Fe^{3+} + 2F^- = FeF_2^+$	$10^{9.09}$	$Zn^{2+} + 3Cl^- = ZnCl_3^-$	$10^{0.53}$
$Fe^{3+} + 3F^- = FeF_3^0$	10^{12}	$Zn^{2+} + 4Cl^- = ZnCl_4^{2-}$	$10^{0.20}$
$Fe^{3+} + Cl^- = FeCl^{2+}$	$10^{1.42}$	$Zn^{2+} + SO_4^{2-} = ZnSO_4^0$	$10^{2.38}$
$FeCl^{2+} + Cl^- = FeCl_2^+$	$10^{0.66}$	$Zn^{2+} + CO_3^{2-} = ZnCO_3^0$	10^5
$FeCl_2 + Cl^- = FeCl_3^0$	10^1	$Cd^{2+} + F^- = CdF^+$	$10^{0.46}$
$Fe^{2+} + SO_4^{2-} = FeSO_4^0$	$10^{2.7}$	$Cd^{2+} + Cl^- = CdCl^+$	10^2
$Fe^{3+} + SO_4^{2-} = FeSO_4^+$	$10^{4.15}$	$Cd^{2+} + 2Cl^- = CdCl_2^0$	$10^{2.70}$
$Cu^{2+} + F^- = CuF^+$	$10^{1.23}$	$Cd^{2+} + 3Cl^- = CdCl_3^-$	$10^{2.11}$
$Cu^{2+} + Cl^- = CuCl^+$	10^0	$Cd^{2+} + SO_4^{2-} = CdSO_4^0$	$10^{2.29}$
$Cu^{2+} + SO_4^{2-} = CuSO_4^0$	$10^{2.3}$	$Al^{3+} + F^- = AlF^{2+}$	$10^{6.13}$
$Cu^{2+} + CO_3^{2-} = CuCO_3^0$	$10^{6.77}$	$Al^{3+} + 2F^- = AlF_2^+$	$10^{11.15}$
$Cu^{2+} + 2CO_3^{2-} = Cu(CO_3)_2^{2-}$	$10^{10.01}$	$Al^{3+} + 3F^- = AlF_3^0$	10^{15}
$Zn^{2+} + F^- = ZnF^+$	$10^{1.26}$	$Al^{3+} + SO_4^{2-} = AlSO_4^+$	$10^{2.04}$

Solution:
The equations and ion pair formation constants can be taken from Table 17.4. *The ion pair estimates are:*

$$ZnCO_3^0 = (Zn^{2+})(CO_3^{2+})(10^5)$$

$$(ZnCO_3^0) = (10^{-6.81})(10^{-4})(10^{5.0}) = 10^{-5.81}\ M = 101\ \mu g/L\ as\ Zn$$

$$ZnSO_4^0 = (Zn^{2+})(SO_4)(10^{2.38})$$

$$(ZnSO_4^0) = (10^{-6.81})(10^{-4})(10^{2.38}) = 10^{-8.43}\ M = 0.24\ \mu g/L\ as\ Zn.$$

Continuing in like manner, the other ion pair concentrations are

$$ZnF^- = 0.018\ \mu g/L\ as\ Zn$$

$$ZnCl^+ = 0.027 \mu g/L\ as\ Zn$$

$$ZnCl^0 = 0.041\ \mu gL\ as\ Zn$$

$$ZnCl_3^- = 0.0311\ \mu g/L\ as\ Zn$$

$$ZnCl_4^{2-} = 0.016\ \mu g/L\ as\ Zn.$$

The total zinc concentration will be 111.4 µg/L, an increase in soluble Zn^{2+} concentration of 101.4 µg/L. Only the $ZnCO_3^0$ ion pair influenced the concentration of total zinc enough to matter. The total soluble zinc concentration was increased 11-fold.

In Ex. 17.4, it was assumed that the Zn^{2+} concentration of 10 µg/L was the result of the equilibrium of Zn^{2+} with its mineral source. The combination of Zn^{2+} with anions to form ion pairs would have disrupted the Zn^{2+}-mineral equilibrium allowing more Zn^{2+} to enter the water until ion pairs were in equilibrium with Zn^{2+}. If the water had been a laboratory concoction made to contain 10 µg/L of Zn^{2+}, the Zn^{2+} would have been divided among the various ion pairs and Zn^{2+}. The result would have been a lower Zn^{2+} concentration than initially added, and lower concentrations of ion pairs than calculated. The upshot of this is that ion pairs can either increase the total dissolved concentration of dissolved metal as shown in Ex. 17.3 or not affect the total concentration while reducing the concentration of the free metal ion.

It also should be observed in Ex. 17.4 that only one ion pair $(ZnCO_3^0)$ contributed significantly to total dissolved Zn^{2+} concentration. Only one or two ion pairs usually are important in metal ion pair formation.

Complex Ions

The Brønsted theory of acids and bases holds that an acid is a compound able to donate a proton (hydrogen ion) and a base is a compound able to accept a proton. Nitric acid is a typical Brønsted acid:

$$HNO_3 \rightarrow H^+ + NO_3^-, \tag{17.4}$$

while the hydroxide ion is an example of a Brønsted base,

$$H^+ + OH^- = H_2O. \tag{17.5}$$

According to Lewis acid and base theory, a Lewis acid can accept a pair of electrons while a Lewis base can donate a pair of electrons (Fig. 17.2). The Lewis theory is much more general than the Brønsted theory. Examples of Lewis acid-base reactions important in water quality are

Fig. 17.2 Simple illustration of a Lewis acid-base reaction

$$Cu^{2+} + 4NH_3 = Cu(NH_3)_4^{2+} \tag{17.6}$$

$$Fe^{2+} + 6CN^- = Fe(CN)_6^{3-} \tag{17.7}$$

$$PbCl_2 + 2Cl^- = PbCl_4^{2-} \tag{17.8}$$

$$Ag^+ + 2NH_3 = Ag(NH_3)_2. \tag{17.9}$$

Lewis acid-base reactions occur in natural waters, and they are important in waters with high concentrations of ammonia, cyanide, and certain other pollutants. Reactions in Eqs. 17.6, 17.7, 17.8, and 17.9 reveal that potentially toxic trace elements can be put into solution by reaction with other ions according to Lewis acid-base theory.

Hydrolysis of Metal Ions

Hydrolysis also involves the Lewis acid-base concept, and metal ions hydrolyze in water causing an acid reaction and formation of a soluble, metal-hydroxide complex or hydrolysis product as illustrated for the cupric ion (Cu^{2+}):

$$Cu^{2+} + H_2O = CuOH^+ + H^+. \tag{17.10}$$

The hydrolysis may alternatively be written as:

$$Cu^{2+} + OH^- - CuOH^+. \tag{17.11}$$

In both representations, H^+ and OH^- are related to the dissociation of water ($H_2O = H^+ + OH^-$) and H^+ increases relative to OH^- causing pH to decrease.

The mass action expression of Eq. 17.11 multiplied by the mass action form of the reaction for dissociation of water gives the mass action form of Eq. 17.10:

$$\frac{(CuOH^+)}{(Cu^{2+})(OH^-)} \times (H^+)(OH^-) = \frac{(CuOH^-)(H^+)}{(Cu^{2+})}.$$

This exercise confirms that the two reactions have the same relationship to water and hydrogen ion concentration (or pH)—they are simply two ways of expressing the same reaction.

The amphoteric property of trace metals deserves some further comment. Another way of writing the reactions of zinc oxide to illustrate its amphoterism is:

$$ZnO + HCl = Zn^{2+} + Cl^- + H_2O \tag{17.12}$$

and

$$ZnO + 2NaOH = 2Na^+ + Zn(OH)_4^{2-}. \qquad (17.13)$$

Aluminum hydroxide [$Al(OH)_3$] is an important amphoteric compound in water quality. The basic reaction of aluminum hydroxide is:

$$Al(OH)_3 + 3H^+ = Al^{3+} + 3H_2O. \qquad (17.14)$$

The acidic reaction is:

$$Al(OH)_3 + OH^- = Al(OH)_4^-. \qquad (17.15)$$

The metal hydroxides formed by hydrolysis usually are soluble, and a list of metal hydrolysis reactions and their reaction constants is given (Table 17.5). The following example illustrates the calculation of a hydrolysis product.

Ex. 17.5 *A water of pH 3 contains 3.47 μg/L ($10^{-7.21}$ M) ferric iron. The concentrations of soluble iron in hydroxides will be estimated.*
  Solution:
  From Table 17.5,

$$\frac{(FeOH^{2+})}{(Fe^{3+})(OH^-)} = 10^{11.17}$$

$$(FeOH^{2+}) = (10^{-7.21})(10^{-11})(10^{11.17}) = 10^{-7.04} \ M \ or \ 5.09 \ μg/L.$$

By like manner, $Fe(OH)^{2+}$ and $Fe(OH)_4^-$ would have concentrations of 4.6 μg/L and $10^{-17.1}$ M (<0.001 mg Fe/L), respectively. The concentration of dissolved iron in hydroxides would total about 9.69 μg/L—3 times the concentration of free ferric iron.

Table 17.5 Equations and hydrolysis constants (K_h) for selected metal ion hydrolysis reactions

Reaction	K_h	Reaction	K_h
$Fe^{3+} + OH^- = FeOH^{2+}$	$10^{11.17}$	$Cu^{2+} + 4OH^- = Cu(OH)_4^{2-}$	$10^{16.1}$
$Fe^{3+} + 2OH^- = Fe(OH)_2^+$	$10^{22.13}$	$Zn^{2+} + OH^- = ZnOH^+$	$10^{5.04}$
$Fe^{3+} + 4OH^- = Fe(OH)_4^-$	$10^{34.11}$	$Zn^{2+} + 3OH^- = Zn(OH)_3^-$	$10^{13.9}$
$Cu^{2+} + OH^- = CuOH^+$	10^6	$Zn^{2+} + 4OH^- = Zn(OH)_4^{2-}$	$10^{15.1}$
$2Cu^{2+} + 2OH^- = Cu_2(OH)_2^{2+}$	10^{17}	$Cd^{2+} + OH^- = CdOH^+$	$10^{3.8}$
$Cu^{2+} + 3OH^- = Cu(OH)_3^-$	$10^{15.2}$		

Organic Complexes or Chelates

Certain organic molecules contain one or more pairs of electrons that can be shared with metal ions. These molecules are known as ligands or chelating agents and a ligand-metal ion complex often is called a chelated metal. According to Pagenkopf (1978), humic and fulvic acids are the most commonly occurring natural ligands in water. As discussed in Chap. 11, humic and fulvic acids and other naturally occurring dissolved humic substances are large, complex molecules with many functional groups, some of which can bind with metals (Figs. 11.1, 11.2, and 11.3). Equilibrium constants for reactions between metals and natural, organic ligands are not known, and salicylic acid is used below to illustrate the formation of a chelated metal complex:

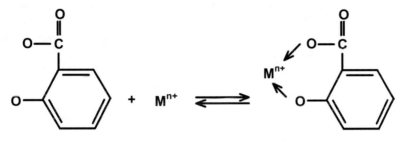

where M^{n+} is the metal ion.

Commercial preparations of metals for use in fertilizers, algicides, and various other products often are chelated with ligands for which equilibrium constants of the ligand-metal ion reactions are known. Triethanolamine (HTEA) is commonly used to chelate copper for use as an algicide. According to Sillén and Martell (1971), triethanolamine (HTEA) dissociates into hydrogen ion and triethanolamine ion (TEA$^-$) as follows:

$$HTEA = H^+ + TEA^- \quad K = 10^{-8.08}. \tag{17.16}$$

The ionized form of HTEA can form complexes with metal ions as illustrated below for copper:

$$Cu^{2+} + TEA^- = CuTEA^+ \quad K = 10^{4.44} \tag{17.17}$$

$$Cu^{2+} + TEA^- + OH^- = CuTEAOH^0 \quad K = 10^{11.9} \tag{17.18}$$

$$Cu^{2+} + TEA^- + 2OH^- = CuTEA(OH)_2^- \quad K = 10^{18.2}. \tag{17.19}$$

Ex. 17.6 *A solution 10^{-4} M in triethanolamine contains $10^{-8.25}$ M (0.357 µg/L) Cu^{2+} at equilibrium at pH 8.0. The concentration of chelated copper will be estimated.*

Solution:
The triethanolamine concentration is given, so the TEA$^-$ concentration available to chelate copper will be computed. From Eq. 17.16,

$$\frac{(H^+)(TEA^-)}{(HTEA)} = 10^{-8.08}$$

$$\frac{(TEA^-)}{(HTEA)} = \frac{10^{-8.08}}{10^{-8}} = 10^{-0.08} = 0.83$$

$$\%TEA^- = \frac{0.83}{1.83} \times 100 = 45.4\%.$$

The   HTEA   concentration   is   0.0001   M,   and   TEA$^-$ =   (0.0001) (0.454) = 0.0000454 M = 10$^{-4.34}$ M.

Equations 17.17, 17.18, and 17.19 allow computations of concentrations of triethanolamine-chelated copper complexes:

$$Cu^{2+} + TEA^- = CuTEA^+$$

$$(CuTEA^+) = (Cu^{2+})(TEA^-)\, K = (10^{-8.35})(10^{-4.34})(10^{4.44})$$

$$= 10^{-8.25}\, M\,(0.357\, \mu g\, Cu^{2+}/L).$$

Following the same procedure, the copper concentrations in CuTEAOH^0 and CuTEA(OH)$_2^-$ are 10^{-6} M (63.6 μg/L) and 10$^{-6.49}$ M (20.6 μg/L), respectively. The total concentration of chelated copper is 84.6 μg/L or two orders of magnitude greater than the Cu^{2+} concentration.

Role of Metal Hydroxides, Ion Pairs, and Chelates in Solubility

Equilibrium concentrations of free trace metal ions in natural water bodies are low even where the controlling minerals occur in abundance. Free metal ions are in equilibrium with their controlling minerals, but they also are in equilibrium with ion pairs and inorganic and organic complexes as illustrated in Fig. 17.3 for copper ion. The presence of dissolved copper associations does not alter the concentration of Cu^{2+} at equilibrium with CuO. The dissolved associations usually result in more dissolved copper than would be expected from the K$_{sp}$ calculation for CuO, and the same situation is true for other metals. At pH 7, the equilibrium concentration of Fe^{3+} is calculated from the K$_{sp}$ of one of the ferric iron minerals listed in Table 17.1 is undetectable by analytical methods, but the dissolved iron concentrations in natural waters are usually 100–500 μg/L. Waters that contain large amounts of humic substances may have higher total dissolved iron concentrations.

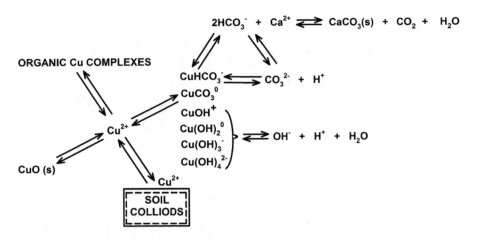

Fig. 17.3 Copper equilibria in a soil-water system containing free calcium carbonate

There is an equilibrium between metal associations and free metal ions just as between free ions and their mineral forms:

$$\text{Mineral form} \rightleftharpoons \text{Free ion} \rightleftharpoons \text{Metal Associations.} \qquad (17.20)$$

Free metal ions removed from solution will be replaced by further dissolution of their mineral sources (if present in the sediment). The dissociation of other dissolved forms of metals can supply free ions for use by aquatic plants much quicker than can the solid mineral forms that occur in the bottom sediment. The various dissolved metal associations tend to buffer metal ion concentrations in water. Moreover, plants can use free ions as sources of mineral nutrients or they can absorb the complexed forms. The toxicity of metals appears to be related primarily to free ion concentrations. Although some of the other metal ion associations have a degree of toxicity, addition of a chelating agent often will reduce toxicity in presence of elevation concentrations of metal ions.

Micronutrients

Nearly all plants require iron, manganese, zinc, copper, boron, selenium, cobalt, and molybdenum in small amounts, and some plants apparently benefit physiologically from traces of chromium, cadmium, nickel, and vanadium. The required trace elements for animals are iron, manganese, zinc, copper, fluoride, iodine, selenium, cadmium, cobalt, and molybdenum, while there appears to be physiological benefits from small quantities of boron, arsenic, and nickel.

Micronutrients usually are divided between metals and nonmetals. The metals include iron, manganese, zinc, copper, cadmium, chromium, nickel, vanadium, cobalt, and molybdenum. The nonmetals are boron, fluoride, iodine, selenium, and arsenic (Pais and Jones 1997).

Iron (Fe)

The two main iron ores are hematite (F_2O_3) and magnetite (Fe_3O_4). Of course, iron occurs in other oxides, hydroxides, sulfides, sulfates, arsenates, and carbonates. Iron oxides and hydroxides are abundant in soil.

Many enzymes such as peroxidases, catalases, and cytochrome oxidases important in cellular energy transformations contain iron. Iron proteins (ferredoxins) are involved in phytophosphorylation in photosynthesis. Hemoglobin in the blood of animals contains an iron porphyrin ring.

Iron concentrations seldom are great enough to be harmful to aquatic life. Nevertheless, it precipitates in water and the resulting floc can be harmful in fish hatcheries by settling on eggs and by occluding the gills of fish. Precipitates of iron on other biological surfaces also can be harmful.

The solubility of iron compounds under aerobic conditions is controlled mainly by pH, and except at very low pH, the oxides and hydroxides found in soil are of low solubilities. But, iron is abundant in nature and dissolved ion associations increase iron concentrations, and freshwaters contain 250 µg/L to 1000 µg/L and even more in water heavily stained by humic substances. Polluted acidic waters can have very high iron concentrations. Ocean water has an average of 10 µg/L of iron mostly in the form of $Fe(OH)_3$.

Iron has been shown to limit phytoplankton productivity in freshwater bodies (Hyenstrand et al. 2000; Vrede and Tranvik 2006) and in the ocean (Nadis 1998). There have been suggested that fertilization of the ocean with iron could increase its productivity and thereby increase both global fish supply and counteract global warming through carbon sequestration.

Respiration in sediment uses dissolved oxygen from the pore water faster than oxygen enters by diffusion and infiltration. This leads to anaerobic conditions other than in the layer near the sediment-water interface. Ferric iron and other oxidized substances in anaerobic sediment act as oxidizing agents by accepting electrons:

$$Fe^{3+} + e^- = Fe^{2+}. \tag{17.21}$$

Some types of bacteria under anaerobic conditions can use ferric iron as a substitute for molecular oxygen as an acceptor of electrons and hydrogen ions resulting from respiration. The general process of iron reduction by bacteria is illustrated below:

$$\text{Organic matter} \rightarrow \text{fragments such as } CH_3COOH \tag{17.22}$$

$$CH_3COOH + 6H_2O \rightarrow 2CO_2 + 8H^+ \tag{17.23}$$

$$8Fe(OH)_3 + 8H^+ \rightarrow 8Fe(OH)_2 + 8H_2O \tag{17.24}$$

$$Fe(OH)_2 = Fe^{2+} + 2OH^-. \tag{17.25}$$

Anaerobic respiration increases the Fe^{2+} concentration in sediment pore water and oxygen-depleted hypolimnetic water. Ferrous iron concentrations begin to increase at redox potentials of 0.2–0.3 volt. Concentrations up to 5–10 mg/L of ferrous iron are common, and concentrations several times this have been measured in bottom soil of fertilized fish ponds.

Water from some wells may contain high concentrations of iron because of low redox potential in aquifers. Ferric hydroxide tends to control iron concentration in shallow aquifers where redox potential usually is comparatively high, and concentrations of ferrous iron usually will be low (Hem 1985). Deep wells often draw water from aquifers where there is a low redox potential and iron concentrations may be controlled by iron sulfide,

$$FeS_2 = Fe^{2+} + 2S^{2-} \qquad K = 10^{-26}. \qquad (17.26)$$

Groundwater usually contains sulfate, and sulfate-reducing bacteria reduce sulfate to sulfide in respiration. Ferrous iron also is present from anaerobic respiration, and iron precipitates as insoluble ferrous sulfide. Waters from deep aquifers usually contain very little iron, but they may yield waters high in sulfide concentration. In aquifers of intermediate depth, redox potential is intermediate between that of shallow and deep aquifers. Ferrous carbonate (siderite) tends to be the mineral controlling iron concentration in aquifers of intermediate depth:

$$FeCO_3 + H^+ = Fe^{2+} + HCO_3^- \qquad K = 10^{-0.3}. \qquad (17.27)$$

Very high concentrations of iron (up to 100 mg/L or more) may occur in waters from aquifers where siderite is the iron-controlling mineral.

Organic matter decomposition in waterlogged soil and in sediment causes low redox potential and high concentrations of ferrous iron often result. The pH usually will be between 6 and 6.5 in such situations (Ex. 17.7).

Ex. 17.7 *The pH will be estimated for a sediment in which the pore water contains 20 mg/L ($10^{-3.47}$ M) Fe^{2+} and 61 mg/L (10^{-3} M) HCO_3^-.*
  Solution:
  From Eq. 17.27,

$$\frac{\left(Fe^{2+}\right)\left(HCO_3^-\right)}{\left(H^+\right)} = 10^{-0.3}$$

$$\left(H^+\right) = \frac{\left(Fe^{2+}\right)\left(HCO_3^-\right)}{10^{-0.3}} = \frac{\left(10^{-3.47}\right)\left(10^{-3}\right)}{10^{-0.3}} = 10^{-6.17}$$

$$pH = 6.17.$$

When water containing ferrous iron contacts oxygen, precipitation of iron compounds occur as illustrated below:

$$4Fe(HCO_3)_2 + 2H_2O + O_2 = 4Fe(OH)_3 \downarrow + 8CO_2. \qquad (17.28)$$

Ferric hydroxide precipitation from water following oxygenation can stain plumbing fixtures and kitchen utensils, and clothes laundered in such water may be permanently stained.

Iron bacteria such as *Leptothrix ochracea* and *Spirophyllum ferrugineum* obtain energy for the synthesis of organic compounds (chemoautotrophy) from oxidation of ferrous salts as follows:

$$4FeCO_3 + O_2 + 6H_2O = 4Fe(OH)_3 \downarrow + 4CO_2. \qquad (17.29)$$

These bacteria tend to form slimy mats known as ochre where waters containing a high concentration of iron flow or seep onto the land surface and absorbs oxygen from the air.

Manganese (Mn)

The principal ore of manganese is manganese dioxide (MnO_2). There are other oxides and hydroxides of manganese, and a great many other manganese compounds in nature. Soils usually contain manganese at a much lower concentration than at which iron is present.

Manganese is a nutrient, because it is a constituent of enzymes or acts as an enzyme activator. This element is particularly important in the action of antioxidants, and catalyzes the photolysis reaction that releases oxygen in photosynthesis. Manganese, like iron, seldom causes toxicity, but elevated concentrations of manganese have been reported to be toxic to benthic invertebrates (Pinsino et al. 2012). Also, like iron, manganese can precipitate in water having various mechanical effects on organisms. It also can stain plumbing fixtures and cause laundry stains.

The factors governing the solubility of manganese are similar to those discussed for iron under both aerobic and anaerobic conditions. The concentration of Mn^{2+} in water is usually very low, and it occurs mainly in combination with other dissolved substances. Manganese concentrations usually are considerably lower than those of iron, because of its lower abundance than iron in nature. Maximum concentrations of manganese typically will be below 100 μg/L in unpolluted natural waters. The average manganese concentration in the ocean is given as 2 μg/L.

Copper (Cu)

The main ore of copper is an iron-copper sulfide ($CuFeS_2$) known as chalcopyrite, and it also is present in other sulfide deposits, sandstones, and shales. Copper is present in soil mainly as copper adsorbed on clay and organic matter.

Copper is a cofactor in many metalloenzymes including those that catalyze RNA and DNA synthesis, melanin production, electron transfers in respiration, collagen and elastin formation. Plants require copper for chlorophyll synthesis, root metabolism, and lignification.

Copper has a long history as a fungicide and algicide (Moore and Kellerman 1905). Its use as an algicide is important in water quality, because copper is widely used to control taste and odor microorganisms in water supply reservoirs and to combat phytoplankton blooms in lakes and ponds. The treatment rate usually is 0.2–2.0 mg/L of copper sulfate (about 50–400 µg/L as Cu^{2+}) with the higher treatment concentrations being made at greater alkalinity. A popular treatment rate for copper sulfate is a concentration equivalent to 1% of the total alkalinity concentration.

The mineral controlling copper solubility in water bodies often is said to be tenorite (CuO) at pH ≥ 7 and malachite $Cu_2(OH)_2CO_3$ at pH <7. Solubility increases as pH decreases for both of these minerals. The equilibrium Cu^{2+} concentration at pH 7 is 14.2 µg/L for tenorite and 7.6 µg/L for malachite, while at pH 6, the concentrations are 1.42 mg/L and 0.76 mg/L, respectively. This suggests that malachite possibly is the controlling copper compound at all pH values.

Calculations of copper ion pairs as done for zinc in Ex. 17.4 reveal that $CuCO_3^0$ is the main ion pair. It is possible to illustrate the relationship between Cu^{2+} and $CuCO_3^0$ in water of different pH values and total alkalinity concentrations by following the method illustrated in Ex. 17.8.

Ex. 17.8 *Concentrations of Cu^{2+} and $CuCO_3^0$ at pH 8 and 50 mg/L total alkalinity will be calculated.*

Solution:

At pH 8, the dissolution of CuO is:

$$\frac{(Cu^{2+})}{(10^{-8})^2} = 10^{7.35} : Cu^{2+} = 10^{-8.65} \, M \, (0.142 \, µg/L)$$

The expression for $CuCO_3^0$ formation is:

$$Cu^{2+} + CO_3^{2-} = CuCO_3^0 \qquad (K_f = 10^{6.77}).$$

Carbonate can be calculated from the expression for dissociation of bicarbonate ($HCO_3^- = H^+ + CO_3^{2-}$) for which $K = 10^{-10.33}$. At pH 8 and 50 mg/L alkalinity (61 mg/L HCO_3^- or 10^{-3} M) carbonate has a calculated concentration of $10^{-5.33}$ M. The $CuCO_3^0$ ion pair concentration is:

$$CuCO_3^0 = (Cu^{2+})(CO_3^{2+})(K_f) = (10^{-8.65})(10^{-5.33})(10^{6.77})$$
$$= 10^{-7.21} \, M \, or \, 3.91 \, µg/L \, of \, CuCO_3^0.$$

Concentrations of Cu^{2+} and $CuCO_3^0$ were calculated for different pH values and alkalinity concentrations (Tables 17.6 and 17.7) using the procedure of Ex. 17.8. When the pH was fixed at 8 and alkalinity varied, the Cu^{2+} concentration was the

Table 17.6 Concentrations of Cu^{2+} and $CuCO_3^0$ at different total alkalinity concentrations and pH 8

Total alkalinity	Concentration (µg/L)			
	Cu^{2+}	$CuCO_3^0$	ΣCu	$CuCO_3/\Sigma Cu$
25	0.142	1.96	2.10	0.93
50	0.142	3.91	4.05	0.97
100	0.142	7.82	7.96	0.98
200	0.142	15.60	15.74	0.99

Table 17.7 Concentrations of Cu^{2+} and $CuCO_3^0$ at 50 mg/L total alkalinity and different pH values

pH	Concentration (µg/L)			
	Cu^{2+}	$CuCO_3^0$	ΣCu	$CuCO_3/\Sigma Cu$
6	1,422	392	1,814	0.22
7	14.22	39.2	53.4	0.73
8	0.142	3.92	4.07	0.96
9	0.00142	0.0392	0.041	0.96

same at all four alkalinities, but the $CuCO_3^0$ and sum of the two copper forms (ΣCu) increased with greater alkalinity (Table 17.6). The proportion of $CuCO_3^0$ to ΣCu was 0.93–0.99. When alkalinity was fixed and the pH varied (Table 17.7), Cu^{2+} and ΣCu concentration dropped drastically with pH. The $CuCO_3^0$ concentration decreased with pH in spite of increasing CO_3^{2-} concentration, because there was less copper to form ion pairs as the pH rose. The proportion of $CuCO_3^0$ was less than that of Cu^{2+} at pH 6, slightly more than Cu^{2+} at pH 7, but it accounted for nearly all of the dissolved copper at higher pH.

Concentrations of dissolved copper in 1500 rivers in the United States averaged 15 µg/L with maximum concentration of 28 µg/L (Kopp and Kroner 1967). The copper concentration of the ocean is 3.0 µg/L and most is in dissolved $CuCO_3^0$. Polluted waters of two rivers in Turkey contained 120 µg/L to 1.37 mg/L of dissolved copper (Seker and Kutler 2014). Ning et al. (2011) reported copper concentrations of 38 µg/L up to 14.6 mg/L in surface waters of a mining area in China.

Zinc (Zn)

Zinc is more abundant than copper in the earth's crust. It occurs in sulfides, oxides, silicates, and carbonates. The common zinc ore is Smithsonite which has a high $ZnCO_3$ content, but it usually is mixed with ZnS and often associated with silver. Zinc ion also is adsorbed on clay minerals and organic matter in soil at low concentrations.

Zinc, like copper, is involved biochemically in metalloenzymes, and its action in stabilizing certain molecules and membranes is especially important. Zinc also is involved in chlorophyll synthesis. Zinc is not as toxic as copper, and it is not as effective as a fungicide and algicide.

Zinc carbonate ($K_{sp} = 10^{-9.84}$) may be the mineral controlling zinc concentration in many water bodies, and the equilibrium concentration would be 86 µg/L. The dissolved zinc concentration seldom is this great. Kopp and Kroner (1967) reported an average of 64 µg/L with a maximum of 1.18 mg/L in rivers of the United States. Ocean water has an average zinc concentration of 10 µg/L.

Polluted water can be much higher in zinc concentration. In a mining area in the United States, zinc concentrations of 175–659 µg/L were found (Besser and Leib 2007). Ning et al. (2011) reported surface water zinc concentrations in a mining area of China ranging from 60 µg/L–3.72 mg/L with an average of 1.01 mg/L.

Cadmium (Cd)

Cadmium occurs in carbonate and hydroxide form with other trace metals, but relatively pure deposits of cadmium compounds are rare. Cadmium for commercial use usually is extracted as an impurity during zinc processing. It also occurs in most soil at low concentrations.

Cadmium is a nutrient for some planktonic, marine diatoms (Lee et al. 1995), in which it has a role in increasing photosynthesis by a beneficial influence on carbonic anhydrase activity (Lane and Morel 2000). Cadmium is toxic to animals at elevated concentrations, but at normal body concentrations it is physiologically beneficial for unknown reasons.

The solubility expression for cadmium carbonate is

$$CdCO_3 = Cd^{2+} + CO_3^{2-} \qquad K_{sp} = 10^{-12.85}. \qquad (17.30)$$

The solubility from K_{sp} is 42.2 µg/L cadmium, and the actual concentration could be greater because of the action of carbon dioxide on carbonates. In natural waters, cadmium forms ion pairs ($CdCl^+$, $CdCl_2^0$, $CdCl_3^-$, and $CdCl_4^{2-}$), the hydrolysis product ($CdOH^+$), and chelates. Nevertheless, because of its low abundance in the earth's crust, cadmium concentrations often are undetectable in natural waters. River water in the United States with measurable cadmium usually had less than 1 µg/L (Kopp and Kroner 1967), and the highest concentration found was 120 µg/L. Concentrations above 1 µg/L were thought the result of pollution. The highest concentration in surface water of a mining area in China was 194 µg/L (Ning et al. 2011). Cadmium has an average concentration in the ocean is of 0.11 µg/L and mainly as $CdCl_2$.

Cobalt (Co)

There are many minerals from which cobalt may be extracted to include arsenides, sulfate salts, sulfides and oxides, and residual weathered rocks. It also is present at low concentration in soil.

Cobalt is a cofactor for several enzymes, the most notable being vitamin B12 that controls red blood cell production in animals. This element also is needed as a cofactor in nitrogen fixation by blue-green algae and certain bacteria.

Cobalt has many properties in common with iron, but the redox potential for the cobaltic form (Co^{3+}) is greater than that for the occurrence of ferric iron. As a result, the cobaltous (Co^{2+}) form is common in aquatic environments. According to Hem (1985), the mineral controlling cobalt concentration in natural water likely is cobalt carbonate ($K_{sp} = 10^{-12.85}$). The cobalt concentration in equilibrium with cobalt carbonate would be about 24 μg/L, and probably greater because of the action of dissolved carbon dioxide. Because of the low abundance of cobalt minerals in the earth's crust, concentrations of cobalt usually are less than expected for cobalt carbonate equilibrium in water. In waters with detectable cobalt, concentrations usually are <1 μg/L (Baralkiewicz and Siepak 1999). Cobalt concentrations above 20 μg/L have been reported in highly mineralized water, and concentrations up to 80 μg/L were reported in mining areas (Essumang 2009). Seawater averages 0.5 μg/L cobalt, and about half of the cobalt in seawater is the free ion and the rest is in hydrolysis products and ion pairs (Ćosović et al. 1982).

Molybdenum (Mo)

The main ore of molybdenum is molybdenite (MoS_2). Small concentrations of molybdenum are present in various minerals, and it also is found in soil.

Molybdenum is an essential constituent of several enzymes that play a role in nitrogen fixation by microorganisms and nitrate reduction by plants. In animals, molybdenum is a cofactor in enzymes influencing oxidation of purines and aldehydes, protein synthesis, and metabolism of several nutritive elements.

Inland waters contain small concentrations of molybdate (MoO_4^{2-}). According to Hem (1970), the average concentration of molybdenum was 0.35 μg/L for North America rivers, and 100 μg/L was observed in a reservoir in Colorado. Groundwater waters often contain 10 μg/L molybdenum or more. The average concentration of molybdenum in seawater is 10 μg/L.

Chromium (Cr)

Chromium is rather common in the earth's crust occurring in chromites such as iron chromium oxide ($FeCr_2O_3$) which is its principal ore. Chromium occurs in the soils and water mainly in trivalent (Cr^{3+}) and hexavalent (Cr^{6+}) forms.

Trivalent chromium is an essential nutrient for humans, livestock, and presumably other animals, because it is necessary in sugar and fat metabolism (Anderson 1997; Lindemann et al. 2009). At elevated concentrations, chromium also is toxic, and trivalent chromium exposure through air and drinking water has been linked to mutations and cancer.

Concentrations of chromium in rivers of the United States ranged from below detection to 112 µg/L with an average of 9.7 µg/L for water in which it could be detected. Samples with particularly elevated concentrations were polluted with industrial effluents. Chromium concentrations up to 60 µg/L were reported in wells in arid areas of southern California (Izbicki et al. 2008), and Ning et al. (2011) found a maximum surface water concentration of 75 µg/L in a mining area in China. There is an average of 0.05 µg/L of chromium in the ocean.

Vanadium (V)

Vanadium sulfide (VS_4) and carnotite $K_2(UO_2)_2(VO_4)_2 \cdot 3H_2O$ are the main ores of this element, but it also is extracted from Venezuelan crude oil. It is present in soils at very low concentration.

Vanadium has been identified as an essential element for the green algae *Scenedesmus obliquus* (Arnon and Wessel 1953), and for some marine macroalgae (Fries 1982). However, it is apparently not essential for higher plants or for animals.

Vanadium concentration is inland waters usually are very low. In some arid areas with rocks of elevated vanadium concentration, river water may contain over 1 mg/L of vanadium (Livingstone 1963). The ocean contains about 2 µg/L of vanadium.

Nickel (Ni)

Nickel occurs in nature as oxides, sulfides, and silicates. The geochemistry of nickel is similar to that of iron.

Nickel is a cofactor at very low concentration in certain metalloenzyme functions. It has been recognized for several decades as an essential nutrient (Spears 1984). These studies have demonstrated that if nickel is absent in the diet, organisms do not grow as well as when it is present in small amounts.

The ionic form of nickel in water is Ni^{2+}, and it forms soluble hydroxides. Nickel carbonate may be considered the controlling mineral:

$$NiCO_3 = Ni^{2+} + CO_3^{2-} \qquad K_{sp} = 10^{-8.2}. \qquad (17.31)$$

Based on K_{sp}, a concentration of 4.66 mg/L would be possible at equilibrium, but such a high concentrations is not found in surface waters. Rivers in the United States contained up to 56 µg/L of nickel (Koop and Kroner 1967), and in another study, Durum and Haffty (1961) reported 10 µg/L as the average for North American rivers. Khan et al. (2011) reported 650 µg/L of nickel in the Shah Alum River in Afghanistan. Ocean water contains around 2 µg/L nickel mostly as $NiCO_3$.

Boron (B)

Boron is a metalloid having properties of both metals and nonmetals. Boron occurs in relatively insoluble aluminosilicate minerals into which it substituted for silicon and aluminum. It also occurs in arid areas as borax ($Na_2B_4O_7 \cdot 10H_2O$), ulexite ($NaCaB_5O_9 \cdot 8H_2O$) and other borate minerals which are mined to make boron products (Kochkodan et al. 2015). Its geochemical cycle is much like that of chloride.

Plants require small amounts of boron, and there have been suggestions that boron may be essential in animals. Boric acid forms bonds with molecules of pectins in plant cell walls and glycolipids in bacterial cell walls to make the cell walls more stable. High boron concentrations in irrigation water are toxic to plants.

Boron behaves as a Lewis acid (Kochkodan et al. 2015) for which the first dissociation is:

$$B(OH)_3 + H_2O = B(OH_4)^- + H^+ \qquad K = 10^{-9.24}. \qquad (17.32)$$

At pH below 9.24, undissociated boric acid comprises more than 50% of the total dissolved boron. For example, at pH 8, only 5% of boric acid is dissociated. The other two dissociations with K values of $10^{-12.74}$ and $10^{-13.80}$, respectively, are seldom of water quality interest. Traditionally, boric acid was considered to dissociate as a Brønsted acid and its dissociation often is still written as follows:

$$H_3BO_3 = H^+ + H_2BO_3^- \qquad K = 10^{-9.24}. \qquad (17.33)$$

Boron concentration in freshwater bodies in the southeastern US seldom exceeds 100 µg/L (Boyd and Walley 1972). A more extensive survey of 1546 river and lake samples throughout the United States gave an average boron concentration of 100 µg/L with a maximum of 5 mg/L (Kopp and Kroner 1970). Much greater concentrations can be found in saline inland waters of arid regions; concentrations above 100 mg/L were reported (Livingstone 1963). Boron occurs in the ocean as B $(OH)_3$ and $B(OH)^{4-}$, and its concentration averages about 4.5 mg/L.

Fluoride (F)

Fluoride, like chlorine, bromine, and iodine, is a highly electronegative element of the halogen group. Calcium fluoride (CaF_2) is a common fluoride bearing mineral, and fluoride also is contained in apatite and some other common minerals. Fluoride compounds are less abundant in the earth's crust and of lower solubility than are chloride compounds. As a result, most fluoride in the earth's crust is bound in rocks while the sea is a huge reservoir of dissolved chloride.

Fluoride is a nutrient at optimal levels, because it is important to the integrity of bone and teeth in humans and other animals. According to Palmer and Gilbert (2012), fluoridation of municipal water supplies for the prevention of dental cavities is likely the most effective public dental health measure in existence. However, high concentrations of fluoride in drinking water can cause bone disease and mottling of tooth enamel. The current recommended upper limit for fluoride in drinking water for avoiding health effects is 4 mg/L, and no negative cosmetic effects on teeth have been reported for fluoride concentrations less than 2 mg/L. The World Health Organization (WHO) suggested an upper limit of 1.5 mg/L for fluoride concentrations in drinking water (https://www.who.int/water_sanitation_health/dwq/chemicals/fluoride.pdf). In the United States, the recommended fluoride concentration in drinking water was 0.7–1.2 mg/L for many years, but the maximum concentration limit was reduced to 0.7 mg/L in 2015 (https://www.medicalnewstoday.com/articles/154164.php). There is a segment of society opposing fluoridation of drinking water, but the available evidence does not support this concern.

The solubility of calcium fluoride may be expressed as follows:

$$CaF_2 = Ca^{2+} + 2F^- \qquad K = 10^{-10.60}. \qquad (17.34)$$

The concentration of fluoride based on K_{sp} from calcium fluoride is 125 µg/L. It forms ion pairs with the major cations and with many metals in water, and this would increase concentration above that of the fluoride ion. Most freshwaters contain less than 100 µg/L of fluoride. Puntoriero et al. (2014) reported fluoride concentrations ranging from 0.15 µg/L to 1.37 mg/L in lakes of different continents. Ocean water typically contains around 1.4 mg/L of fluoride present as F^-, MgF^+, and CaF^+.

Iodine (I₂)

Iodine is present as iodide (I^-) and in combination with other elements in the earth's crust. Iodide is obtained commercially by extracting it from natural brines or brines from oil wells. Iodine (I_2) also is extracted from seaweed.

Iodine is important for proper function of the thyroid gland in producing thyroid hormones needed for normal growth, development, and metabolism in humans and other vertebrates. In areas of the world where soils and waters are deficient in iodine, those who rely heavily on local sources of food may develop iodine-deficiency disorders—the most famous of which is enlargement of the thyroid gland (goiter) producing a prominent swelling of the neck. The diet is an important source of iodine, and in the past when foods were produced locally, the incidence of goiter was high in areas where soils and crops produced in them were low in iodide. Dietary iodine supplements such as iodized salt were and still are a preventive measure for goiter. Since at least the 1960s, most people in developed countries obtain food from many different places, and dietary iodine deficiency has become less common.

Being a halogen, iodide's chemistry is similar to that of chloride and fluoride. According to Hem (1985), rainwater typically contains 1–3 µg/L of iodine, while the range for river water is 3–42 µg/L. The British Geological Survey (2000) reported that potable groundwater typically contained 1–70 µg/L iodide with extreme values up to 400 µg/L. Seawater has an average iodine concentration of 60 µg/L mainly as IO_3^-.

Selenium (Se)

Selenium occurs in the earth's crust mainly as elemental selenium, ferric selenite, and calcium selenite. The chemistry of selenium is similar to that of sulfur, but it is much less common than sulfur. The ion, SeO_3^{2-} is the most stable form of selenium in water.

Selenium is a component of the amino acids selenocysteine and selenomethionine. It also is a cofactor for certain peroxidases and reductases. Selenium accumulates in the food chain, and in spite of being an essential element for aquatic animals, selenium pollution can have serious consequences on aquatic ecosystems (Hamilton 2004). Those who drink water of high selenium concentration over many years may experience hair and fingernail loss, numbness in extremities, and circulation problems.

Selenium concentrations in groundwater and surface water seldom exceed 1 µg/L (Hem 1985), but concentrations of 0.06–400 µg/L have been reported. There is one report of 6 mg/L of selenium in groundwater (World Health Organization 2011). The average concentration of selenium in ocean water is 4 µg/L.

Arsenic (As)

Arsenic occurs in minerals such as arsenopyrite (AsFeS), realgar (AsS), and orpiment (As_2S_3) from which it is released by weathering as arsenate. Small percentages of arsenate have substituted for phosphate in apatite (rock phosphate). Arsenic compounds tend to be of low solubility, and they are seldom abundant in soils or bottom sediments. The chemistry of arsenic and phosphorus is similar. Arsenic is strongly absorbed on clay minerals and increases in solubility when redox potential is low. Even the dissociation constants for the three step ionization of arsenic acid (H_3AsO_4) and orthophosphoric acid (H_3PO_4) are similar. Compare Eqs. 17.35, 17.36, and 17.37 with Eqs. 14.1, 14.2, and 14.3).

$$H_3AsO_4 = H^+ + H_2AsO_4^- \qquad K = 10^{-2.2} \qquad (17.35)$$

$$H_2AsO_4^- = H^+ + HAsO_4^{2-} \qquad K = 10^{-7} \qquad (17.36)$$

$$HAsO_4^{2-} = H^+ + AsO_4^{3-} \qquad K = 10^{-11.5.} \qquad (17.37)$$

The different arsenate ions also react similarly to phosphate ions, and arsenate cannot be distinguished from phosphate in the usual methods for determining phosphate concentration.

Arsenic is considered a micronutrient because it apparently is a factor in methionine metabolism by certain animals (Uthus 1992). Arsenic is important in medical history because it was once used to treat sleeping sickness and syphilis. It is still used in treatment of acute promyelocytic leukemia.

Arsenic is most famous as a murder weapon. Arsenic trioxide is colorless, tasteless, and virtually impossible to detect in food and drink, and it became known as the "poison of kings." Of course, it has been used for murder in all levels of society. Arsenicals also are used as insecticides, fungicides, and wood preservatives. The United States Army even has a chemical weapon that is an arsenical (Firth 2013).

Arsenate is the main form of arsenic in water, and the concentration range for its soluble inorganic form is similar to phosphorus. Of course, arsenic is not accumulated in plankton and detritus to the extent that phosphorus is. According to Kopp (1969), arsenic concentrations in river water in the United States ranged from 5 to 336 µg/L with an average of 64 µg/L. Arsenic concentrations as high as 187 µg/L were found in surface waters of a mining area in China (Ning et al. 2011). On average, ocean water contains 3 µg/L of arsenic.

Natural arsenic contamination of groundwater is a serious problem in a few regions of the world. The most highly publicized incidence is in the lower Ganges River basin in nine districts of West Bengal, India and 42 districts in Bangladesh where arsenic levels in drinking water exceed the World Health Organization maximum permissible limit of 50 µg/L (Chowdhury et al. 2000). In Bangladesh, about 21 million people use well waters with more than 50 µg/L arsenic, and some well waters have over 1 mg/L. Arsenic poisoning symptoms include lesions, keratosis, conjunctival congestion, edema of feet, and liver and spleen enlargement. In advanced cases, cancers affecting lungs, uterus, bladder, and genitourinary tract often are seen. According to Chowdhury et al. (2000), of 11,180 people examined who had been drinking water with over 50 µg/L of arsenic, about 25% had arsenical skin lesions. An estimated 100 million people are at risk of arsenic poisoning in West Bengal and Bangladesh.

Elevated concentrations of arsenic in public water supplies increase cancer risk, and many countries have been imposing stricter arsenic standards for drinking waters. In the United States, the standard was 50 µg/L for many years, but in 2006, it was reduced to 10 µg/L. Of course, many municipal water supplies have not yet complied with the new standard.

Nonessential Trace Elements

Nonessential elements are of concern in water quality mainly because, like micronutrients, they can be toxic to aquatic organisms at elevated concentrations. Trace elements also can be toxic through drinking water for humans, livestock, and other animals.

Aluminum (Al)

Aluminum is present in many silicate rocks and in deposits of aluminum oxides and hydroxides, and only oxygen and silicon exceed aluminum in abundance in the earth's crust. Deposits of bauxite, a mixture of gibbsite [$Al(OH)_3$] and boehmite

[AlO(OH)], and certain other aluminum oxides and hydroxides are mined and processed to produce aluminum metal. Aluminum oxides and hydroxides are abundant in soils and especially in acidic soils.

Although abundant in nature, aluminum is at low concentration in all except very acidic waters. This can be seen by calculation of the equilibrium Al^{3+} concentrations for solubility of gibbsite at different pH levels:

$$Al(OH)_3 + 3H^+ = Al^{3+} + 3H_2O \qquad K = 10^9. \qquad (17.38)$$

The equilibrium aluminum (Al^{3+}) concentrations decline as follows: pH 4, 27 mg/L; pH 5, 27 μg/L; pH 6, 0.027 μg/L; pH 7, 0.000027 μg/L.

Aluminum forms several soluble hydroxide complexes in dilute aqueous solution, several of which are polynuclear. There are various ways of writing the equations for the formation of these complexes, and experts do not completely agree on the formulas. The equations and equilibrium constants for aluminum hydroxide complexes given below were from Hem and Roberson (1967) and Sillén and Martell (1964):

$$Al^{3+} + H_2O = AlOH^{2+} + H^+ \qquad K = 10^{-5.02} \qquad (17.39)$$

$$2Al^{3+} + 2H_2O = Al_2(OH)_2^{4+} + 2H^+ \qquad K = 10^{-6.3} \qquad (17.40)$$

$$7Al^{3+} + 17H_2O = Al_7(OH)_{17}^{4+} + 17H^+ \qquad K = 10^{-48.8} \qquad (17.41)$$

$$13Al^{3+} + 34H_2O = Al_{13}(OH)_{34}^{5+} + 34H^+ \qquad K = 10^{-97.4}. \qquad (17.42)$$

The solubility of aluminum is greatly increased by its formation of hydroxide complexes as illustrated in Ex. 17.9.

Ex. 17.9 *The concentrations of Al^{3+} and inorganic complexes of aluminum will be calculated for a water-gibbsite system at equilibrium at pH 5, where Al^{3+} concentration is 10^{-6} M (27 μg/L).*

Solution:
Using Eqs. 17.39, 17.40, 17.41, and 17.42, we obtain

$$\left(AlOH^{2+}\right) = \frac{\left(Al^{3+}\right)\left(10^{-5}\right)}{H^+} = \frac{\left(10^{-6}\right)\left(10^{-5}\right)}{10^{-5}} = 10^{-6} \; M \; or \; 27 \; μg/L \; of \; Al^{3+}$$

$$\left[Al_2\left(OH\right)_2^{4+}\right] = \frac{\left(Al^{3+}\right)^2 \left(10^{-6.3}\right)}{\left(H^+\right)^2} = \frac{\left(10^{-6}\right)^2 \left(10^{-6.3}\right)}{\left(10^{-5}\right)^2}$$

$$= 10^{-8.3} \; M \; or \; 0.14 \; μg/L \; of \; Al^{3+}$$

$$\left[Al_7\left(OH\right)_{17}^{4+}\right] = \frac{\left(Al^{3+}\right)^7\left(10^{-48.8}\right)}{\left(H^+\right)^{17}} = \frac{\left(10^{-6}\right)^7\left(10^{-48.8}\right)}{\left(10^{-5}\right)^{17}}$$

$$= 10^{-5.8} \ M \ or \ 42 \ \mu g/L \ of \ Al^{3+}$$

$$\left[Al_{13}\left(OH\right)_{34}^{5+}\right] = \frac{\left(Al^{3+}\right)^{13}\left(10^{-97.4}\right)}{\left(H^+\right)^{34}} = \frac{\left(10^{-6}\right)^{13}\left(10^{-97.4}\right)}{\left(10^{-5}\right)^{34}}$$

$$= 10^{-5.4} \ M \ or \ 107 \ \mu g/L \ of \ Al^{3+}.$$

The aluminum contained in hydrolysis products has a sum of 176.1 μg/L—a 6.52-fold increase over the aluminum ion concentration.

As already mentioned, aluminum hydroxide is amphoteric. It is a base when reacting with hydrogen ion (Eq. 17.38), but it is an acid when forming a soluble aluminum hydrolysis product as shown below:

$$Al(OH)_3 + OH^- = Al(OH)_4^- \qquad K = 10^{1.3}. \qquad (17.43)$$

At pH values above 7 where the basic reaction of gibbsite to release Al^{3+} is essentially nil, aluminum hydroxide can dissolve by reaction with hydroxide to increase aluminum concentration (Ex. 17.10).

Ex. 17.10 *Concentrations of $Al(OH)_4^-$ will be estimated for pH 7, 8, 9, and 10 for a gibbsite-water system at equilibrium.*
Solution:
From Eq. 17.43,

$$\left[Al(OH)_4^-\right] = \left(OH^-\right)\left(10^{1.3}\right).$$

The concentration at pH 7 is

$$\left[Al(OH)_4^-\right] = \left(10^{-7}\right)\left(10^{1.3}\right) = 10^{-5.7} \ M \ or \ 54 \ \mu g/L.$$

Repeating the computation for other pH values, we get: pH 8, 540 μg/L; pH 9, 5.4 mg/L; pH 10, 54 mg/L.

In a simple distilled water-aluminum hydroxide system, dissolved aluminum decreases to a minimum at pH 6.5 and then its concentration rises as pH increases (Fig. 17.4). The $Al(OH)_4^-$ concentration increase from 54 μg/L at pH 7 to 54 mg/L at pH 10. In nature, formation of $Al(OH)_4^-$ depends upon the presence of gibbsite, and there may be little gibbsite in nonacidic sediment, and this will greatly limit the formation of $Al(OH)_4^-$. In spite of its amphoteric property, aluminum seldom has an appreciable concentration in water with pH >5. Of course, $Al(OH)_4^-$ may be important in instances where there is a large input of aluminum from pollution.

Fig. 17.4 Concentrations of aluminum from $Al(OH)_3$ in distilled water at 25 °C and different pH values

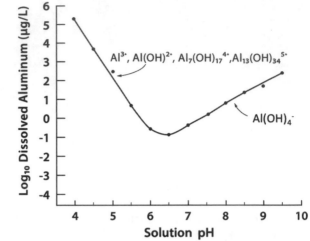

Surface waters in 203 lakes in the Adirondack Mountains in the northeastern United States had an average dissolved aluminum concentration of 40 µg/l, and 168 lakes in Florida (USA) contained an average of 22 µg/L of dissolved aluminum. Unfiltered samples from the lakes had 138 µg/L and 89 µg/L of aluminum (Gensemer and Playle 2010). This reveals that much of the aluminum was in particles suspended in the water. Moreover, much of the dissolved aluminum was contained in hydrolysis products; ion pairs, and organic matter. Stream water in Europe contained 0.1–812 µg/L of aluminum (http://weppi.gtk.fi/publ/foregsatlas/text/Al.pdf). Acidic streams in a mining area in Colorado (USA) contained 77 µg/L–1.38 mg/L of dissolved aluminum (Besser and Leib 2007). It appears accurate to state that most unpolluted, natural waters will contain less than 100 µg/L of aluminum. Ocean water has average aluminum concentration of 10 µg/L where it is present mainly as $Al(OH)_4{}^-$ and $Al(OH)_3$.

Antimony (Sb)

Antimony has properties somewhat like arsenic, and like arsenic, it tends to concentrate in sulfide ores. As a result, it is often found along with copper, lead, gold, and silver. The principle ore minerals are Stibnite (Sb_2S_3) and Jamesonite ($Pb_4FeSb_6S_{14}$). Antimony also occurs in soils at low concentrations. In aerobic waters it is usually in pentavalent, hydrated form $[Sb(OH)_5{}^0]$ or other arrangements with oxygen. It also forms antimonic acid $[Sb(OH)_5{}^0]$ which acts as a monoprotic acid (Accornero et al. 2008) as follows:

$$Sb(OH)_5^0 + H_2O = Sb(OH)_6^- + H^+ \qquad K = 10^{-2.85}. \qquad (17.44)$$

Most natural waters have antimony concentrations <1 µg/L, but in mining areas concentrations may be 100 times or more greater, while ocean water averages 0.33–0.5 µg/L antimony.

Barium (Ba)

Barium occurs in the earth's crust as barite ($BaSO_4$) and witherite ($BaCO_3$). The solubilities of these compounds can be expressed as

$$BaSO_4 = Ba^{2+} + SO_4^{2-} \quad K = 10^{-9.97} \tag{17.45}$$

$$BaCO_3 = Ba^{2+} + CO_3^{2-} \quad K = 10^{-8.56}. \tag{17.46}$$

Based on calculations using K_{sp}, barite and witherite are rather insoluble, and equilibrium Ba^{2+} concentrations would be 5.47 mg/L and 5.34 mg/L, respectively. Witherite solubility is increased by the action of carbon dioxide. Depending upon sulfate and carbonate concentrations in natural waters, rather large amounts of Ba^{2+} and its hydrolysis product $BaOH^+$ could exist in equilibrium with barium sulfate or barium carbonate.

Barium concentrations in natural waters are generally far below equilibrium concentrations. Surface waters and public water supplies in the United States had average barium concentrations of 43 µg/L and 45 µg/L, respectively (Durum and Haffty 1961; Koop and Kroner 1967), but concentrations up to 3 mg/L were found. Seawater contains an average of 30 µg/L barium.

Beryllium (Be)

Beryllium occurs in geological formations as silicates and hydroxyl silicates. Its ores include beryl, penakite, Bertrandite, and several others which contain 6–45% BeO depending upon the ore. Beryllium also is found in soil at very low concentration.

Beryllium hydroxide [$Be(OH)_2$] is very insoluble in water ($K_{sp} = 10^{-21.84}$), but beryllium oxide (BeO) will dissolve when exposed to acid. Because beryllium is rare and its minerals are not highly soluble, concentrations in natural waters are low. River waters in the United States contained 0.01–112 µg/L of beryllium with an average of 9.7 µg/L (Kopp 1969). Korečková-Sysalová (1997) reported that beryllium concentrations usually were 0.01–1 µg/L in the Czech Republic. However, higher concentrations may result from pollution. Seawater has an average of 0.05 µg/L of beryllium.

Bismuth (Bi)

The most common bismuth ores are bismuthinite (Bi_2S_3) and bismite (B_2O_3). This element also occurs in some other minerals and in soils at low concentration. The Bi_2S_3 form is highly insoluble ($K_{sp} = 10^{-97}$), but the oxide form will dissolve in presence of acid:

$$Bi_2CO_3 + 6H^+ = 2Bi^{3+} + 3H_2O. \qquad (17.47)$$

Bismuth concentrations in natural waters are poorly documented. The ionic form is Bi^{3+}, and its concentration range has been given from undetectable to 0.26 µg/L in mining areas (http://weppi.gtk.fi/publ/foregsatlas/text/Bi.pdf). The average bismuth concentration of the ocean is given as 0.02 µg/L.

A review of bismuth concentrations in surface water (Filella 2010) included inland water concentrations of 1–250 µg/L and seawater concentrations of 0.02–0.04 µg/L. The conclusion of the review was that the existing data on dissolved bismuth concentration in surface waters are unreliable because of analytical limitations.

Bromine (Br)

Bromine occurs in soils and waters mostly as bromide (Br^-). The main commercial sources of bromide are oil well brines and Dead Sea brine from which it is extracted (Wisniak 2002).

Bromide is not highly toxic, but it can form trihalomethanes by reaction with dissolved organic matter (Magazinovic et al. 2004). Because it can form trihalomethanes, bromide has been considered a potential risk in drinking water. Of course, bromine (Br_2) is highly toxic and has been used to disinfect drinking water as an alternative to chlorine.

Magazinovic et al. (2004) summarized bromide concentrations reported in inland surface waters: 6–170 µg/L (USA); 9–760 mg/L (Germany); 24–200 µg/L (France); 30–70 µg/L (UK, Spain, and France); 4–76 µg/L (Sweden); 2.0 mg/L (Israel); 30 µg/L-4.3 mg/L (Australia). The higher concentrations were in arid areas, e.g., Israel and some places in Australia. Bromide is a major constituent of seawater at an average concentration of 65–80 mg/L.

Cyanide (CN)

Cyanide is not a trace element. Rather, it is a trace compound containing one or more carbon atoms triple bonded to nitrogen ($-C \equiv N$), and it occurs naturally in many plants including food plants (Jones 1998). Hydrogen cyanide and cyanides are used in plastic and dye production, occur in exhaust from fuel combustion, and are widely used in gold and silver mining and in metallurgical processes. Cyanide ions enter natural waters mainly in wastewater. Cyanide is highly toxic to all forms of life. It is a rapidly acting poison which mainly interferes with cellular respiration.

The forms of cyanide that enters water include $Fe_4[Fe(CN_6)]_3$, $Na_4Fe(CN)_6$, $K_4Fe(CN)_6$, NaCN and KCN. The more complex compounds break down into $Fe(CN)_6^{3-}$ and $Fe(CN)_6^{4-}$, and all forms ultimately result in CN^- (Jaszczak et al. 2017). The CN^- ion can be associated with hydrocyanic acid:

$$HCN = H^+ + CN^- \qquad K = 10^{-9.31}. \qquad (17.48)$$

At pH 9.31 and less, most of the cyanide will be in HCN. Cyanide reacts with trace metals according to the Lewis acid-base theory forming both strong soluble complexes $Fe(CN)_6^{4-}$, $Au(CN)^{2-}$, $Co(CN)_6^{4-}$, and $Ni(CN)_4^{2-}$, and weak soluble complexes, $CdCN^-$, $Ag(CN)_2^-$, $Cu(CN)_3^{2-}$, and $Zn(CN)_4^{2-}$ (Jaszczak et al. 2017). Concentrations of cyanide in surface water ranged from 0.77 to 5.11 µg/L, drinking water contained 0.6 µg/L or less, and wastewater varied from 0.04 to 1.2 µg/L in electroplating waste to 540 mg/L in waste from gold extraction (Jaszczak et al. 2017).

Lead (Pb)

Lead occurs in several minerals as sulfides, sulfates, and carbonates. The principal ores of lead are galena (PbS), cerussite ($PbCO_3$), and anglesite ($PbSO_4$). It also enters the environment through pollution. Soils usually have low concentrations of lead.

Because of the common use of lead and its harmful effects, it has been the focus of considerable environmental and public health concerns. Lead is toxic to plants and animals, and humans may take in lead through contact and in drinking water. It accumulates in the body, and excessive exposure can have several effects, the most serious being adverse effects on the nervous system and retarded mental development in children.

The equilibrium concentration for $PbSO_4$ and $PbCO_3$ in water are 32.8 mg/L and 82 µg/L, respectively. Most other lead compounds are less soluble. Lead concentration in rivers of the world were 1–10 µg/L (Livingstone 1963), and Kopp and Kroner (1967) found a maximum lead concentration of 140 µg/L in rivers of the United States. Hard rock mining and ore processing are major sources of lead pollution. Ning et al. (2011) reported an average concentration of 69 µg/L (range = 1.8–434 µg/L) in surface waters of a mining area in China. Average concentration of lead in the ocean is given as 4 µg/L; it is mostly as the $PbCO_3$ ion pair.

Mercury (Hg)

Mercury is of low abundance in geological formations, but it is found in many minerals and is contained in coal. The common ore is cinnabar (HgS).

Mercury compounds are not appreciably soluble as illustrated with the following K_{sp} values: Hg_2Cl_2, $10^{-25.4}$; HgS, $10^{-51.8}$; $HgCO_3$, $10^{-16.4}$. Total mercury concentrations for baseline streams not receiving mercury pollution in Alaska (USA) were 0.001 µg/L total mercury and 0.0004 µg/L methyl mercury. These concentrations increased 2,000–300,000 times for total mercury and 6–350 times for methyl mercury in mercury mining areas (Wentz et al. 2014). In mercury mining areas in Texas (USA), 1.1–9.7 µg/L total mercury and 0.03–0.61 µg/L of methyl mercury were measured (Gary et al. 2015). Ning et al. (2011) also found high concentrations of total mercury in surface waters of a mining area in China. The average concentration was 61 µg/L with a range of 0.01–827 µg/L. Ocean water has an average mercury concentration of 0.03–0.15 µg/L.

There are several sources of mercury pollution in addition to mercury mining. These include goldmining, combustion of coal, production of nonferrous metals, and cement production. Mercury can enter water bodies in effluents or from atmospheric deposition   (https://www.epa.gov/international-cooperation/mercury-emissions-global-context).

Goldmining—especially small-scale and artisanal goldmining—is the greatest global source of mercury contamination. Liquid mercury is mixed with the crushed ore in a 1:20 weight ratio to form a mercury-gold concentrate. The concentrate and excess mercury are removed, but the extracted ore which still contains some mercury is released into the environment. Additional mercury is used to further concentrate the mercury-gold amalgam, and the amalgam is heated in the open to distill the mercury and isolate the gold. Mercury enters the environment during extraction of the ore, the concentration step, and during vaporization of the mercury from the amalgam. This practice is widespread, but particularly in several South American, Asian, and African countries accounting for about 1400 t of mercury pollution and around 40% of global mercury pollution (Esdaile and Chalker 2018).

Silver (Ag)

Silver is rather rare geologically, but it occasionally is found at greater abundance in deposits of argentite (Ag_2S) and chlorargyrite ($AgCl$). Much silver also is obtained along with lead from the lead ore galena for which the formula often is written as (Pb, Ag, Sb, Cu)S. This formula is not exact, because the proportions of the four metals vary among different sources of galena. Most silver compounds are insoluble with K_{sp} values of 10^{-10} to 10^{-50}, but a few like silver sulfate ($K_{sp} = 10^{-5}$) are more soluble.

The solubility of silver oxide (Ag_2O) is so low as to prevent high concentrations of silver at high pH, and chloride concentration is usually adequate to precipitate silver at low pH and avoid high concentrations. This can be seen from the equations below:

$$\text{High pH} : Ag^+ + 2OH^- = Ag_2O \downarrow \qquad (17.49)$$

$$\text{Low pH} : Ag^+ + 2Cl^- = AgCl \downarrow . \tag{17.50}$$

Monovalent silver (Ag^+) is the common form in water. The mean silver concentration of rivers in the United States was 0.09 µg/L (Durum and Haffty 1961). Flegal et al. (1997) reported that surface water subjected to pollution may contain up to 0.3 µg/L or more. Ocean water contains about 0.05 µg/L silver.

Strontium (Sr)

The main ores of strontium are celestite ($SrSO_4$) and strontite ($SrCO_3$). Strontium also occurs in various minerals, and limestone usually contains about 0.1% of strontium carbonate.

Strontium carbonate is sometimes found in lake sediment. Using this compound as the controlling mineral, the equilibrium concentration would be about 2.2 mg/L according to K_{sp}. Such high concentrations in inland waters have only been reported in a few arid areas, and the strontium concentration averaged 60 µg/L in North American rivers (Hem 1970). The ocean has an average strontium concentration of 8 mg/L.

Thallium (Tl)

Thallium is found in several minerals such as Crookesite [$Cu_7(Tl,Mg)Se_4$], Hutchinsonite ($Tl,Pb)_2As_5S_9$], and in iron pyrite. For commercial purposes, it usually is recovered from byproducts of lead and zinc refining. It is contained in coal and enters the air in gaseous emissions, but elevated environmental concentrations usually result from ore processing. Thallium is highly toxic to animals and humans.

Thallium exists in water primarily as Tl^+, but it forms hydrolysis products and other soluble combined forms. Thallium oxide is rather insoluble and is probably the controlling mineral. Concentrations in surface waters were reported to average 0.01–1 µg/L in uncontaminated surface water and up to 96 µg/L in surface water of mining regions (Frattini 2005). Thallium concentrations in rivers of Poland ranged from 5 to 17 µg/L, a lake in Macedonia contained 0.5 µg/L, and contaminated sites of the Huron and Raisin rivers in Michigan (USA) contained 21 µg/L and 2.62 mg/L, respectively, according to a review by Karbowska (2016). The average thallium concentration in ocean water is around 0.014 µg/L.

Tin (Sn)

The main ores mined for tin extraction are cassiterite which is a high tin oxide (SnO_2) content and tin sulfides such as teallite ($PbSnS_2$). Tin occurs in many other minerals in low concentrations, and most soils contain 2–3 mg/kg of tin (Howe and Watts 2003).

Tin is not considered by most authorities to be an essential nutrient. Nevertheless, many health advocates claim that tin supports hair growth and enhances reflexes. Deficiency symptoms such as certain types of baldness and less hemoglobin synthesis have been claimed.

Tin is not highly soluble, but it is often at detectable concentrations in water. Tin occurs in natural waters as Sn^{2+} and Sn^{4+}, with the Sn^{2+} form favored by low dissolved oxygen concentration. Quadrivalent tin hydrolyzes and precipitates as $Sn(OH)_4$, but divalent tin forms soluble hydroxides [$SnOH^+$, $Sn(OH)_2^0$, $Sn(OH_3)^-$, $Sn_2(OH)_2^{2+}$, and $Sn(OH)_4^{2+}$]. Livingstone (1963) reported that some rivers contained up to 100 µg/L of tin, but most rivers contained <2.5 µg/L. Howe and Watts (2003) reported the following: rivers in Maine (USA) had an average tin concentration 0f 0.03 µg/L; Canadian rivers had tin concentrations of 1–37 µg/L; a river in Turkey usually contained 0.004 µg/L tin, but near its mouth, pollution increased the tin concentration to 0.7 µg/L; Lake Michigan (USA) contained 0.08–0.5 µg/L tin. Ocean water contains on average 3 µg/L tin.

Uranium (U)

Uranium is most commonly associated with nuclear fuels and weapons. There are several types of uranium deposits that provide its ores to include uraninite (UO_2 + UO_3), Brannerite [$U(TiFe)_2O_2$], and Carnotite ($K_2O \cdot 2U_2O_3 \cdot V_2O_5 \cdot 3H_2O$). It also is found in rock phosphate and a few other minerals. Uranium is the heaviest, natural element that occurs in abundance; its atomic weight is 238. Natural uranium has three major isotopes: ^{238}U, ^{235}U, ^{234}U, and ^{238}U is most abundant (99.28%). The isotopes are unstable and emit gamma particles eventually decaying into lead. Uranium exists in several valence states, and the hexavalent form UO_2^{2+} is the most stable in water (https://periodic.lanl.gov/926shtml).

Uranium was once used as a coloring agent in ceramic glazes and glass, but this has been discontinued because of its potential harmful radiation. It now is used mainly as a nuclear fuel and in nuclear weapons. It is highly toxic both chemically and radiologically.

Uranium is leached from geological formations and soils and pollution from ore processing. The chemistry of uranium in aqueous solution, unlike most other trace elements, has been thoroughly investigated because of its importance in nuclear applications and its chemical and radiological hazards (Thoenen and Hummel 2007). Uranium forms many complexes, but in water, soluble hydroxides and carbonates are predominant, e.g., $UO_2(OH)_2^+$, $UO_2(CO_3)_2^{2-}$, $UO_2(CO_3)_3^{4-}$, and $UO_2(CO_3)(OH)_3^-$. The worldwide average for uranium in natural waters is about 0.5 µg/L (range = 0.02–6 µg/L), but extreme values in excess of 20 µg/L have resulted from pollution (IRSN 2012). Examples of extreme uranium concentrations are 61.3 µg/L in a lake and 57.8 µg/L in a river in Russia reported by Shiraishi et al. (1994). Seawater contains around 3 µg/L uranium.

Toxicity of Trace Elements

The toxicity of trace elements in aquatic ecosystems is a major concern. The pH has a great effect on solubility of trace elements. The form of the trace metal in water (free ions versus ion pairs, hydrolysis products or chelated ions) has a great influence on toxicity. Trace metals enter fish mainly through the gills. Calcium and magnesium ions—the sources of hardness in water—interfere with the transport of trace metals across the gill and into the blood of aquatic animals. Trace elements are transported across the gills by an active carrier mechanism, and a high abundance of calcium (and presumably magnesium) in the water compete for absorption sites on the carrier mechanism lessening the amount of trace metal that can enter the fish. As a result, a higher concentration of a trace metal is necessary to cause toxicity in hard water than in soft water (Howarth and Sprague 1978).

One approach to estimating safe concentrations of trace elements (and other toxins) is development of the criterion maximum concentration (CMC) and the criterion continuous concentration (CCC) which are based on models made from existing toxicity data (Mu et al. 2014). The CMC and CCC values accepted are conservative, because they are based on the values for the most sensitive species in the database. Another method for metals is the Biotic Ligand Model (BLM) which predicts effects in waters of different chemical concentrations. The BLM incorporates 10 water quality variables: temperature, pH, dissolved organic carbon, calcium, magnesium, sodium, potassium, sulfate, chloride, and alkalinity to estimate the safe concentration (USEPA 2007). A list of the CMC and CCC concentrations for trace metals are given in Table 17.8.

Possible Health Effects of Trace Elements

Standards for concentrations of trace elements in drinking water are provided in Table 17.9. The World Health Organization and governments in most countries have developed recommended limits on concentrations of trace elements in drinking water. These limits are based upon the accumulation of experience in the public health aspects of water supply, observed effects of trace elements on animals in laboratory studies, and epidemiological studies relating exposure to different concentrations of trace elements in drinking water to health problems.

Table 17.8 Maximum allowable toxicant concentrations for short-term contact (MATC) and the probable no effect concentrations (PNEC) of trace elements to aquatic life

Element	Freshwater (μg/L) MATC	PNEC	Marine water (μg/L) MATC	PNEC
Aluminum[a]	–	125	–	–
Antimony[b]	9,000	610	–	–
Arsenic[c]	340	150	69	36
Barium[b]	Unwarranted		Unwarranted	
Beryllium[c]	130	5.3	–	–
Bismuth	No information found			
Boron[b,d]		750	–	–
Bromide[e]		7,800 μg/L		
Cadmium[c]	1.8	0.72	33	7.9
Chromium III[c]	570	74	–	–
Chromium IV[c]	16	11	1,100	50
Cobalt[f]	–	4.0	–	–
Copper[g]	13	9	4.8	3.1
Cyanide[c]	22	5.2	1.0	1.0
Fluoride[h]	–	500	–	–
Iodide[i]	Possibly toxic to invertebrates		–	–
Iron[c]	–	1,000	–	–
Lead[c]	65	2.5	201	8.1
Manganese[j]	–	600	Unwarranted	
Mercury[c]	1.4	0.77	1.8	0.94
Molybdenum[k]	–	36.1	–	3.85
Nickel[c]	470	52	74	8.2
Selenium (lentic)[c,l]	–	1.5	–	–
(lotic)		3.1	290	71
Silver[c]	3.2	–	1.9	–
Strontium	Apparently no limit warranted			
Tin (tributyl tin)[c]	0.46	0.072	0.42	0.0074
Uranium[m]	–	5	–	–
Vanadium[n]	–	50	–	–
Zinc[c]	120	120	90	81

[a]Cardwell et al. (2018)
[b]USEPA (1986)

[c]USEPA (2018) these are the USEPA criterion maximum concentration (CMC) and criterion continuous concentration (CCC) recommendations (https://www.epa.gov/wqc/national-recommended-water-quality-criteria-aquatic-life-criteria-table)

[d]Boron standard is only related to water use for irrigation

[e]Canton et al. (1983)

[f]Nagpal (2004)

[g]USEPA (2004)

[h]Camargo (2003)

[i]Laveroch et al. (1995)

[j]Reimer (1988)

[k]Heijerick and Carey (2017)

[l]USEPA (2016)

[m]Sheppard et al. (2005)

[n]Schiffer and Karsten (2017)

Table 17.9 Acceptable limits for trace elements in drinking water

Trace element	WHO guideline[a]	USEPA standard[b]	Possible health effects of elevated concentrations in drinking water
Aluminum	None	0.05–0.2 mg/L	Possible increase in risk of Alzheimer's disease
Antimony	20 μg/L	6 μg/L	Vomiting; stomach ulcers
Arsenic	10 μg/L	10 μg/L	Skin lesions; circulatory problems; greater cancer risk
Barium	1.3 mg/L	2 mg/L	Increased blood pressure
Beryllium	12 μg/L	4 μg/L	Intestinal lesions
Bismuth	None	None	Kidney and liver damage
Boron	2.4 mg/L	None	Possible testicular lesions
Bromide	None	None	Possible cancer risk by trihalomethanes
Bromate	None	0.01 μg/L	No consensus of effects
Cadmium	3 μg/L	0.005 μg/L	Kidney damage
Chromium	50 μg/L	100 μg/L	Allergic dermatitis, increased cancer risk
Copper	2 mg/L	1 mg/L	Liver and kidney damage
Cyanide	None	0.2 mg/L	Thyroid problems
Fluoride	1.5 mg/L	4 mg/L	Bone problems; mottled teeth
Iron	None	0.3 mg/L[c]	No health effects, but taste, odor, stains
Lead	70 μg/L	0.0 μg/L	Developmental problems in children; kidney problems; high blood pressure
Manganese	None	None	No health effects, but taste, odor, stains
Mercury	6 μg/L	2 μg/L	Impaired neurological development in fetuses, infants, and children; neurological issues in adults; kidney damage
Molybdenum	None	None	Diarrhea; liver and kidney lesions
Nickel	70 μg/L	None	Dermatitis; intestinal upset; increased red blood cell counts and protein in urine
Selenium	40 μg/L	50 μg/L	Nail and hair loss; circulating problems
Silver	None	0.1 μg/L[c]	Possible change in skin color (argyria)
Strontium	None	None	Impairs bone and tooth development in children
Thallium	None	2 μg/L	Enzyme disruption; stomach and intestinal ulcers; neurological problems
Tin	None	None	No reported problems found
Uranium	30 μg/L	0.03 μg/L	Kidney damage; increased cancer risk
Zinc	None	5 mg/L	Anemia, pancreatic damage

World Health Organization (WHO) guidelines and United States Environmental Protection Agency (USEPA) guidelines primary standards and secondary regulations
[a]https://apps.who.int/iris/bitstream/10665/254637/1/9789241549950-eng.pdf
[b]https://www.epa.gov/sites/production/files/2018-03/documents/dwtable2018.pdf
[c]Secondary regulations. These are non-mandatory and intended as guidelines for water supply system management

References

Accornero M, Marini L, Lelli M (2008) The dissociation constant of antimonic acid at 10-40°C. J Solution Chem 37:785–800

Anderson RA (1997) Chromium as an essential nutrient for humans. Regul Toxicol Pharmacol 26:535–541

Arnon DI, Wessel G (1953) Vanadium as an essential element in green plants. Nature 172:1039–1040

Baralkiewicz D, Siepak J (1999) Chromium, nickel, and cobalt in environmental samples and existing legal norms. Pol J Environ Stud 8:201–208

Besser JM, Leib KJ (2007) Toxicity of metals in water and sediment to aquatic biota. In: Church SE, von Guerard P, Finger SE (eds) Integrated investigations of environmental effects of historical mining in the Animas River Watershed, San Juan County, Colorado. U.S. Geological Survey, Washington, p. 839–849

Boyd CE, Walley WW (1972) Studies of the biogeochemistry of boron. I. Concentrations in surface waters, rainfall, and aquatic plants. Am Midl Nat 88(1):1–14

British Geological Survey (2000) Iodine. Water quality fact sheet, London

Camargo JA (2003) Fluoride toxicity to aquatic organisms: a review. Chemosphere 50:251–264

Canton JH, Webster PW, Mathijssen-Speikman EA (1983) Study on the toxicity of sodium bromide to different freshwater organisms. Food Chem Toxicol 21:369–378

Cardwell AS, Adams WJ, Gensemer RW, Nordheim E, Santore RC, Ryan AC, Stubblefield WA (2018) Chronic toxicity of aluminum, at pH 6, to freshwater organisms: empirical data for the development of international regulatory standards/criteria. Environ Toxicol Chem 37:36–48

Chowdhury UK, Biswas BK, Chowdhury TR, Samanta G, Mandal BK, Basu GC, Cahnda CR, Lodh D, Saha KC, Murkherfee SK, Roy S, Kalir S, Quamruzzaman Q, Chakraborti D (2000) Groundwater arsenic contamination in Bangladesh and West Bengal, India. Environ Health Perspect 108:393–397

Ćosović B, Degobbis D, Bilinski H, Branica M (1982) Inorganic cobalt species in seawater. Geochim Cosmochim Acta 46:151–158

Deverel SJ, Goldberg S, Fujii R (2012) Chemistry of trace elements in soils and groundwater. In: Wallender WW, Tanji KK (eds) ASCE manual and reports on Engineering practice No 71 Agricultural salinity assessment and management, 2nd edn. ASCE, Reston, pp 89–137

Durum WH, Haffty J (1961) Occurrence of minor elements in water. United States Geological Survey Circular 445, United States Government Printing Office, Washington

Esdaile LJ, Chalker JM (2018) The mercury problem in artisanal and small-scale gold mining. Chem Eur J 24:6905–6916

Essumang DK (2009) Levels of cobalt and silver in water sources in a mining area in Ghana. Int J Biol Chem Sci 3:1437–1444

Filella M (2010) How reliable are environmental data on "orphan elements?" The case of bismuth concentrations in surface waters. J Environ Monit 12:90–109

Firth J (2013) Arsenic—the 'poison of kings' and the 'saviour of syphilis'. J Mil Vet Health 21:11–17

Flegal AR, Patterson CC (1985) Concentrations of thallium in seawater. Mar Chem 15:327–331

Flegal AR, Rivera-Durarte SA, Sanudo-Wilhelmy SA (1997) Silver contamination in aquatic environments. Rev Environ Contam Toxicol 148:139–162

Frattini P (2005) Thallium properties and behaviour—a literature study. Geological survey of Finland. http://tupa.gtk.fi/raportti/arkisto/s41_0000_2005_2.pdf

Fries L (1982) Vanadium an essential element for some marine macroalgae. Planta 154:393–396

Gaillardet J, Viers J, Duprèe B (2003) Trace elements in river waters. In: Turekian K, Holland H (eds) Treatise on geochemistry. Elsevier, Amsterdam, pp 5–9

Gary JE, Theodorakos PM, Fey DL, Krabbenhoft DP (2015) Mercury concentrations and distribution in soil, water, mine waste leachates, and air in and around mercury mines in the Big Bend region, Texas, USA. Environ Geochem Health 37:35–48

Gensemer RW, Playle RC (2010) The bioavailability and toxicity of aluminum in aquatic environments. Crit Rev Environ Sci Tech 29:315–450

Goldberg ED (1963) The oceans as a chemical system. In: Hill MN (ed) Composition of sea water, comparative and descriptive oceanography, Vol II. The sea. Wiley, New York

Goldman CR (1972) The role of minor nutrients in limiting the productivity of aquatic ecosystems. In Likens GE (ed) Nutrients and eutrophication: the limiting-nutrients controversy. Lim Ocean Spec Sym 1:21–33

Guo T, Delaune RD, Patrick WH (1997) The effect of sediment redox chemistry on solubility/ chemically active forms of selected metals in bottom sediment receiving produced water discharge. Spill Sci Tech Bull 4:165–175

Hamilton SJ (2004) Review of selenium toxicity in the aquatic food chain. Sci Total Environ 326:1–31

Heijerick DG, Carey S (2017) The toxicity of molybdate to freshwater and marine organisms. III. Generating additional chronic toxicity data for the refinement of safe environmental exposure concentrations in the US and Europe. Sci Total Environ 609:420–428

Hem JD (1970) Study and interpretation of the chemical characteristics of natural water. Water-supply paper 1473, United States Geological Survey, United States Government Printing Office, Washington

Hem JD (1985) Study and interpretation of the chemical characteristics of natural water. Water-supply paper 2254, United States Geological Survey, United States Government Printing Office, Washington

Hem JD, Roberson CE (1967) Form and stability of aluminum hydroxide complexes in dilute solution. Water-supply paper 1827-A, United States Geological Survey, United States Government Printing Office, Washington

Howarth RS, Sprague JB (1978) Copper lethality to rainbow trout in waters of various hardness and pH. Water Res 12:455–462

Howe P, Watts P (2003) Tin and inorganic tin compounds. Concise International Chemical Assessment Document 65, World Health Organization, Geneva

Hyenstrand P, Rydin E, Gunnerhed M (2000) Response of pelagic cyanobacteria to iron additions—enclosure experiments from Lake Erken. J Plankton Res 22:1113–1126

IRSN (Institut de Radioprotection et de Sûretè Nuclèaire) (2012) Natural uranium in the environment. https://www.irsn.fr/EN/Research/publications-documentation/radionuclides-sheets/environment/Pages/Natural-uranium-environment.aspx

Izbicki JA, Ball JW, Bullen TD, Sutley SJ (2008) Chromium, chromium isotopes and selected trace elements, western Mojave Desert, USA. Appl Geochem 23:1325–1352

Jaszczak E, Palkowska Z, Narkowicz S, Namieśnik J (2017) Cyanides in the environment— analysis—problems—challenges. Environ Sci Pollut Res 24:15929–15948

Jones DA (1998) Why are so many plant foods cyanogenic? Phytochemistry 47:155–162

Karbowska B (2016) Presence of thallium in the environment: sources of contaminations, distribution, and monitoring methods. Environ Monit Assess 188:640

Khan T, Mohammad S, Khan B, Khan H (2011) Investigating the levels of heavy metals in surface water of Shah Alam River (a tributary of River Kabul, Khyber Pakhtunkhwa). Asian J Earth Sci 44:71–79

Kochkodan V, Darwish NB, Hilal N (2015) The chemistry of boron in water. In: Kabay N, Hilal N, Bryak M (eds) Boron separation processes. Elsevier, The Netherlands, pp 35–62

Kopp JF (1969) The occurrence of trace elements in water. In: Hemphill DD (ed) Proceedings of the Third Annual Conference on Trace Substances in Environmental Health. University of Missouri, Columbia, pp 59–79

Kopp JF, Kroner RC (1967) Trace metals in waters of the United States. A five year summary of trace metals in rivers and lakes of the United States (October 1, 1962 to September 30, 1967). United States Department of the Interior, Federal Water Pollution Control Administration, Cincinnati

Kopp JF, Kroner RC (1970) Trace metals in waters of the United States. Report PB-215680. Federal Water Pollution Control Administration, Cincinnati

Korečková-Sysalová J (1997) Determination of beryllium in natural waters using atomic absorption spectrometry with tantalum-coated graphite tube. Int J Environ Anal Chem 68:397–404

Lane TW, Morel FMM (2000) A biological function for cadmium in marine diatoms. Proc Natl Acad Sci 97:4627–4631

Laveroch MJ, Stephenson M, Macdonald CR (1995) Toxicity of iodine, iodide, and iodate to *Daphnia magna* and rainbow trout (*Oncorhynchus mykiss*). Arch Environ Con Toxicol 29 (3):344–350

Lee JG, Roberts SB, Morel FMM (1995) Cadmium: a nutrient for the marine diatom *Thelassiosira weissflogii*. Limnol Oceanogr 40:1056–1063

Lindemann MD, Cho JH, Wang MQ (2009) Chromium—an essential mineral. Rev Colom de Cien Pec 22:339–445

Livingstone DA (1963) Chemical composition of rivers and lakes. Professional Paper 440-G, United States Geological Survey, United States Government Printing Office, Washington

Magazinovic RS, Nicholson BC, Mulcahy DE, Davey DE (2004) Bromide levels in natural waters: its relationship to chloride and total dissolved solids and the implications for water treatment. Chemosphere 57:329–335

McBride MB (1989) Reactions controlling heavy metal solubility in soils. In: Stewart BA (ed) Advances in soil science. Springer, New York, pp 1–56

McNevin AA, Boyd CE (2004) Copper concentrations in channel catfish, *Ictalurus punctatus*, ponds treated with copper sulfate. J World Aquacult Soc 35:16–24

Moore GT, Kellerman KF (1905) Copper as an algicide and disinfectant in water supplies. Bull Bur Ind 76:19–55

Mu Y, Wu F, Chen C, Liu Y, Zhao X, Liao H, Giesy JP (2014) Predicting criteria continuous exposure concentrations of 34 metals or metalloids by use of quantitative ion character-activity relationships-species sensitivity distributions (QICAR-SSD) model. Environ Pollut 188:50–55

Nadis S (1998) Fertilizing the sea. Sci Am 177:33

Nagpal NK (2004) Technical report-water quality guidelines for cobalt. Ministry of Water, Land, and Air Protection, Victoria

Ning L, Liyuan Y, Jirui D, Xugui P (2011) Heavy metal pollution in surface water of Linglong gold mining area, China. Procedia Environ Sci 10:914–917

Pagenkopf GK (1978) Introduction to natural water chemistry. Marcel Dekker, Inc., New York

Pais I, Jones JB Jr (1997) The handbook of trace elements. Saint Lucie Press, Boca Raton

Palmer CA, Gilbert JA (2012) Position of the Academy of Nutrition and Dietetics: the impact of fluoride on health. J Acad Nutr Diet 112:1443–1453

Pinsino A, Matranga V, Roccheri MC (2012) Manganese: a new emerging contaminant in the environment. In: Srivastava J (ed) Environmental contamination. InTech Europe, Rijeka, pp 17–36

Puntoriero ML, Volpedo AJ, Fernandez-Cirelli A (2014) Arsenic, fluoride, and vanadium in surface water (Chasicó Lake, Argentina). Front Environ Sci 2:1–5

Reimer PS (1988) Environmental effects of manganese and proposed freshwater guidelines to protect aquatic life in British Columbia. Thesis, University of British Columbia

Ryan D (1992) Minor elements in seawater. In: Millero FJ (ed) Chemical oceanography. CRC Press, Boca Raton, pp 89–119

Schiffer S, Karsten L (2017) Estimation of vanadium water quality benchmarks for protection of aquatic life with reference to the Athabasca Oil Sands region using species sensitivity distributions. Environ Toxicol Chem 36:3034–3044

Seker S, Kutler B (2014) Determination of copper (Cu) levels for rivers in Tunceli, Turkey. World Environ 4:168–171

Sheppard SC, Sheppard NI, Gallerand MO, Sanipelli B (2005) Deviation of ecotoxicity thresholds for uranium. J Environ Radioact 79:55–83

Sillén LG, Martell AE (1964) Stability constants for metal-ion complexes. Special Publication 17, Chemical Society, London

Sillén LG, Martell AE (1971) Stability constants of metal-ion complexes. Special Publication 25, Chemical Society, London

Shiraishi K, Igarashi Y, Yamamoto M, Nakajima T, Los IP, Zelensky AV, Buzinny MZ (1994) Concentrations of thorium and uranium in freshwater samples collected in the former USSR. J Radioanal Nucl Chem 185:157–165

Spears JW (1984) Nickel as a "newer trace element" in the nutrition of domestic animals. J Anim Sci 59:823–835

Stralberg E, Varskog ATS, Raaum A, Varskog P (2003) Naturally occurring radio-nuclides in the marine environment—an overview by current knowledge with emphasis on the North Sea area. Norse Decom AS, Kjeller, Norway

Thoenen T, Hummel (2007) The PSI/Nogra chemical thermodynamic database (Update of the Nagra/PSI TDB 01/01: data selection for uranium. Paul Scherrer Institute

Turekian KK (1968) Oceans. Prentice-Hall, Englewood Cliffs

USEPA (1986) Quality criteria for water. EPA 440/S-86-001. USEPA Office of Water, Washington

USEPA (2004) National recommended water quality criteria. USEPA Office of Water, Washington

USEPA (2007) Aquatic life ambient freshwater quality criteria: copper, EPA 822-R-07-001. http://www.epa.gov/waterscience/criteria/copper/index.htm

USEPA (2016) Aquatic life ambient water quality criterion for selenium—freshwater. USEPA Office of Water, Washington

USEPA (2018) Aquatic life criteria and methods for toxics. https://www.epa.gov/wqc/aquatic-life-criteria-and-methods-toxics

Uthus EO (1992) Evidence for arsenic essentiality. Environ Geochem Health 14:55–58

Vrede T, Tranvik LJ (2006) Ion constraints on planktonic primary production in oligotrophic lakes. Ecosystems 9:1094–1105

Wentz DA, Brigham ME, Chasar LC, Lutz MA, Krabbenholf DP (2014) Mercury in the nation's steams—levels, trends, and implications. Circular 1395, US Geological Survey, Washington

Wisniak J (2002) The history of bromine from discovery to commodity. Indian J Chem Technol 9:262–271

World Health Organization (2011) Selenium in drinking water. WHO/HSE/WSH/10.01/14

Water Quality Protection

18

Abstract

Impaired water quality can result from natural causes, but the most common cause is anthropogenic pollution. Soil erosion leads to turbidity and sedimentation in water bodies. Organic wastes impart a high oxygen demand often culminating in low dissolved oxygen concentrations and nitrogen and phosphorus in effluents cause eutrophication. Pesticides, synthetic organic chemicals and heavy metals from industry, and pharmaceutical compounds and their degradation products can be toxic to aquatic animals or have other adverse effects on them. Toxins in drinking water can lead to several serious illnesses to include cancer in humans. Water bodies also may be contaminated with biological agents that cause aquatic animal and human diseases. Elevated sulfur dioxide and carbon dioxide concentrations in the atmosphere as a result of air pollution can influence water quality. Wetland destruction must be considered in a discussion of water pollution, because functional wetlands are important for natural water purification. Water quality regulations are important for avoiding conflicts among water users, minimizing public health risks of certain chemical and biological pollutants, protecting the environment, and preventing conditions that lessen the recreational and aesthetic value of water bodies. Most countries have developed water quality regulations with which effluents must comply. These permits typically have limits on concentrations of pollutants in effluents, and many times, there are limits on quantities of pollutants that may be discharged. There is a growing tendency to develop total maximum daily pollutant loads (TMDL) that specify the total quantities of selected pollutants that can be discharged into a stream or other water body by all permit holders. Standards with which municipal water supply operations must comply also have been developed to protect drinking water quality and protect public health.

© Springer Nature Switzerland AG 2020
C. E. Boyd, *Water Quality*, https://doi.org/10.1007/978-3-030-23335-8_18

Introduction

From the beginning, human population grew slowly for several millennia, few areas were heavily populated, and natural ecosystems were capable of sustainably supplying resources and services to support society. The demand placed on ecosystems by humans did not cause significant damage to overall ecosystem structure and function other than in some highly populated areas.

As people learned how to exert a degree of control over their environment and to produce food by agriculture, the population gradually grew and spread over the land. The population was around 30 million by 2000 BC, and it reached 400–500 million by 1500 AD. Growing knowledge of science and technology led to the industrial revolution which began in Europe in the mid-1700s, spread quickly to North America, and eventually to most of the world.

Since the industrial revolution began, food supply, housing, health care, and other conditions improved allowing a greater annual ratio of births to deaths. The result was a population explosion with exponential growth since the mid-1800s (Fig. 18.1). The expanding human population has placed a huge demand on the world's ecosystems for water, food and fiber, and other resources as well as taxing their waste assimilation capacity. One of the major impacts of the growing human population has been to increase pollution loads that have often caused the quality of aquatic ecosystems and water supplies to deteriorate.

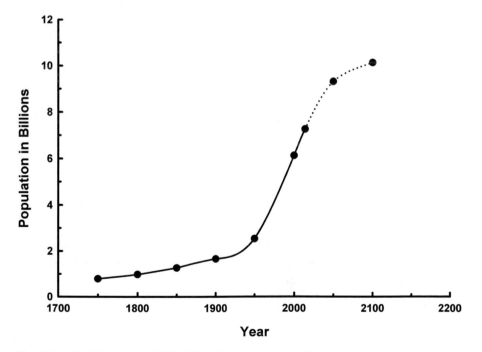

Fig. 18.1 World population 1750–2014 with projection to 2100

The major sources of pollution and their effects on natural aquatic ecosystems and water use by humans will be discussed in this chapter.

Types of Water Pollution

Discharges that cause water pollution are separated into two broad categories: point sources and nonpoint sources. Well-defined effluent streams discharged in both wet and dry weather via pipes, channels, or other conduits are point sources. Common point sources of pollution are industrial operations and municipal wastewater treatment plants. Urban and suburban runoff from streets, parking lots, lawns, etc., enters storm sewers, ditches, and similar conduits that discharge untreated into water bodies. Runoff from farmland, construction sites, etc., also enters water bodies by overland flow and atmospheric deposition falls directly into water bodies. Storm sewer flow, overland runoff, and atmospheric deposition are nonpoint source discharges.

There are several types of pollutants. Organic waste impose an oxygen demand when microorganisms decompose it. Organic matter is a major contaminant in domestic and municipal wastewater, animal feedlot effluents, and discharges from food processing and paper manufacturing.

Turbidity from suspended solids is unsightly and interferes with light penetration and aquatic plant growth. Solids that settle from water create sediment deposits that may suffocate benthic organisms. Sediment also reduces water depth, and shallow water favors growth of rooted aquatic macrophytes. The oxygen demand of sediment with a high organic matter content can cause anaerobic conditions in shallow areas. Suspended solids in municipal, industrial, and feedlot effluents tend to be highly organic, while effluents from agricultural land, logging operations, construction sites, and surface mining have a high proportion of inorganic, suspended solids.

Nutrient pollution results primarily from nitrogen and phosphorus in runoff and effluents. Municipal wastewater and other effluents with high concentrations of organic matter also tend to have large concentrations of nitrogen and phosphorus. Runoff from residential lawns and from cropland and pastureland also contain elevated concentrations of nitrogen and phosphorus from fertilizers and contributes to eutrophication.

Chemicals used for domestic, industrial, and agricultural purposes find their way into water bodies from normal operations or in leaks from storage depots or seepage from waste disposal sites. Toxic substances may be directly harmful to aquatic life or they may cause toxicity through bioconcentration and be toxic to organisms in the food chain. Bioconcentration also presents a potential food safety hazard for consumers of aquatic products. Toxins also can be present in domestic drinking water or in waters used for agricultural purposes.

Runoff from surface mining and seepage from underground mines are well known sources of acidification in natural waters. Combustion of fossil fuels contaminates the air with sulfur dioxide, nitrous oxide, and other compounds that oxidize to form mineral acids. Rainfall in heavily populated or industrialized areas is a major source of acidification. Nitrification also can be a significant source of acidity in surface waters. On the other hand, some effluents may be alkaline and cause an excessive pH in receiving waters.

Many natural waters contain residues and degradation products of pharmaceutical chemicals. Pharmaceuticals often enter water by home disposal of unwanted medicine and other health products into the sewer system. Some are actually toxic to aquatic life, while others act more subtly by having negative physiological or genetic effects upon prolonged exposure. Pharmaceutical residues also can enter the water supply for humans.

Contamination of waters with disease organisms of human origin is still a major concern in many developing nations. If human fecal material enters water, the risk of disease spread through drinking water is greatly increased.

Many industrial processes generate waste heat that may be disposed of by transfer to water. Heated effluents may raise temperatures in streams or other water bodies to cause serious ecological perturbations. Desalination of seawater necessary to supplement water supply in some countries also causes pollution. The discharge water from reverse osmosis plants is of higher salinity than coastal waters. Distillation plants discharge thermally polluted cooling water, and metals from heat exchangers enter the cooling water.

Accidents may result in spills of potentially toxic chemicals or other substances. A highway or rail accident can result in a cargo being inadvertently spilled into a watercourse, or a ship accident can spill crude oil or other substances into the ocean or into coastal and inland waters. The most famous cases of crude oil pollution probably are the Exxon Valdez oil spill that occurred when an Exxon tanker struck a reef in Prince William Sound, Alaska in 1989, and the BP Deepwater Horizon oil spill that resulted from an accident on an off-shore oil drilling platform in the Gulf of Mexico in 2010.

Water quality can be impaired through natural processes without human intervention. In some coastal areas, soils contain iron pyrite that oxidizes in dry weather to produce sulfuric acid which leaches in rainy weather to cause acidification of surface water. Some groundwaters may be unfit for domestic or other uses because of high iron or manganese concentrations. The inhabitants of areas in Bangladesh and India are at risk of poisoning by naturally-occurring arsenic in groundwater (see Chap. 17). Salinization has made freshwaters in some areas too salty for domestic and agricultural purposes.

Considerable attention is given to amounts of suspended solids, biochemical oxygen demand, nitrogen, phosphorus, and dissolved metals in effluents. In the United States, about 50% of the total point source amounts of these five pollutants are from the 10 industries listed in Table 18.1. However, point source pollution contributes less than 0.2% of total suspended solids, 5% of phosphorus, 10% of nitrogen, and 20% of oxygen demand. Nonpoint source discharges are the most serious pollution threats to water bodies. Agriculture contributes more than 50% of nonpoint source pollution.

Table 18.1 Major sources of point source pollution

Source	Major pollutants				
	TSS	BOD$_5$	N	P	DM
Municipal sewage plants	+	+	+	+	+
Power plants	+				+
Pulp and paper mills	+	+			
Feedlots	+	+	+	+	
Metallurgical industries	+	+			
Organic chemical production	+	+			
Food and beverage industry	+	+	+	+	
Textile production	+	+			
Mining	+				
Seafood processing	+	+	+	+	

TSS = total suspended solids, BOD$_5$ = 5-day biochemical oxygen demand, N = nitrogen, P = phosphorus, DM = dissolved metals

Erosion

Soil particles contribute the largest quantity of suspended solids entering most water bodies, and their main source is soil erosion. Falling raindrops dislodge soil particles, and the energy of flowing water further erodes the land surface and keeps particles in suspension during transport. Factors opposing erosion are the resistance of soil to dispersion and movement, slow moving runoff because of gentle slope, vegetation that intercepts rainfall, vegetative cover to shield soil from direct raindrop impacts, roots to hold the soil in place, and organic litter from vegetation to protect the soil from direct contact with flowing water.

Erosion usually is considered to be one of three types: raindrop erosion, sheet erosion, or gully erosion. Raindrop erosion dislodges soil particles and splashes them into the air. Usually, the dislodged particles are splashed into the air many times, and because they are separated from the soil mass, they are readily transported in runoff. Sheet erosion refers to the removal of a thin layer of soil from the surface of gently sloping land. True sheet erosion does not occur, but flowing water erodes many tiny rills in surface soil to cause more or less uniform erosion of the land surface. These rills are not seen in cultivated fields because they are removed by tillage. Gully erosion produces much larger channels than rills and these channels are visible on the landscape.

The universal soil loss equation (USLE) is used widely to estimate soil loss by erosion. The initial efforts to predict soil erosion by mathematical procedures of Zingg (1940) and Smith (1941) led to further research on the topic, and the first complete version of the USLE was published in 1965 (Wischeier and Smith 1965). The equation has been slightly revised over time, and the present form is

$$A = (R)(K)(LS)(C)(P) \tag{18.1}$$

where A = soil loss, R = a rainfall and runoff factor, K = soil erodibility factor, LS = slope factor (length and steepness), C = crop and cover management factor, and P = conservation practice factor. The instructions and tabular and graphical material for obtaining the necessary factors for solving the USLE are too lengthy to include here, but there are many online sources including calculators for solving the equation.

Some typical soil loss rates for different land uses are listed (Table 18.2). Land surface disruption facilitates erosion; construction, logging and mining sites typically have very high rates of soil loss. Deforestation is a major concern both because of reduction in forest area and because of the serious erosion that follows. Row cropland also has a high erosion potential. The lowest rates of erosion are for watersheds that are forested or completely covered with grass. Erosion of streambeds and shorelines also can be important sources of suspended solids in water bodies.

A portion of the soil particles dislodged from watersheds by erosion remain suspended in runoff when it enters streams and other water bodies. Suspended solids in waters create turbidity making the water less appealing to the eye, and less enjoyable for watersports. Turbidity also reduces light penetration into the water, and diminishes primary productivity. Moreover, suspended solids often must be removed from water to allow its use for human and industrial water supply adding to the cost of water treatment.

When turbulence in water carrying suspended solids is reduced, sedimentation occurs. Sediment creates deposits of coarse particles in areas where turbid water enters water bodies and finer particles over the entire bottom. Elevated sedimentation rates have several undesirable consequences. They make water bodies shallower, and this may lead to greater growth of rooted aquatic macrophytes. Shallower water bodies have less volume, and this may have negative ecological effects as well as reducing the volume of water that can be stored for flood control or human uses. Sediment also destroys breeding areas for fish and other species, and it can smother fish eggs and benthic communities.

Erosion and sedimentation are, of course, natural processes that have been operating since the earth was created. The morphology of the earth's surface is the result of millennia of erosion and sedimentation and other geological processes. However, natural processes tend to operate slowly allowing living organisms time to adapt. The problem today is that rates of erosion and sedimentation have been greatly accelerated by human activities, and many negative impacts are resulting.

Table 18.2 Typical amounts of soil loss for different land uses (USEPA 1973 and Magleby et al. 1995)

Land use	Soil loss by erosion (t/ha/yr)
Forest	1.8
Pasture, range, and grassland	2.5
Cropland	7.5
Forest logging areas	17.0
Construction sites	68.0

Biological Oxygen Demand

Bacteria and other saprophytes in aquatic ecosystems remove dissolved oxygen for use in decomposing organic matter. The effect of addition of organic matter in pollutants on dissolved oxygen concentration depends upon the capacity of a water body to assimilate organic matter relative to the amount of organic matter introduced. A given organic matter load might not influence dissolved oxygen concentrations in a large body of water, but the same load might cause oxygen depletion in a smaller body of water. A rapidly flowing stream reaerates more rapidly than a sluggish stream of the same cross-sectional area, and therefore can assimilate a greater organic matter input than a sluggish stream. The oxygen demand of wastewater is usually estimated as the biochemical oxygen demand (BOD).

The standard 5-day BOD (BOD_5) determination provides an estimate of the pollutional strength of wastewaters (Eaton et al. 2005). In the BOD procedure, an aliquot of wastewater is typically diluted with inorganic nutrient solution and a bacterial seed added. The inorganic nutrients and bacterial seed are necessary to prevent a shortage of bacteria and inorganic nutrients that might result from dilution. Because of the possibility of oxygen demand from the bacterial seed, a blank consisting of the same quantity of bacterial seed used in the sample is introduced into nutrient solution and carried through the same incubation as the sample. Samples are incubated in the dark to prevent photosynthetic oxygen production. The incubation is continued in the dark for 5 days at 20 °C. At the beginning and end of incubation, the dissolved oxygen concentration is measured in blank and sample to permit estimation of BOD_5 as illustrated in Ex. 18.1.

Ex. 18.1 In a BOD_5 analysis, the sample is diluted 20 times. The initial dissolved oxygen concentration is 9.01 mg/L in sample and blank. After 5 days of incubation, the dissolved oxygen concentration is 8.80 mg/L in the blank and 4.25 mg/L in the sample. The BOD will be calculated.

Solution:
The oxygen loss caused by the bacterial seed is the blank BOD,

$$Blank\ BOD = Initial\ DO - Blank\ DO$$

$$or\ (9.01 - 8.80)mg/L = 0.21\ mg/L.$$

The oxygen consumption by the sample is

$$(Initial\ DO - Final\ DO) - Blank\ BOD$$

$$or\ (9.01 - 4.25) - 0.21 = 4.55\ mg/L.$$

The sample BOD is the oxygen consumption by the sample multiplied by a correction factor equal to the number of times the sample was diluted—20 times in this case.

$$BOD = 4.55 \times 20 = 91 \; mg \; L.$$

A formula for estimating BOD is

$$BOD \; (mg/L) = (I_{DO} - F_{DO})_s - (I_{DO} - F_{DO})_b \times D \qquad (18.2)$$

where I_{DO} and F_{DO} = initial and final DO concentrations in sample bottle and blank bottle, respectively, and subscript s = sample, subscript b = blank, and D = the dilution factor.

The BOD of a sample represents the amount of dissolved oxygen that will be used up in decomposing the readily-oxidizable organic matter. In samples with a lot of phytoplankton, a large portion of the BOD will represent phytoplankton respiration. By knowing the volume of an effluent and its BOD concentration, the oxygen demand of the effluent can be estimated by multiplying the BOD (mg/L or g/m^3) by the daily input of effluent (m^3/day).

Organic matter in a sample does not decompose completely in 5 days as shown in Fig. 18.2, and it would require many years for complete degradation of all the organic matter. The rate of oxygen loss from a sample (expression of BOD) usually is exceedingly slow after 30 days, and the BOD$_{30}$ is a good indicator of the ultimate BOD (BOD$_u$) of a sample.

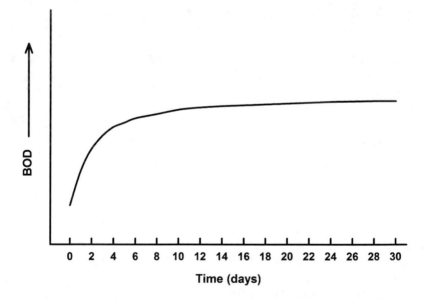

Fig. 18.2 A typical expression of carbonaceous biochemical oxygen demand (BOD) over a 30-day period

Many wastewaters contain appreciable ammonia nitrogen. The oxidation of ammonia to nitrate by bacteria (nitrification) consumes two moles of oxygen for each mole of ammonia nitrogen (Chap. 13) contributing an oxygen demand of 4.57 mg/L for each 1 mg/L of ammonia N. The oxygen demand of ammonia nitrogen in a sample can be estimated from the total ammonia nitrogen concentration by multiplying this concentration by the factor 4.57.

In samples that are diluted several fold for BOD analysis, the abundance of nitrifying organisms is greatly diluted, and it takes more than 5 days for the nitrifiers to build up a population great enough to cause significant nitrification. The typical influence of nitrification on BOD in a highly diluted sample is compared to that of a sample that is not diluted or only diluted a few times (Fig. 18.3). The BOD resulting exclusively from organic matter decomposition (carbonaceous BOD), can be determined by adding a nitrification inhibitor such as 2-chloro-6-(trichloromethyl) pyridine (TCMP) to the sample. If it is desired to determine both carbonaceous BOD and nitrogenous BOD, one portion of the sample is treated with nitrification inhibitor and another portion is not. The oxygen demand of ammonia N is obtained by subtracting the results of the nitrification-inhibited portion from the uninhibited one.

The BOD of natural waters usually is in the range of 1–10 mg/L. Effluents from municipal and industrial sources have much greater BOD (Table 18.3). Relatively small daily inputs of some of these effluents can impose a large oxygen demand on receiving water bodies.

The typical response of streams to BOD loads is a dissolved oxygen concentration sag (in extreme cases, dissolved oxygen depletion) downstream from the effluent outfall (Fig. 18.4). The distance downstream before dissolved oxygen concentration returns to normal depends upon the amount of BOD added and the rate of stream reaeration (see Chap. 7). The rate of change of the oxygen deficit with time at a location in a stream is equal to the rate of deoxygenation caused by the BOD load minus the rate of stream reaeration (Vesilind et al. 1994). Mathematical models based on this concept are used to predict dissolved oxygen concentrations at

Fig. 18.3 Illustration of the expression of biochemical oxygen demand (BOD) over time in water samples that were either greatly diluted or slightly diluted with nutrient solution

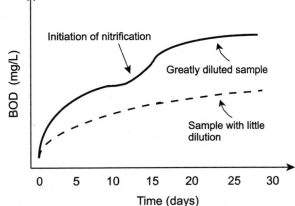

Table 18.3 Typical concentrations of 5-day biological oxygen demand (BOD_5) in various effluents (van der Leeden et al. 1990; Boyd and Tucker 2014)

Effluent	BOD_5 (mg/L)
Pond aquaculture effluent	10–30
Domestic sewage	100–300
Laundry	300–1000
Milk processing	300–2000
Canneries	300–4000
Beet sugar refining	450–2000
Brewery	500–1200
Meat packing	600–2000
Grain distilling	1500–20,000

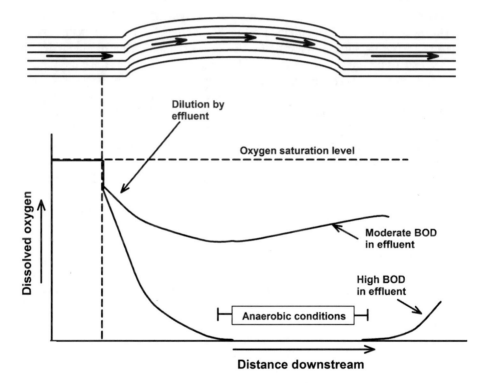

Fig. 18.4 Oxygen sag curve below an effluent outfall in a stream

different distances downstream from effluent outfalls. Discharge of effluents into lakes, estuaries, or the ocean also can depress dissolved oxygen concentrations in the vicinity of the outfall. The severity of this effect depends both on the BOD load and the extent to which the effluent is transported away from the outfall by water currents.

Organic wastes typically contain nitrogen and phosphorus, resulting in ammonia and phosphate being released along with carbon dioxide during decomposition. Carbon dioxide, ammonia, and phosphorus concentrations tend to increase in

Fig. 18.5 Changes in total ammonia nitrogen (TAN), nitrate-nitrogen (NO_3^--N), and nitrite-nitrogen (NO_2^--N) in a stream downstream of a major sewage outfall

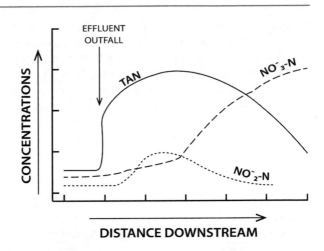

streams downstream of effluent outfalls, or in the vicinity of outfalls into lakes, estuaries, and the sea. Solids in effluents settle in the vicinity of outfalls and oxygen depletion may occur in sediments.

The typical pattern in nitrogen concentrations downstream from outfalls is an initial increase in organic nitrogen. The organic nitrogen concentration then declines and total ammonia nitrogen increases as a result of decomposition of organic matter from effluent. Nitrite may also increase because of low dissolved oxygen concentration. Finally, downstream of the oxygen sag, nitrate increases and ammonia nitrogen decreases because of nitrification (Fig. 18.5).

Biological Pollution

The main concern over biological pollution in water is the introduction of human pathogens in untreated sewage. The major waterborne diseases are gastroenteritis, typhoid, bacillary dysentery, cholera, infectious hepatitis, amebic dysentery, and giardiasis. Of course, in tropical nations, additional diseases also may be spread via the drinking water supply. The role of coliform organisms as indicators of contamination of water with human feces was discussed in Chap. 12.

Diseases of aquatic animals also can be of concern in water quality management. For example, fish and shrimp diseases are common in pond aquaculture. When the effluents from a pond containing diseased animals are released into natural waters, disease can spread (Boyd and Clay 1998).

Groundwater Pollution

Groundwater from wells is the source of drinking water for many people, and it is not unusual for well water to be consumed without treatment. Water in aquifers usually originates from rainfall that seeps downward until it reaches an impermeable layer.

Groundwater may accumulate pollutants, because wastes are sometimes disposed in or on the soil. Manure of farm animals is disposed by application as fertilizer to fields and pastures, sewage is used for irrigation in some parts of the world, septic tanks seep into the soil, waste is buried in landfills, and waste is sometimes put into dumps on the land surface. Tanks that hold fuel and other chemicals may be buried in the ground or placed on the land surface, and leakage or spillage is not uncommon. Pesticides are applied to fields, and a portion reaches the soil surface and move downward with infiltrating water.

In spite of the many opportunities for contamination, groundwater is protected from pollution to a large extent by the natural purifying action of the soil, soil microorganisms, and geological formations with which it comes in contact. The soil is composed of a mixture of particles of sand, silt, clay, and organic matter, and the underlying geological formations also contain layers of sand, silt, clay, gravel, and limestone and other fractured rocks. Water infiltrates slowly and bacteria in the soil decompose some of the dissolved organic material. Bacteria can even degrade or alter the molecular structure of agricultural and industrial chemicals to make them less toxic or render them harmless. More importantly, the soil and underlying geological formations act as a porous medium to filter harmful bacteria and other small particles from the infiltrating water. Clay, in particular, has a great ability to adsorb substances from infiltrating water. Bollenbach (1975) summarized data on the movement of bacterial and chemical pollution in infiltrating water. Coliform bacteria moved through the first 1.5 m at the same rate as the water, and a few moved about 5 m in 3 days. After 2 months, a few coliforms had seeped to a depth of 10 m. Chemical pollution, however, was found to move 3 m in 4 days and to ultimately seep to a depth of 30 m. Some studies in more permeable soil showed greater rates of movement of bacteria and pollutants, but the medium through which water must seep to reach the permanently saturated groundwater zone is highly effective in filtering out particles such as bacteria and in adsorbing chemicals.

Wells themselves provide a conduit between the land surface and aquifers that can result in pollution of groundwater. A sanitary seal should be placed between the well casing and the borehole to prevent water from flowing downward through this space. Dug wells are still used in some parts of the world, and means for preventing surface runoff from entering these wells after rains should be provided. Abandoned wells should be sealed for human safety and to prevent them from being conduits for groundwater contamination.

In spite of the purification of water by natural process as it seeps downward, groundwater in many aquifers is contaminated with bacteria and chemicals. This is particularly true in areas where wastes have been disposed in the soil or on its surface, and pollutants have infiltrated into shallow aquifers.

Toxins and Human Risk Assessment

Many chemicals used by agriculture, industry, and households can be toxic to humans and other organisms. There is a large number of potentially toxic chemicals to include petroleum products, inorganic substances (ammonia and heavy metals), pesticides and other agricultural chemicals, industrial chemicals, and pharmaceuticals. Evaluation of the toxicity of waterborne toxins to aquatic organisms is difficult, because the toxicity of a substance depends upon its concentration, its degradation rate, and environmental conditions. Mortality may be rapid (acute) if a high concentration of the toxin is introduced or slow (chronic) if a lower concentration is maintained in the water. The lowest concentration at which mortality can be detected is the threshold toxic concentration. The threshold concentration also may be defined as the lowest concentration necessary to elicit some response other than death. Responses may include failure to reproduce, lesions, aberrant physiological activity, susceptibility to disease, or behavioral changes. The exposure time necessary for a toxin to produce some undesirable effect on organisms decreases with increasing concentration.

Organisms usually must absorb a certain amount of a toxin before the threshold body burden necessary to produce a toxic effect occurs. The total body burden at a particular instant for a toxin is described by the following equation:

$$TBB = (DI + R) - DL \qquad (18.3)$$

where TBB = total body burden, DI = daily intake of toxin, R = residual of toxin in body before exposure, and DL = daily loss of toxin from body by metabolism or excretion. Toxicity occurs when the body burden reaches the threshold level. If the toxin disappears from the water, organisms will eliminate the toxin, but the rate of loss usually declines as the total body burden decreases.

Bioaccumulation occurs when an organism accumulates a toxin in specific organs or tissues. Many pesticides are fat-soluble and tend to accumulate in fatty tissues. The term bioconcentration is used to describe the phenomenon in which a toxic substance accumulates at greater and greater concentrations as it passes through the food chain. A toxin introduced into water may be bioaccumulated by plankton. Fish eating the plankton may store this toxin in their fat and have higher body burdens than did the plankton. Birds feeding on the fish may further concentrate the toxin until a toxic body burden is reached. Bioaccumulation and bioconcentration of toxic substances by aquatic food organisms is also a human food safety concern.

Toxicity may increase as a result of synergism of two toxic compounds. Toxicity normally increases with increasing water temperature. In the case of metals, the free ion usually is the most toxic form, and the toxicity of a metal will be less in water with high concentrations of humic substances that complex metals than in clear water. Other water quality factors such as pH, alkalinity, hardness, and dissolved oxygen concentration can affect the toxicity of substances.

Toxicity tests are not conducted on human subjects, and humans are seldom exposed to a high enough concentration of a particular toxin in water to elicit an immediate response of lesions, sickness, or death. Still, over a long time, exposure to pollutants in water can adversely influence an individual's health. The USEPA has developed a system of evaluating human risk from pollutants that is used in developing drinking water quality standards that are protective of human health.

A lengthy discussion of the epidemiological techniques for assigning human health risks to pollutants is beyond the scope of this book, but a brief illustration of the general procedure is in order. The USEPA uses the concept of unit risk. For pollutants in water, a unit risk is the risk incurred by exposure to 10^{-9} g/L of the substance. A unit lifetime risk is the risk associated with exposure to 10^{-9} g/L of the pollutant for 70 years. In the case of carcinogens in water, the risk is given in terms of latent cancer fatalities. Such data must be obtained from complex epidemiological studies that have been done for various compounds, and it is presented as the number of latent cancer fatalities (LCF) per 100,000 people.

The unit annual risk (UAR) for a compound is

$$UAR = \frac{LCF/yr}{10^{-9} \, g/L} \tag{18.5}$$

or unit lifetime risk (ULR)

$$ULR = \frac{LCF}{(10^{-9} \, g/L)(70 \, yr)}. \tag{18.6}$$

The ULR of a hypothetical compound is calculated in Ex. 18.2.

Ex. 18.2 *A community has been drinking water for 10 years that contains 10^{-7} g/L of a compound known to be a carcinogen with a LCF of 0.2/100,000. The risk of these people developing cancer will be estimated using Eq. 18.6.*
  Solution:

$$ULR = \frac{(10^{-7} \, g/L)(0.2 \, LCF)(10 \, yr)}{(100,000)(10^{-9} \, g/L)(70 \, yr)} = 2.86 \times 10^{-5} \, LCF.$$

About three cancer deaths per 100,000 people could be expected as a result of the community drinking the water for 10 years. Of course, individuals in the community may also develop cancer from other causes beside drinking water contaminated with the carcinogen.

Toxicity Tests and Their Interpretation

Toxicity tests are important tools of aquatic toxicology for determining effects, including death, of different concentrations of toxins. Threshold concentrations for different responses can be estimated, and the influence of exposure time and water quality conditions on toxicity can be evaluated. Toxicity tests often are important in establishing safe concentrations of pollutants for natural waters. Some species are more sensitive than others, and the overall risk of waterborne toxins to ecosystems is extremely difficult to establish with certainty.

In acute toxicity tests, aquatic organisms are exposed for specific time periods to a concentration range of a toxicant under carefully controlled and standardized conditions in the laboratory. The mortality at each concentration is determined, and the resulting data are helpful in assessing toxicity under field conditions. Toxicity studies may be conducted as static tests in which water with toxicant is placed in chambers and organisms introduced. There may or may not be water or toxicant renewal during the exposure period. The duration of static tests seldom exceed 96 hours and sometimes is shorter. Toxicity studies also may be conducted as flow-through trials in which fresh toxicant solution is continuously flushed through the test chambers. Animals may be fed in flow-through tests, and animals may be exposed to a toxicant for weeks or months. There are many sources of information on toxicity test methodology; an excellent one is the *Standard Methods for the Examination of Water and Wastewater* (Eaton et al. 2005).

The most common way of analyzing results of acute toxicity tests is to calculate the percentage survival (or mortality) at each test concentration, and plot percentage survival on the ordinate against toxicant concentrations on the abscissa. Semi-log paper is normally used for preparing the graph of concentration versus mortality because the relationship is logarithmic. The concentration of toxicant that caused 50% mortality can be estimated from the graph by direct interpolation or by aid of regression analysis. The concentration of the toxicant necessary to kill 50% of the test animals during the time that organisms were exposed to the toxicant (exposure time) is called the lethal concentration 50 (LC50). The exposure time of animals to toxicants usually is specified by placing the number of hours of exposure before LC50, e.g., 24-hr LC50, 48 hr-LC50, or 96-hr LC50. The graphical estimation of the LC50 from the results of a toxicity test is illustrated (Fig. 18.6).

In addition to providing the LC50, toxicity testing can reveal the lowest concentration of a substance that causes toxicity or the highest concentration that causes no toxicity. Sometimes tests may be conducted in which the endpoint is some response other than toxicity. For example, in long-term tests, the concentration that inhibits reproduction could be measured, the concentration that produces a particular lesion, or the concentration that elicits a particular physiological or behavioral change might be ascertained.

In many instances the only toxicity data for a substance will be the short-term LC50 for one or possibly a few species at a single temperature and a specific water quality regime. Lists of 96-hr LC50 concentrations of selected inorganic elements, pesticides, and industrial chemicals for freshwater fish (Tables 18.4, 18.5, and 18.6) reveal the wide ranges in toxicity of these classes of substances.

The LC50s for acute mortality (exposures up to 96 hours) will be larger than LC50s for chronic mortality (longer exposure). A plot of LC50 versus exposure time for a given toxicant will show a curvilinear decline in LC50 until the LC50 becomes asymptotic to the abscissa (Fig. 18.7). The asymptotic LC50 is the toxicant concentration above which the LC50 does not decline with greater exposure time.

The concentration of potential toxicants in water bodies should be below the concentration that has adverse effects on growth and reproduction. Information on minimum lethal concentrations and no-effect levels can be obtained from full- or partial-life-cycle tests that are more difficult and expensive to conduct than are short-term tests.

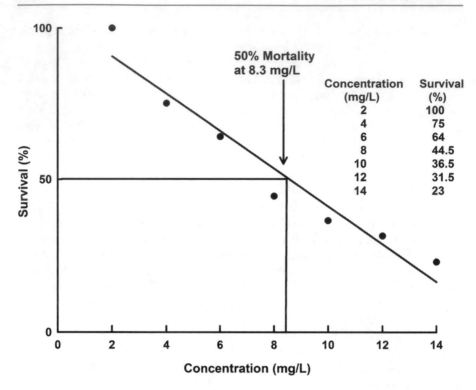

Fig. 18.6 Graphical estimation of the LC50

Table 18.4 Ranges in 96-hr LC50 values for various species of fish exposed to selected inorganic elements

Inorganic substance	96-hr LC50 (mg/L)	Inorganic substance	96-hr LC50 (mg/L)
Aluminum	0.05–0.2	Iron	1–2
Antimony	0.3–5	Lead	0.8–542
Arsenic	0.5–0.8	Manganese	16–2400
Barium	50–100	Mercury	0.01–0.04
Beryllium	0.16–16	Nickel	4–42
Cadmium	0.9–9	Selenium	2.1–28.5
Chromium	56–135	Silver	3.9–13.0
Copper	0.05–2	Zinc	0.43–9.2

In life-cycle or partial-life-cycle chronic toxicity tests, the lowest toxicant concentration at which no effect is observed is called the no observed effect concentration (NOEC) and the lowest concentration that causes an effect is known as the lowest observed effect concentration (LOEC). The highest concentration of a toxicant that should not cause a negative impact on an organism is the maximum allowable toxicant concentration (MATC). The MATC is calculated as the geometric mean of the product of the NOEC and LOEC:

Table 18.5 Acute toxicities of some common pesticides to fish

Trade name	96-hr LC50 (μg/L)	Trade name	96-hr LC50 (μg/L)
Chlorinated hydrocarbon Insecticides		**Pyrethum insecticides**	
DDT	8.6	Permethrin (synthetic pyrethroid)	5.2
Endrin	0.61	Natural pyrethroid	58
Heptachlor	13	**Miscellaneous insecticides**	
Lindane	68	Diflubenzuron	>100,000
Toxaphene	2.4	Dinitrocresol	360
Aldrin	6.2	Methoprene	2900
Organophosphate insecticides		Mirex	>100,000
Diazinon	168	Dimethoate	6000
Ethion	210	**Herbicides**	
Malathion	103	Dicambia	>50,000
Methyl parathion	4380	Dichlobenil	120,000
Ethyl parathion	24	Diquat	245,000
Guthion	1.1	2,4-D (phenoxy herbicide)	7500
TEPP	640	2,4,5-T (phenoxy herbicide)	45,000
Carbamate insecticides		Paraquat	13,000
Carbofuran	240	Simazine	100,000
Carbaryl (Sevin)	6760	**Fungicides**	
Aminocarb	100	Fenaminosulf	85,000
Propoxur	4800	Triphenyltin hydroxide	23
Thiobencarb	1700	Anilazine	320
		Dithianon	130
		Sulfenimide	59

$$\text{MATC} = \sqrt{(\text{NOEC})(\text{LOEC})}. \qquad (18.4)$$

In order to protect aquatic life, the U. S. Environmental Protection Agency (USEPA) establishes limits for pollutant concentrations that should protect aquatic communities rather than just individual species from harmful chemicals. The procedure for establishing these criteria for a pollutant will not be presented, but criteria are based on the response of very sensitive species and take into account differences in water quality that affect the toxicity of a substance. The procedures for estimating CMCs and CCCs may be found at http://water.epa.gov/learn/training/standardsacademy/aquatic_page3.cfm. The criterion maximum concentration (CMC) is the highest concentration of a substance in ambient water to which an aquatic community can be exposed briefly (1 hour every 4 years) without resulting in an undesirable effect—an acute criterion. The criterion continuous concentration (CCC) is the highest concentration to which an aquatic community can be continuously exposed without an undesirable effect resulting—a chronic criterion.

Table 18.6 Acute toxicities of some common industrial chemicals to fish

Compound	96-hr LC50 (mg/L)
Acrylonitrile	7.55
Benzidine	2.5
Linear alkylate sulfonates and alkyl benzene sulfonates	0.2–10
Oil dispersants	>1000
Dichlorobenzidine	0.5
Diphenylhydrazine	0.027–4.10
Hexachlorobutadiene	0.009–0.326
Hexachlorocyclopentadiene	0.007
Benzene	<5.30
Chlorinated benzenes	0.16
Chlorinated phenols	0.004–0.023
2,4-Dimethylphenol	2.12
Dinitrotoluenes	0.33–0.66
Ethylbenzene	0.43–14
Nitrobenzenes	6.68–117
Nitrophenols	0.23
Phenol	10
Toluene	6.3–240
Nitrosamines	5.85

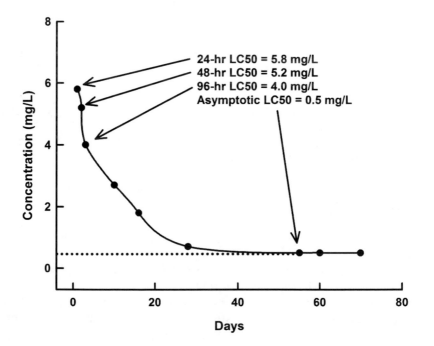

24-hr LC50 = 5.8 mg/L
48-hr LC50 = 5.2 mg/L
96-hr LC50 = 4.0 mg/L
Asymptotic LC50 = 0.5 mg/L

Fig. 18.7 Illustration of the asymptotic LC50

When no information is available other than the 96-hr LC50, the MATC often is estimated by multiplying the LC50 by an application factor. An application factor of 0.05 sometimes is suggested for common toxins such as carbon dioxide, nitrite, ammonia, or hydrogen sulfide. If the 96-hour LC50 for un-ionized ammonia to a species is 1.2 mg/L, the MATC would be $1.2 \times 0.05 = 0.060$ mg/L (60 µg/L). Smaller application factors usually are selected for more toxic substances; they range from 0.01 to 0.001 or even less. In the case of trace metals, pesticides, and industrial chemicals, it is common to use an application factor of 0.01. A pesticide with a 96-hr LC50 of 100 µg/L would have a MATC of 1 µg/L. Needless to say, there is a risk associated with the use of the application factor, but it is often the only means of estimating the safe concentration of a given toxicant.

The toxicity of most substances will increase with temperature. Where LC50 values are available for only a single temperature, it is prudent to assume that the toxicity will double with a 10 °C increase in temperature, i.e., $Q_{10} = 2.0$. If the 96-hr LC50 for a pesticide at 20 °C is 0.2 mg/L, at 25 °C the 96-hour LC50 could be expected to be about 0.15 mg/L. In natural ecosystems, the concentration of a toxicant will seldom be as constant as in toxicity tests. The toxin will almost never be delivered at a constant rate, and various processes will gradually or rapidly remove the toxicant from the water.

Natural environmental conditions are much different than those in toxicity tests. The toxicity of a substance may change in response to water quality conditions. To illustrate, the toxicity of nitrite to fish is much greater when dissolved oxygen concentration is low than when it is high. Animals in poor physiological condition because of environmental stress are more susceptible to most toxicants than healthy animals living in high quality water.

More than one toxicant may be present in the same water, and the two toxicants may act synergistically to produce greater effects than either will produce alone. There also can be antagonistic effects between toxicants in which the mixture of two toxicants is less toxic than either of the toxicants alone.

Different sizes or life-stages of animals may have different tolerances to toxicants. Fish fingerlings usually are more susceptible to toxins than are larger fish. Coldwater species typically are more sensitive to toxins than are warmwater species.

Wetland Destruction

Wetlands play an important role in water quality protection. Although wetlands are a major feature of many landscapes, a widely acceptable definition of a wetland is difficult to formulate. Cowardin et al. (1979) indicated that wetlands are covered by shallow water or the water table is at or near the land surface. According to Mitsch and Gosselink (1993) the international definition of a wetland follows: "wetlands, areas of marsh, fen, peatland or water, whether natural or artificial, permanent or temporary, with water that is static or flowing, fresh, brackish, or salt including areas of marine water, the depth of which at low tide does not exceed 6 meters." The legal definition of a wetland in the United States is as follows: "the term wetlands means those areas that are inundated or saturated by surface or groundwater at a frequency and duration sufficient to support, and that under normal circumstances do support, a prevalence of vegetation typically adapted for life in saturated soil conditions. Wetlands generally include swamps, marshes, bogs, and similar areas."

As an analogy to emphasize the role of wetlands in purifying water, Mitsch and Gosselink (1993) called wetlands "the kidneys of the landscape," but wetlands also provide many other ecological services. Wetlands are highly productive ecosystems. They are nurseries and feeding grounds for aquatic animals, waterfowl, and other bird life. They also are important habitats for amphibians and some reptiles. Wetlands act as sediment and nutrient traps and provide flood protection. Riparian vegetation is a buffer zone that filters runoff and reduces the input of suspended solids and nutrients into streams. Riparian vegetation also helps control bank erosion. Coastal mangrove forests and other marine wetlands provide a similar filtration system at mouths and deltas of rivers, and mangroves reduce wave and storm damage to the coastline.

The major cause of freshwater wetland loss is conversion of wetlands to agricultural land. It is estimated that 26,000,000 ha of farmland in the United States were obtained by draining wetlands (Dahl 1990). Over half of the original mangrove wetlands that occur in coastal areas of the world have been destroyed or converted to other uses (Massaut 1999). Coastal pond aquaculture was a major driver of mangrove loss between 1970 and 2000, but its role in mangrove loss is declining greatly because of better regulations. Wetlands are today protected by law in the United States and many other countries, and their conversion to agricultural land, aquacultural ponds, and municipal and industrial areas has greatly declined. However, in some parts of the world, wetland destruction continues.

Water Quality Regulations and Pollution Control

The usefulness of water declines as water quality deteriorates, and high quality water is in greater demand and has more value than low quality water. The concepts of water quality and water quantity developed simultaneously during human history, but until recently, few quantitative means of assessing water quality were available. Because of the importance of water in human affairs and the relative scarcity of water in many regions, conflicts over water use rights and water quality have often occurred. Water rights issues historically focused on water quantity and especially on the quantity of surface water. Disputes over water rights have been settled by sundry methods, but mainly warfare, feuds, court decisions, and mutual agreements.

In the United States and most developed nations, water rights disputes have traditionally been settled through legal actions based on the principles of common law. In common law, court rulings are based on precedents set in previous court cases of similar nature, or if there is no precedent, the court must make a decision that will become precedent. According to Vesilind et al. (1994), common law has provided a reasonable way to deal with surface water disputes through the theories of riparian doctrine, prior appropriations doctrine, the principle of reasonable use, and the concept of prescriptive rights. Riparian doctrine holds that ownership of the land beneath or adjoining a body of surface water includes the right to use this water. Prior appropriations take the approach that water use rights are based on a "first-come, first-serve basis" and land ownership does not necessarily grant control over the use of the water associated with the land. The principle of reasonable use states that the riparian owner is entitled to make reasonable use of the water, but the court may take into account the needs of others downstream. The concept of prescriptive rights basically allows the upstream user to abuse water quantity and water quality provided the downstream riparian owner does not use the water. Thus, by lack of use, the downstream owner gives up his water rights.

Doctrines, principles, and concepts of water rights used in common law do not provide clear guidance about issues related to water quality. Of course, if a water user impairs the quality of water that another party has a right to use, the injured party may initiate a court case to seek relief. It takes years to set precedents, and current issues often are settled by ancient precedents. This does not allow good use of new technology and knowledge, and it docs not provide adequate protection of environmental quality and public health. Modern governments establish rules or regulations related to water quality and water pollution abatement through governmental mandates known as statutory laws. Statutory laws about water quality allow a government to exert a degree of control over the quality of waters within its boundaries, minimize disputes over water quality, protect aquatic ecosystems, and guard public health.

In the United States, federal legislation known as the Clean Water Act was passed in 1965 to provide a uniform series of procedures to deal with water pollution. The US Environmental Protection Agency (USEPA) has the responsibility for enforcing the regulations and laws mandated by the Clean Water Act, but the day to day enforcement of the Clean Water Act has been relegated mainly to the individual states. The Clean Water Act applies to governmental facilities, municipalities, industries, and private individuals. Implementation and enforcement of the Clean Water Act is obviously a tremendous task, and it has yet to be applied to all activities that affect water quality in the United States. Implementation and enforcement of the act are continuing, and water quality in the United States has improved greatly and is continuing to improve as a result. Most countries have a system of water quality legislation, but in developing nations, water quality legislation often is either poorly structured or not adequately implemented and enforced.

Effluent Discharge Permits

The initial step in water pollution control is to require a permit for discharge of pollutants from a point source into natural waters. The Clean Water Act in the United States requires a National Pollutant Discharge Elimination System (NPDES) permit for every pollutant discharge. The holder of a NPDES permit has the right to discharge effluents containing specified concentrations or amounts of pollutants at a given point usually for a 5-year period. The NPDES does not apply to storm runoff, discharges into water treatment systems, and some other waters. Individual states and municipalities must establish regulations for discharges not covered by NPDES.

Pollution discharge permits typically contain effluent limitations, monitoring requirements, and reporting schedules. They also may contain other features such as use of best management practices (BMPs) to prevent or reduce the release of pollutants or to clean up spills. Permits also specify conditions related to operation of treatment systems, record keeping, inspection and entry, etc. In addition to effluent discharge permits, legally enforceable standards are imposed on drinking water quality in most countries.

Effluent Limitations

The reason for limits on the discharge of pollutants into natural waters is to avoid water quality degradation. Prohibiting discharge is rarely a possibility, because many activities that cause water pollution are essential for society. Complete removal of pollutants by treatment of effluents before final discharge usually is both technologically and economically impossible. As a compromise, regulatory agencies usually try to achieve a balance in water quality permits to allow an activity to continue but with limits on pollutants in effluents that are technologically achievable, affordable, and protective of water quality.

Use Classification for Water Bodies

The task of assigning effluent limitations in effluent permits can be facilitated as it is in the United States by classifying streams and other water bodies according to their anticipated maximum beneficial use such as public drinking water supplies, propagation of fish and wildlife, recreational activities, industrial and agricultural water sources, navigation, and others. Each water use requires a certain level of water quality. Water bodies can be classified into use categories and each use category assigned water quality standards. A stream classified for agricultural and industrial use will have lower water quality standards than a stream designated for fish and wildlife. Stream classification force water users to limit or treat discharges to prevent streams from violating their water quality standards. Use classification also can be applied to other types of water bodies and water sources.

A stream classification system usually has both quantitative and qualitative standards as illustrated in Table 18.7. The US Clean Water Act requires that each state develop a stream classification system with criteria and standards for each use category. The standards developed by each state must attain the Clean Water Act's goal of fishable, swimmable water wherever possible and should prevent further stream degradation.

Most streams already were receiving pollution and degraded below their pristine condition before the initiation of stream classification, but stream classification and

Table 18.7 Summary of Alabama Stream Classification System with quantitative water quality standards. Some details related to Tennessee and Cahaba River basins and coastal waters not included

Classification	Wastewater effluent limits (mg/L)	Dissolved oxygen (mg/L)	Bacteria (cfu/100 mL)	Qualitative narrative criteria
Outstanding national resource waters	(No discharge permitted into these waters)			No discharge permitted.
Outstanding Alabama water	DO: 6.0 NH$_3$: 3.0 BOD$_5$: 15.0	5.5	200	Must meet all toxicity requirements, not affect propagation, palatability of fish/shellfish, or affect aesthetic values.
Swimming		5.0	200	Must be safe for water contact, be free from toxicity, not affect fish palatability, not affect aesthetic value or impair waters for this use.
Shellfish harvesting		5.0	FDA regulations	Must be free from toxicity, not affect fish/shellfish palatability, not affect aesthetic value or impair waters for this use.
Public water supply		5.0	2000/4000 June-Sept: 200	Must be safe for water supply, free from toxicity, not have adverse aesthetic values for this use.
Fish and wildlife		5.0	1000-2000 June-Sept: 200	Must not exhibit toxicity to aquatic life of propagation, impair fish palatability, or affect aesthetic values for this use.
Agricultural and industrial water supply		3.0		Must not impair agricultural irrigation, livestock watering, industrial cooling, industrial water supply, fish survival, or interfere with downstream uses. Does not protect fishing, recreational use, or use as a drinking water supply.
Industrial operations		3.0		Must not impair use as industrial cooling and process water. Does not protect use as fishing, recreation, water supply for drinking or food processing.

All classifications into which discharge is permitted require pH 6–8.5, maximum increase in temperature to 90 °F (32 °C) with maximum rise of 5 °F (2.35 °C), and turbidity increase not to exceed 50 NTU

effluent standards can prevent water quality in a stream or a reach of a stream from degrading further and may cause water quality to improve. The system can be used to force improvements in surface water quality by giving streams a higher classification or disallowing the lowest classification.

Stream classification can fail to provide the intended benefits of protecting public health and the aquatic environment for several reasons. The stream standards may not be adequate to protect water quality, or the effluent standards may not be strict enough to prevent waste from causing the receiving water to violate its standard. Moreover, there are many cases where standards simply are not enforced.

The way in which governments foster economic development also can affect water quality protection. Two general approaches are used in setting water quality standards: (1) stipulation and (2) a policy of minimum degradation of water quality. Where the major goal of a government is to encourage economic development, subsidies can be stipulated to industry. One possible subsidy to industry is to classify steams according to low standards. From an environmental standpoint, it would be better for governments to subsidize wastewater treatment rather than to lower stream classification or relax discharge standards (Tchobanoglous and Schroeder 1985).

Water Quality Standards

Once streams have been classified, the major focus is on limiting pollutant levels in effluents to assure water quality in the receiving stream does not violate the standards of its use classification. The writer of an effluent permit has the responsibility to protect the environment while not unduly penalizing municipalities, industries, or other users. The criteria and standards in effluent permits are selected based on experience, technical attainability, economic attainability, bioassays and other tests, ability to reliably measure the criteria, evidence of public health effects, educated guess or judgment, mathematical models, and legal enforceability (Tchobanoglous and Schroeder 1985). Considerable experience has been accumulated in setting water quality standards, and many water quality guidelines have been published, e.g., guidelines for drinking water, for protection of aquatic ecosystems, for irrigation water, for livestock watering, and for recreational waters. Nevertheless, the permit writer must decide upon the safe limits of pollutants in effluents in order to achieve the goals of the guidelines. Permits must be renewed at intervals, and obvious flaws in permits may be negotiated between the permit holder and the permitting agency. No permit is perfect, but discharge permits with water quality criteria and standards are important for protecting water quality.

Concentration Based Standards

The simplest standards in effluent permits have criteria regarding permissible concentrations of selected water quality variables. Examples of concentration-based criteria in a water quality standard for an effluent follow:

Criteria	Standard
pH	6–9
Dissolved oxygen	5 mg/L or above
5-day biochemical oxygen demand	30 mg/L or less
Total suspended solids	25 mg/L or less

Standards of this type can prevent adverse effects on water quality in the mixing zone where the effluent mixes with the receiving water. They also put a limit on the concentration of pollutants to avoid future increases in concentration. Effluent discharge permit holders can dilute wastewater to assure compliance with concentration limits in standards. Dilution of effluents may allow compliance with standards without lessening the loads of pollutants entering natural waters. Some effluent discharge permits with concentration limits may also impose a limit on discharge volume to avoid the possibility of compliance through dilution.

Load Based Standards

The load of a pollutant is calculated by multiplying effluent volume by pollutant concentration:

$$L_x = (V)(C_x)(10^{-3})$$

where L_x = maximum load of pollutant x (kg/day), V = effluent volume (m³/day), C_x = concentration of pollutant x (g/m³), 10^{-3} = kg/g.

A load-based standard could limit BOD_5 to no more than 100 kg/day. Such a simple standard is unacceptable alone, because a small discharge could have a very high BOD_5 and not exceed the daily load standard. This could lead to dissolved oxygen depletion in the mixing zone. To avoid this possibility, a concentration limit usually is specified along with the load limit for a pollutant. The load standard might limit BOD_5 load to 100 kg/day, but a concentration standard prohibiting a maximum daily BOD_5 concentration above 30 mg/L might be added.

The weakness of load standards lies in the fact that the permissible effluent load for the receiving water is seldom known. Nevertheless, limiting the load can prevent the permit holder from increasing the load of a pollutant over time. A degree of control over pollution loads can be imposed with the combination of a concentration limit standard and an effluent volume standard.

Delta Based Standards

A delta standard specifies the maximum allowable increase (delta) for one or more variables. A TSS standard could require that TSS concentration in effluent cannot exceed the TSS concentration of receiving water by more than 10 mg/L. In other cases, the standard might give the permissible effluent concentration of a variable as an increase above the expected seasonal average (ambient) concentration of the receiving water, e.g., the turbidity of the discharge must not exceed the ambient turbidity by more than 10 NTU. The delta standard differs from a concentration limit standard by including a relationship to the quality of the receiving water.

Total Maximum Daily Loads

The shortcomings of effluent standards discussed above have led the United States and some other countries to establish total maximum daily loads (TMDLs) for priority pollutants in receiving waters. This approach consists of calculating the maximum amount of a pollutant from all sources (natural or anthropogenic) that can be allowed without causing a water body to violate its classification standards. In some situations, the TMDL of a pollutant must be allocated among the different sources of the pollutant. Suppose a stream reach has a TMDL for phosphorus of 500 kg/day, several industries discharge into the stream reach, and natural sources of phosphorus are 100 kg/day. The maximum amount of phosphorus permissible in effluents is 400 kg/day, but this amount normally would not be allowed in order to have a safety factor. A safety factor of 1.5 would lower the TMDL of this example to 267 kg/day. This load would have to be allocated among the different industries, and in some cases, the current load of a pollutant might already exceed the TMDL leading to stricter limits in permits. Use of TMDLs allows industries to trade in pollution loads. If an industry does not need its entire assigned TMDL for a pollutant, it could sell the unneeded portion of its TMDL allocation to another industry that cannot meet its TMDL for the particular pollutant.

Toxic Chemical Standards

Water pollution control agencies often publish lists of acceptable concentrations of pollutants that may be used as guidelines for establishing limits in permits. It is difficult to establish standards for toxic chemicals, and metals in particular, in effluent permits because their toxicities vary with water quality conditions. To avoid this problem, toxicity-based limitations may be established through effluent toxicity testing. The toxicity tests are conducted by exposing certain species of aquatic organisms to the effluent in question. A permit may require that toxicity testing be done to prove that the effluent is not acutely toxic to organisms in the receiving water body at the time of its discharge.

Biological Standards

Discharge permits often contain standards for coliform organisms that can be indicators of fecal pollution and certain other microorganisms of human health concern. There is a growing tendency to include biological standards based on biocriteria associated with the flora and fauna of the receiving water body. Biocriteria may be used to supplement the traditional water quality criteria and standards or used as an alternative where traditional methods have not been effective. Development of biocriteria requires a reference condition (minimal impact) for each use classification, measurement of community structure and function in reference to water quality to establish biocriteria, and a protocol for determining if community structure and function has been impaired. Biocriteria are much more difficult to assess than chemical and physical criteria and standards.

Waste Treatment and Best Management Practices

Pollution control involves technological, political, legislative, regulatory, enforcement, business, ethical educational, and other issues. Most governments have developed regulations of point source effluents to impose concentration limits, load limits, or both. Treatment of point source pollution allows compliance with the standards in the discharge regulations. Activities that discharge nonpoint source pollution are typically required to operate in a manner that minimizes pollution, and inspections are made by the appropriate authority to verify that pollution is being controlled.

Point source effluent can be directed through treatment facilities to lessen concentrations and loads of pollutants. Some treatment techniques in common use are filtration and sedimentation to remove solids, activated sludge basins with aeration to rapidly oxidize organic matter, precipitation of phosphorus with ferrous chloride, aerobic reactors for nitrification, anaerobic reactors for denitrification, air-stripping of ammonia, outdoor wastewater stabilization ponds, neutralization, and precipitation of metals by chemical treatment or pH manipulation. Removal of certain substances from industrial effluents may require development of specific treatment technologies. Some effluents that contain biological pollutants must be disinfected—often by chlorination—before final discharge.

Nonpoint source effluents are not confined in a conduit, and conventional treatment techniques cannot generally be applied. The most common means of reducing nonpoint source pollution is through the use of practices that lessen the amounts of pollutants entering runoff. Such practices are called best management practices (BMPs). Sometimes BMPs and qualitative standards also are included in standards for point source discharges.

Agriculture is the largest single source of nonpoint pollution, and BMPs have long been used for controlling agricultural pollution. There are three classes of agricultural BMPs: erosion control, nutrient management, and integrated pest management. Some examples of erosion control BMPs in agriculture are cover crops, no-till farming, conservation tillage, grass-lined ditches, terraces in fields on sloping land, etc. Storm water runoff management in municipalities also may consist of a system of BMPs, and BMPs are used to lessen pollution from logging, construction, and mining operations.

Monitoring and Enforcement of Discharge Permits

Many times it is not possible for effluent discharge permit holders to comply immediately with the conditions of the permit, and a schedule for compliance may be specified. Most governments depend largely upon self-monitoring to document compliance with permit standards. The permit will specify minimum monitoring requirements including frequency of sampling, type of sample, variables to be analyzed, and reporting schedule. Permits normally require the permit holder to immediately report when discharges are not in compliance.

Enforcement of water discharge permits normally involves administrative actions, because the permit holder must report compliance or noncompliance with the permit on a scheduled basis. In theory, the permit holder should realize when compliance is not being achieved and work to correct the problem. In the United States, federal and state enforcement of the Clean Water Act can be in the form of administrative actions or judicial actions. The administrative actions may take many forms depending upon the type of violation. An order to comply may be issued along with a compliance schedule. For serious violations, the administrative order may include a fine or other penalty. Failure to comply with administrative orders can lead to criminal prosecution. Both civil and criminal judicial enforcement is possible if administrative orders do not solve problems related to permits. The penalties resulting from judicial enforcement are more severe than those resulting from administrative orders.

Parties who feel that they have been injured by water discharges may bring civil lawsuits against permit holders if the government is not "diligently prosecuting" the violation (Gallagher and Miller 1996). The form of enforcement of water discharge permits will vary greatly among nations.

Drinking Water Standards

Many countries, individual states or provinces, and international agencies have made standards for drinking water quality. The current National Drinking Water Regulations for the United States may be found at (https://www.epa.gov/sites/production/files/2018-03/documents/dwtable2018.pdf). These standards come from the Office of Groundwater and Drinking Water of the USEPA, and the primary standards are legally enforceable standards that apply to public water systems. The purpose of these primary standards is to protect drinking water quality by limiting the levels of specific contaminants that can adversely affect public health and are known or anticipated to occur in public water supply systems. The Secondary Drinking Water Regulations for the United States are non-enforceable guidelines regulating contaminants that may cause cosmetic effects such as skin or tooth discoloration and aesthetic effects such as taste, odor, or color in drinking water.

Table 18.8 Guidelines for protection of freshwater aquatic ecosystems. (Source: Australian and New Zealand Environmental and Conservation Council 1992)

Variable	Concentration/level	Variable	Concentration/level
Physico-chemical		**Industrial organic (µg/L)**	
Color and clarity	<10% change in compensation depth	Hexachlorobutadiene	0.1
Dissolved oxygen	>6 mg/L	Benzene	300
pH	6.5–9.0	Phenol	50
Salinity	<1000 mg/L	Toluene	300
Suspended particulate matter and turbidity	<10% change in seasonal average	Acrolein	0.2
Nutrients	Site specific[a]	Di-n-butylphthalate	4
Temperature	<2 °C increase	Di(2-ethylexyl) phthalate	0.6
Inorganic (µg/L)		Other phthalate esters	0.2
Aluminum	<5 (pH <6.5)	Polychlorinated biphenyls	0.001
	<100 (pH >6.5)	Polycyclic aromatic hydrocarbons	3
Antimony	30	**Pesticides (µg/L)**	
Arsenic	50	Aldrin	0.01
Beryllium	4	Chlordane	0.004
Cadmium	0.2–2.0 (depends on hardness)	Chlorpyrifos	0.001
Chromium (total)	10	DDE	0.014
Chromium III	---	DDT	0.001
Chromium VI	---	Demeton	0.1
Copper	2–5 (depends on hardness)	Dieldrin	0.002
Cyanide	5	Endosulfan	0.01
Iron (Fe^{3+})	1000	Endrin	0.003
Lead	1–5 (depends on hardness)	Guthion	0.01
Mercury	0.1	Heptachlor	0.01
Nickel	15–150 (depends on hardness)	Lindane (BHC)	0.003
Selenium	5	Malathion	0.07
Silver	0.1	Methoxychlor	0.04
Sulfide	2	Mirex	0.001
Thallium	4	Parathion	0.004
Tin (tributyltin)	0.008	Thoxaphene	0.008
Zinc	5–50 (depends on hardness)		

[a]See URL in caption for instructions

Water Quality Guidelines

There are many lists of guidelines or criteria for public water supplies, fish and wildlife, agriculture, recreation and aesthetics, and industry. Guidelines recommended for protection of aquatic ecosystems in Australia, and New Zealand are provided in Table 18.8. Such guidelines usually are not legally enforceable, but they show acceptable concentrations and may be the basis for criteria in legally enforceable effluent permits.

Water quality guidelines for agriculture can help farmers to protect their crops and livestock from damage by poor quality water. Guidelines for industrial waters may be very important in assuring adequate water quality for various processes.

Conclusions

Humans must use water for many purposes, and water quality often is impaired through use. The demand for water is increasing because of the rapidly growing human population, and water quality deterioration has become a serious issue in many countries. Measures to conserve both the quantity and quality of water must be imposed or the world will face serious water shortages in the future. Some countries have developed rather elaborate systems of water quality regulations for maintaining or improving the quality of their waters. Other countries have done little to protect water quality, and serious water quality problems are occurring. It is urgent for all countries to develop water quality regulations and to enforce them seriously. It is equally important to educate the public about the importance of protecting our limited and fragile water supplies for future use.

References

Australian and New Zealand Environment and Conservation Council (1992) Australian water quality guidelines for fresh and marine waters. Australian and New Zealand Environment and Conservation Council, Canberra

Bollenbach WM Jr (1975) Ground water and wells. Johnson Division, UOP Inc, Saint Paul

Boyd CE, Clay J (1998) Shrimp aquaculture and the environment. Sci Am 278:42–49

Boyd CE, Tucker CS (2014) Handbook for aquaculture water quality. Craftmaster Printers, Auburn

Cowardin LM, Carter V, Golet FC, LaRoe ET (1979) Classification of wetlands and deepwater habitats of the United States. Publication FWS/OBS-79/31, United States Fish and Wildlife Service, Washington

Dahl TE (1990) Wetlands losses in the United States, 1780s to 1980s. United States Department of the Interior, Fish and Wildlife Service, Washington

Eaton AD, Clesceri LS, Greenburg AE (eds) (2005) Standard methods for the examination of water and wastewater. American Public Health Association, Washington

Gallagher LM, Miller LA (1996) Clean water handbook. Government Institutes, Inc., Rockville

Magleby R, Sandretto C, Crosswaite W, Osborn CT (1995) Soil erosion and conservation in the United States. An overview. USDA Economic Research Service Report AIB-178, Washington

Massaut L (1999) Mangrove management and shrimp aquaculture. Department of Fisheries and Allied Aquacultures, Auburn University, Alabama

Mitsch WJ, Gosselink JG (1993) Wetlands. Van Nostrand Reinhold, New York

Smith DD (1941) Interpretation of soil conservation data for field use. Ag Eng 21:59–64

Tchobanoglous G, Schroeder ED (1985) Water quality: characteristics, modeling, modification. Adison-Wesley Publishing Company, Reading

USEPA (1973) Methods for identifying and evaluating the nature and extent of nonpoint sources of pollution. EPA 430/9-73-014, Washington

van der Leeden F, Troise FL, Todd DK (1990) The water encyclopedia. Lewis Publishers, Inc., Chelsea

Vesilind PA, Peirce JJ, Weiner RF (1994) Environmental engineering. Butterworth-Heinemann, Boston

Wischeier WH, Smith DD (1965) Predicting rainfall erosion losses from cropland east of the Rocky Mountains. Agricultural Handbook 282, United States Department of Agriculture, Beltsville, Maryland

Zingg AW (1940) Degree and length of land slope as it affects soil loss in runoff. Ag Eng 22:173–175

Appendix: Some Fundamental Principles and Calculations

Introduction

This appendix provides some very basic principles of chemistry and some information on calculating concentration of substances in solution. Much more detailed information is available in a general inorganic chemistry textbook.

Elements, Atoms, Molecules, and Compounds

Elements are the basic substances of which more complex substances are composed. Anything that has mass (or weight) and occupies space is matter made up of chemical elements such as silicon, aluminum, iron, oxygen, sulfur, copper, etc. Substances more complex than atoms of a single element consist of molecules. A molecule is the smallest entity of a substance that has all the properties of that substance.

Elements are classified broadly as metals, nonmetals, or metalloids. Metals have a metallic luster, are malleable (but often hard), and can conduct electricity and heat. Some examples are iron, zinc, copper, silver, and gold. Nonmetals lack metallic luster, they are not malleable (but tend to be brittle), some are gases, and they do conduct heat and electricity. Metalloids have one or more properties of both metals and nonmetals. Because metalloids can both insulate and conduct heat and electricity, they are known as semiconductors. The best examples of metalloids are boron and silicon, but there are several others.

There also are less inclusive groupings of elements than metals, nonmetals, and metalloids. Alkali metals such as sodium and potassium are highly reactive and have an ionic valence of +1. Alkaline earth metals include calcium, magnesium, and other elements that are moderately reactive and have an ionic valence of +2. Halogens illustrated by chlorine and iodine are highly reactive nonmetals, and they have an ionic valence of −1. This group is unique in that its members can exist in solid, liquid, and gaseous form at temperatures and pressures found on the earth's surface. Because of their toxicity, halogens often are used as disinfectants. Noble gases such as helium, argon, and neon are unreactive except under very special conditions. In all there are 18 groups and subgroups of elements. Most elements are included in the

© Springer Nature Switzerland AG 2020
C. E. Boyd, *Water Quality*, https://doi.org/10.1007/978-3-030-23335-8

groups referred to as alkali metals, alkaline earth metals, transitional metals, halogens, noble gases, and chalcogens (oxygen family). The elements are grouped based on similar properties and reactions. However, in reality, each element has one or more unique properties and reactions.

The most fundamental entity in chemistry and the smallest unit of an element is the atom. An atom consists of a nucleus surrounded by at least one electron. Electrons revolve around the nucleus in one or more orbitals or shells (Fig. A.1). The nucleus is made up of one or more protons, and with the sole exception of hydrogen, one or more neutrons. The protons and neutrons do not necessarily occur in the nucleus in equal numbers. For example, the oxygen nucleus has eight protons and neutrons, sodium has 11 protons and 12 neutrons, potassium has 19 protons and 20 neutrons, copper has 29 electrons and 34 neutrons, and silver has 47 electrons and 61 neutrons.

Protons are positively charged and assigned a charge value of +1 each, electrons possess a negative charge and are assigned a charge value of −1 each, and neutrons are charge neutral. In their normal state, atoms have equal numbers of protons and electrons resulting in them being charge neutral.

Atoms are classified according to the numbers of protons. All atoms with the same number of protons are considered to be the same element. For example, oxygen atoms always have eight protons while chlorine atoms always have 17 protons. The atomic number of an element is the same as the number of protons, e.g., the atomic number of oxygen is 8 while that of chlorine is 17. There are over 100 elements each with an atomic number assigned to it according to the number of protons in its nucleus.

Some atoms of the same element may have one to several neutrons more than do other atoms of the particular element, e.g., carbon atoms may have 6, 7, or 8 neutrons but only 6 protons. These different varieties of the same element are known as isotopes. Moreover, all atoms of the same element may not have the same number of electrons.

When uncharged atoms come close together, one or more electrons may be lost from one atom and gained by the other. This phenomenon results in an imbalance

Fig. A.1 The structure of oxygen, hydrogen, and sodium atoms.

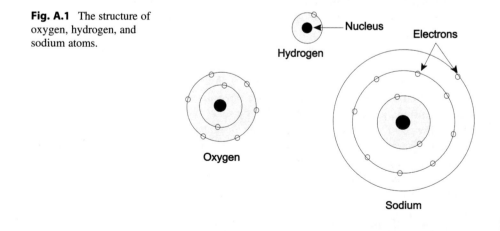

between electrons and protons in each of the two interacting atoms that imposes a negative charge on the one that gained the electron(s) and a positive charge on the one that lost the electron(s). The charge on the atom is equal to the number of electrons gained or lost (-1 or $+1$ charge per electron). Charged atoms are called ions, but the normal atom is uncharged. In the periodic table of chemical elements, an element is assumed uncharged and to have equal numbers of protons and electrons.

The mass of atoms results almost entirely from their neutrons and protons. The masses of the two entities are almost identical; 1.6726×10^{-24} g for one proton and 1.6749×10^{-24} g for one neutron. Thus, their atomic masses usually are considered unity in determining relative atomic masses of elements. An electron has a mass of 9.1×10^{-28} g—nearly 2000 times less than the masses of protons and neutrons. The mass of electrons is omitted in atomic mass calculations for elements. The atomic mass of elements increases as the atomic number increases, because the number of protons and neutrons in the nucleus increases with greater atomic number. The atomic mass of a given element differs among its isotopes because some isotopes have more neutrons than others, while all isotopes of an element have the same number of protons. The loss or gain of electrons by atoms forming ions is not considered to affect atomic mass.

Because of the different natural isotopes of atoms of a particular element, the atomic mass typically listed in the periodic table for elements does not equal to the sum of the masses of the protons and neutrons contained in atoms of these elements. This results because the atomic masses typically reported for elements represent the average atomic masses of their isotopes. For example, copper typically has 29 protons and 34 neutrons, and the atomic mass of the most common isotope would be 63 based on addition of neutrons and protons. However, the atomic mass reported in the periodic table of copper is 63.546. The additional mass results from the effects of averaging the atomic masses of the copper isotopes. Atomic masses of some common elements are provided (Table A.1).

The atomic mass is very important in stoichiometric relationships in reactions of atoms and molecules. The relative molecular mass (or weight) of a molecule is the sum of the atomic masses of the atoms contained in the molecule. Thus, when sodium atoms (relative atomic mass of 22.99) react with chlorine atoms (relative atomic mass of 35.45) to form sodium chloride, the reaction will always be in the proportion of 22.99 sodium to 35.45 chlorine, and the molecular mass of sodium will be the sum of the atomic masses of sodium and chloride or 58.44. Atomic and molecular masses may be reported in any unit of mass (or weight), but the most common is the gram. The mass or weight of atoms (Table A.1) often is referred to as the gram atomic mass (or weight) and the molecular mass of molecules usually is referred to as the gram molecular weight.

Each element is assigned a symbol, e.g., H for hydrogen, O for oxygen, N for nitrogen, S for sulfur, C for carbon, and Ca for calcium. But, because of the large number of elements, it was not possible to have symbols suggestive of the English name for all elements. For example, sodium is Na, tin is Sn, iron is Fe, and gold is Au. The symbols must be memorized or found in reference material. The periodic

Table A.1 Selected atomic weights

Element	Symbol	Atomic weight	Element	Symbol	Atomic weight
Aluminum	Al	26.9815	Magnesium	Mg	24.305
Antimony	Sb	121.76	Manganese	Mn	54.905
Arsenic	As	74.9216	Mercury	Hg	200.59
Barium	Ba	137.327	Molybdenum	Mo	95.94
Beryllium	Be	9.0122	Nickel	Ni	58.6934
Bismuth	Bi	208.9804	Nitrogen	N	14.0067
Boron	B	10.811	Oxygen	O	15.9994
Bromine	Br	79.904	Phosphorus	P	30.9738
Cadmium	Cd	112.411	Platinum	Pt	195.078
Calcium	Ca	40.078	Potassium	K	39.0983
Carbon	C	12.0107	Selenium	Se	78.96
Chlorine	Cl	35.453	Silicon	Si	28.0855
Chromium	Cr	51.996	Silver	Ag	107.8682
Cobalt	Co	58.9332	Sodium	Na	22.9897
Copper	Cu	63.546	Strontium	Sr	87.62
Fluorine	F	18.9984	Sulfur	S	32.065
Gold	Au	196.9665	Thallium	Tl	204.3833
Helium	He	4.0026	Tin	Sn	118.71
Hydrogen	H	1.0079	Tungsten	W	183.84
Iodine	I	126.9045	Uranium	U	238.0289
Iron	Fe	55.845	Vanadium	V	50.9415
Lead	Pb	207.19	Zinc	Zn	65.39
Lithium	Li	6.941			

table of the elements is a convenient listing of the elements in a way that elements with similar chemical properties are grouped together. The periodic table is presented in various formats, but most presentations include, at minimum, each element's symbol, atomic number, and relative atomic mass and indicate the group of elements to which it belongs.

The maximum number of electrons that can occur in a shell of an atom usually is $2n^2$ where n is the number of the shell starting at the shell nearest the nucleus, i.e., two electrons in the first shell, eight in the second shell, etc. Elements with a large atomic number have several shells and some shells may not contain the maximum number of electrons. Chemical reactions among atoms involve only the electrons in the outermost shell of atoms.

The laws of thermodynamics dictate that substances spontaneously change towards their most stable states possible under existing conditions. To be stable, an atom needs two electrons in the inner shell, and at least eight electrons in its outermost shell. Chemical combinations (reactions) occur so that atoms gain or lose electrons to attain stable outer shells.

Nonmetals tend to have outer shells nearly full of electrons. For example, chlorine has seven electrons in its outer shell. If it gains one electron, it will have a stable outer shell, but the acquisition of this electron will give the chlorine atom a charge of -1. Gaining the electron causes a loss of energy and the chlorine ion is more stable than the free atom. Nonmetals tend to capture electrons. The source of electrons for nonmetals is metals which tend to have only a few electrons in their outer shells. Sodium is a metal with one electron in its outer shell. It can lose the single electron and attain a charge of $+1$ and lose energy to become more stable.

If sodium and chlorine are brought together, each sodium atom will give up an electron to each chlorine atom (Fig. A.2). The chlorine atoms will now have fewer protons than electrons and acquire a negative charge while the opposite will be true of sodium atoms. Because opposite charges attract, sodium ions and chlorine ions will combine because of the attraction of unlike charges forming sodium chloride or common salt (Fig. A.2). Bonds between sodium and chlorine in sodium chloride are called ionic bonds. The number of electrons lost or gained by an atom or the charge it acquires when it becomes an ion is the valence.

Some elements like carbon are unable to gain or lose electrons to attain stable outer shells. These elements may share electrons with other atoms. For example, carbon atoms have four electrons in their outer shells and can share electrons with four hydrogen atoms as shown in Fig. A.3. This provides stable outer shells for both

Fig. A.2 Transfer of electron from a sodium atom to a chlorine atom to form an ionic bond in sodium chloride

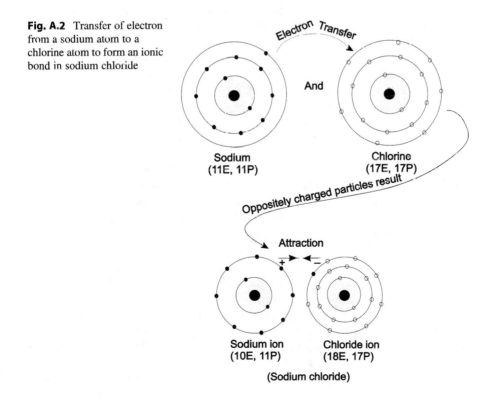

Fig. A.3 Covalent bonding
of hydrogen and carbon atoms
to form methane

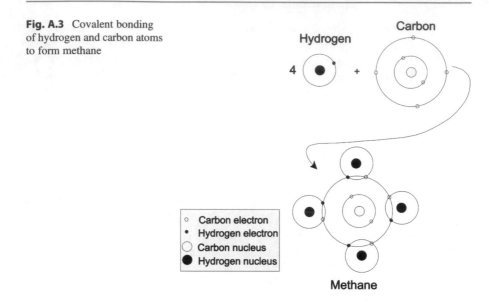

Methane

the carbon and hydrogen atoms. The resulting compound (CH_4) is methane. The chemical bond that connects each hydrogen to carbon in methane is known as a covalent bond.

When two or more elementary substances bond together, the resulting substance is called a compound. A compound has a characteristic composition and unique properties that set it aside from all other compounds. The Law of Definite Proportions holds that all samples of a given compound substance contain the same elements in the same proportions by weight. The smallest part of a substance with all of the properties of that substance is a molecule. A molecule may be composed of a single element as in the case of an elemental substance such as oxygen, sulfur, or iron or it may be composed of two or more elements as in acetic acid (CH_3COOH) which contains carbon, hydrogen, and oxygen. The gram molecular weight or gram atomic mass of a substance contains as many molecules as there are oxygen atoms in 15.9994 g of oxygen. This quantity is Avogadro's number of molecules (6.02×10^{23}), and it is known as a mole.

Elements are represented by symbols, but to represent molecules, the elemental symbols are given numerical subscripts to represent their proportions. Such a notation is called a formula. For example, molecular oxygen and nitrogen have the formulas O_2 and N_2, respectively. Sodium chloride is NaCl and sodium carbonate is Na_2CO_3. The molecular weights of molecules can be determined by summing the atomic weights of all of the constituent elements. If an element has a subscript, its atomic weight must be summed the number of times indicated by the subscript. One molecular weight of a substance in grams is a gram molecular weight or a mole as illustrated in Ex. A.1. The percentage of an element in a compound is determined by dividing the weight of the element by the formula weight of the compound and multiplying by 100 (Ex. A.2).

Ex. A.1 *The molecular weights of O_2 and $CaCO_3$ will be calculated from their formulas.*
   Solution:

O_2: *The atomic weight of O is 15.9994 (Table A.1), but it usually is rounded to 16, thus $16 \times 2 = 32$ g/mole.*
$CaCO_3$: *The atomic weights of Ca, C, and O are 40.08, 12.01, and 16, respectively (Table A.1), thus $40.08 + 12.01 + 3(16) = 100.09$ g/mole.*

The percentage composition of substances can be estimated from their formulas because of the Law of Definite Proportions.

Ex. A.2 *The percentage Cu in $CuSO_4 \cdot 5H_2O$ will be calculated.*
   Solution:
   From Table A.1, *the atomic weights of Cu, S, O, and H are 63.55, 32.06, 16.00 and 1.01, respectively. Thus, the molecular weight of $CuSO_4 \cdot 5H_2O$ is $63.55 + 32.06 + 9(16) + 10(1.01) = 249.7$ g. The percentage Cu is*

$$\frac{Cu}{CuSO_4 \cdot 5H_2O} \times 100 \;\; or \;\; \frac{63.55}{249.71} \times 100 = 25.4\%.$$

Some compounds dissociate into ions in water, e.g. sodium nitrate ($NaNO_3$) dissociates into Na^+ and NO_3^-. The ionic weight of a complex ion such as nitrate is calculated in the same manner as for the molecular weight of a compound. Of course, the weight of Na^+ is the same as the atomic weight of sodium.

Definition of a Solution

A solution consists of a solvent and a solute. A solvent is defined as the medium in which another substance—the solute—dissolves. In an aqueous sodium chlorine solution, water is the solvent and sodium chloride is the solute. Miscibility refers to a solute and solvent mixing in all proportions to form a homogenous mixture or solution. A true solution is by definition a homogenous mixture of two or more components that cannot be separated into their individual ionic or molecular components. Most solutes are only partially miscible in water. If a soluble compound such as sodium chloride is mixed with water in progressively increasing amounts, the concentration of the solute will reach a constant level. Such a solution is said to be saturated, and the amount of the sodium chloride held in solution is the solubility of sodium chloride in water at the particular temperature. Any more sodium chloride added to the saturated solution will settle to the bottom without dissolving.

Solubilities of chemical compounds often are reported as grams of solute in 100 mL of solution. Chemists generally consider substances that will dissolve to the extent of 0.1 g/100 mL in water as soluble. A general list of classes of soluble compounds follows:

- nitrates
- bromides, chlorides, and iodides—except those of lead, silver, and mercury
- sulfates—except those of calcium, strontium, lead, and barium
- carbonates and phosphates of sodium, potassium, and ammonium
- sulfides of alkali and alkaline earth metals and ammonium.

Much of this book is devoted to a discussion of factors controlling concentrations of dissolved matter in natural waters. This dissolved matter in water includes inorganic ions and compounds, organic compounds, and atmospheric gases, and much of water quality depends upon the concentrations of solutes in water.

Methods of Expressing Solute Strength

The solubility of a solute in a saturated solution or its concentration in an unsaturated solution may be expressed in several ways—the most common of which will be described.

Molarity

In a molar solution, the solute strength is expressed in moles of solute per liter, e.g., a 1 molar (1 M) solution contains 1 mole of solute in 1 L of solution. In other words, a 1 M solution of NaCl consists of 1 mole (58.44 g NaCl) in 1 liter. Note: this is not the same as putting 1 mole of NaCl in 1 L of solvent. It is 1 mole of NaCl contained in 1 L of solution. The NaCl is dissolved in the solvent and the solution diluted to 1 l with additional solvent. Calculations for a molar solution are shown in Ex. A.3.

Ex. A.3 *How much Na_2CO_3 must be dissolved and diluted to 1 L to give a 0.25 M solution?*

Solution:
The molecular weight of Na_2CO_3 is.
$2\ Na = 22.99\ g/mol \times 2 = 45.98\ g$
$1\ C = 12.01\ g/mol \times 1 = 12.01\ g$
$3O = 16\ g/mol \times 3 = \dfrac{48.0\ g}{105.99\ g}$

$$105.99\ g/mole \times 0.25\ mol/L = 26.5\ g/L.$$

Thus, 26.5 g Na_2CO_3 diluted to 1 L in distilled water gives 0.25 M Na_2CO_3.

Molality

Not to be confused with a 1 molar solution, a 1 molal solution contains 1 molecular weight of a solute in 1 kg of solvent. The unit of molality is moles of solute per kilogram of solvent, and it is abbreviated as m or *m*.

Ex. A.4 *The preparation of a 0.3 m KCl solution is illustrated.*
   Solution:
   The molecular weight of KCl is.
   K = 39.1 g/mol
   Cl = 35.45 g/mol
   KCl = 74.55 g/mol

$$74.55 \ g/mol \times 0.3 \ mol/kg = 22.36 \ g \ KCl/kg \ solvent.$$

Thus, to make the 0.3 m KCl solution weigh 22.36 kg KCl and dissolve it in 1 kg of distilled water (or other solvent).

Formality

There also is a formal concentration which is calculated as the number of moles of a substance in a liter of solution. The formal solution is denoted by the symbol F and represents the formula weights of solute per liter of solution. The formal solution is seldom used in water quality.

Normality

Because reactions occur on an equivalent weight basis, it is often more convenient to express concentration in equivalents per liter instead of moles per liter. A solution containing 1 gram equivalent weight of solute per liter is a 1 normal (1 N) solution. In cases where the equivalent weight and formula weight of a compound are equal, a 1 M solution and a 1 N solution have identical concentrations of solute. This is the case with HCl and NaCl, but with H_2SO_4 or Na_2SO_4, a 1 M solution would be a 2 N solution.

   The following rules may be used to compute equivalent weights of most reactants: (1) the equivalent weight of acids and bases equal their molecular (formula) weights divided by their number of reactive hydrogen or hydroxyl ions; (2) the equivalent weights of salts equal their molecular weights divided by the product of the number and valence of either cation component (positively charged ion) or anion component (negatively charged ion); (3) the equivalent weights of oxidizing and reducing agents may be determined by dividing their molecular weights by the number of electrons transferred per molecular weight in oxidation-reduction reactions.

Oxidation-reduction reactions usually are more troublesome to students than other types of reactions. An example of an oxidation reduction reaction is provided below in which manganese sulfate reacts with molecular oxygen:

$$2MnSO_4 + 4NaOH + O_2 \rightarrow 2MnO_2 + 2Na_2SO_4 + 2H_2O \qquad (A.1)$$

Manganese in $MnSO_4$ has a valence of +2 and sulfate has a valence of -2. In manganese dioxide the valance of manganese is +4 (notice each of the two oxygens have a valence of -2 making a total valence of -4 for oxygen). Manganese was oxidized, because its valence increased. Molecular oxygen with valence of 0 was reduced to a valence of -2 in manganese dioxide. Each molecule of manganese sulfate (the reducing agent) lost two electrons which were gained by oxygen (the oxidizing agent). The equivalent weight of magnesium sulfate in this reaction is its formula weight divided by two.

Additional examples illustrating calculations of equivalent weights and normalities of acids, bases, and salts are provided in Exs. A.5, A.6, and A.7.

Ex. A.5 *What is the equivalent weight of sodium carbonate when it reacts with hydrochloric acid?*
   Solution:
   The reaction is: $Na_2CO_3 + 2HCl = 2NaCl + CO_2 + H_2O$.
   One sodium carbonate molecule reacts with two hydrochloric acid molecules. Thus, the equivalent weight of sodium carbonate is

$$\frac{Na_2CO_3}{2} = \frac{106}{2} = 53 \ g.$$

Ex. A.6 *What are the equivalent weights of sulfuric acid (H_2SO_4), nitric acid (HNO_3), and aluminum hydroxide [$Al(OH)_3$]?*
   Solution:

(i) *Sulfuric acid has two available hydrogen ions*

$$H_2SO_4 \rightarrow 2H^+ + SO_4^{2-}.$$

Thus, the equivalent weight is

$$\frac{H_2SO_4}{2} = \frac{98}{2} = 49 \ g.$$

(ii) *Nitric acid has one available hydrogen ion*

$$HNO_3 \rightarrow H^+ + NO_3^-.$$

The equivalent weight is

$$\frac{HNO_3}{1} = \frac{63}{1} = 63 \ g.$$

(iii) *Aluminum hydroxide has three available hydroxide ions*

$$Al(OH)_3 = Al^{3+} + 3OH^-.$$

The equivalent weight is

$$\frac{Al(OH)_3}{3} = \frac{77.98}{3} = 25.99 \ g.$$

Ex. A.7 *How much Na_2CO_3 must be dissolved and diluted to 1 L to give a 0.05 N solution?*
   Solution:

$$Na_2CO_3 \rightarrow 2Na^+ + CO_3^{2-}.$$

The formula weight must be divided by 2 because $2Na^+ = 2$, or because CO_3^{2-}
$= -2$. The sign difference ($2Na^+ = +2$ and $CO_3^{2-} = -2$) with respect to valence) does
not matter in calculating equivalent weight.

$$\frac{106 \ g \ Na_2CO_3/mole}{2} = 53 \ g/equiv.$$

$$53 \ g/equiv. \ \times 0.05 \ equiv./L = 2.65 \ g/L.$$

In expressing solute strength for dilute solutions, it is convenient to use milligrams instead of grams. Thus, we have millimoles (mmol), millimolar (mM), millimoles/L (mmol/L that is the same as mM), milliequivalents (meq), and milliequivalents/liter (meq/L). A 0.001 M solution is a 1 mM solution.

Weight per Unit Volume

In water quality, concentrations often are expressed in weight of a substance per liter. The usual procedure is to report milligrams of a substance per liter (Ex. A.8).

Ex. A.8 *What is the concentration of K in a solution that is 0.1 N with respect to KNO_3?*
   Solution:

$$KNO_3 = 101.1 \ g/equiv.$$

$$101.1 \ g/equiv. \ \times 0.1 \ equiv./L = 10.11 g/L.$$

The amount of K in 10.11 g of KNO$_3$ is

$$\frac{39.1}{101.1} \times 10.11 \ g/L = 3.9 \ g/L.$$

$$3.9 \ g/L \times 1,000 \ mg/g = 3,900 \ mg/L.$$

The unit, milligrams per liter, is equivalent to the unit, parts per million (ppm) as shown in Ex. A.9; it is common in water quality to use parts per million (ppm) interchangeably with milligrams per liter.

Ex. A.9 *It will be demonstrated that 1 mg/L = 1 ppm for aqueous solutions.*
    <u>*Solution:*</u>

$$\frac{1 \ mg}{1 \ L} = \frac{1 \ mg}{1 \ kg} = \frac{1 \ mg}{1,000 \ g} = \frac{1 \ mg}{1,000,000 \ mg} = 1 \ ppm.$$

It also is common in water quality to express concentration of minor constituents in micrograms per liter (µg/L); and, of course, 1 µg/L = 0.001 mg/L. To convert milligrams per liter to micrograms per liter, simply move the decimal place to the right three places, e.g., 0.05 mg/L = 50 µg/L, and vice versa. Sometimes micrograms per liter will be expressed as parts per billion (ppb). The rationale for this is easily seen if the calculation in Ex. A.9 is begun with 1 µg/L and we get 1 µg/ 1,000,000,000 µg.

It also is convenient to express the strength of more concentrated solutions in parts per thousand or ppt (Ex. A.10). One part per thousand is equal to 1 g/L because there are 1000 g in 1 L. It also is equal to 1000 mg/L. An alternative way of indicating concentration in parts per thousand is the symbol ‰.

Ex. A.10 *Express the concentration of sodium chloride in a 0.1 M solution as parts per thousand and milligrams per liter.*
    <u>*Solution:*</u>

$$0.1 \ M \times 58.45 \ g \ NaCl/mole = 5.845 \ g/L.$$

The salinity is 5.845 ppt that is equal to 5845 mg/L.

It is easy to convert water quality data in milligrams per liter to molar or normal concentrations (Ex. A.11).

Ex. A.11 *Molarity and normality of 100 mg/L calcium will be calculated.*
  Solution:

(i) $\dfrac{100 \ mg/L}{40.08 \ mg \ Ca/mmol} = 2.495 \ mmol/L \ or \ 0.0025 \ M.$

(ii) *Calcium is divalent, so*

$$\frac{100 \ mg/L}{20.04 \ mg/meq} = 4.99 \ meq/L \ or \ 0.005 \ N.$$

Concentrations of solutions often are in milligrams per liter with no regard for molarity of normality as shown in Ex. A.12.

Ex. A.12 *How much magnesium sulfate ($MgSO_4 \cdot 7H_2O$) must be dissolved and made to 1000 mL final volume and to provide a magnesium concentration of 100 mg/L?*
  Solution:

$$\begin{array}{ccc} x & & 100 \ mg/L \\ MgSO_4 \cdot 7H_2O & = & Mg \\ mw = 246.31 \ g & & mw = 24.31 \ g \end{array}$$

rearranging,

$$x = \frac{24,631}{24.31} = 1,013.2 \ mg/L \ of \ MgSO_4 \cdot 7H_2O.$$

Weight Relationships in Reactions

Stoichiometry is the area of chemistry that considers weight relationships among reactants and products in chemical reactions. The equations for reactions allow weights of reactants and products to be calculated. The reaction of sodium hydroxide to neutralize hydrochloric acid is $NaOH + HCl = NaCl + H_2O$. Each mole of HCl requires 1 mol of NaOH for neutralization. Thus 40.00 g NaOH reacts with 36.46 g of HCl to yield 58.44 g NaCl and 18.02 g H_2O. The weight on each side of the equation remains the same before and after a reaction that goes to completion (there were 76.46 g of reactants and 76.46 g of products in this instance). Calculations usually are more complex than for hydrochloric acid and sodium hydroxide as illustrated in Ex. A.13.

Ex. A.13 *The amount of calcium carbonate necessary to neutralize the acidity for 100 kg of aluminum chloride will be calculated.*
   Solution:
   Al(Cl)$_3$ dissolves releasing Al^{3+}

$$Al(Cl)_3 = Al^{3+} + 3Cl^-$$

Al^{3+} hydrolyzes producing hydrogen ion (H$^+$) which is a source of acidity:

$$Al^{3+} + 3H_2O = Al(OH)_3 \downarrow + 3H^+.$$

Hydrogen ion is neutralized by calcium carbonate (CaCO$_3$):

$$3H^+ + 1.5\ CaCO_3 = 1.5\ Ca^{2+} + 1.5\ CO_2 + 1.5\ H_2O.$$

Weight of Al in 100 kg of AlCl$_3$ is

$$
\begin{array}{ccc}
100\ kg & X & \\
AlCl_3 = & Al^{3+} & + 3Cl^- \\
133.34\ g/mol & 26.98\ g/at.wt. &
\end{array}
$$

Al^{3+} = 20.23 kg Al^{3+}.
   Weight of CaCO$_3$ needed to neutralize the acidity is.
   Al^{3+} = 3H$^+$; therefore, Al^{3+} = 1.5 CaCO$_3$.

$$
\begin{array}{cc}
20.23\ kg & X \\
Al^{3+} = & 1.5\ CaCO_3 \\
26.98\ g/at.wt. & 150\ g
\end{array}
$$

X = 112.5 kg CaCO$_3$.

Two more examples will be provided.

Ex. A.14 *The amount of carbon dioxide required to produce 100 kg organic matter in photosynthesis will be determined.*
   Solution:
   The photosynthesis equation is

$$6CO_2 + 6H_2O = C_6H_{12}O_6 + 6O_2.$$

The relationship of CO$_2$ to carbon in organic matter is CO$_2$ = CH$_2$O.
and

$$
\begin{array}{cc}
X & 100\ kg \\
CO_2 = & CH_2O \\
44\ g/mol & 30\ g/mol
\end{array}
$$

CO$_2$ = 146.7 kg.

Ex. A.15 *The amount of molecular oxygen necessary to oxidize 5 kg of elemental sulfur to sulfuric acid will be computed.*
   Solution:
   The reaction is:

$$S + 1.5O_2 + H_2O = H_2SO_4.$$

The relationship of elemental sulfur to molecular oxygen is S = 1.5 O₂:

$$\frac{5\ kg}{S} = \frac{X}{1.5O_2}$$
$$\frac{}{32} \quad \frac{}{48}$$

$O_2 = 7.5\ kg.$
   Note: In this particular instance, the amount of oxygen required is 1.5 times the amount of sulfur oxidized. This was the result of the coincidence that the molecular weight of O_2 is the same as the atomic weight of S.

Some Shortcuts

Some important water quality variables are radicals such as nitrate, nitrite, ammonia, ammonium, phosphate, sulfate, etc. Sometimes, the concentration will be given as the concentration of the radical, and other times, it will be given as the concentration of the element of interest that is contained in the radical. For example, the concentration of ammonia may be given as 1 mg NH_3/L or it may be given as 1 mg NH_3-N/L. It is possible to convert back and forth between the two methods of presenting concentration. In the case of NH_3-N, use of the factor N/NH_3 (14/17 or 0.824): 1 mg NH_3/L × 14/17 = 0.82 mg NH_3-N/L. It follows that 1 mg NH_3-N/L ÷ 14/17 = 1.21 mg NH_3/L. Similar reasoning may be used to convert between

Table A.2 Factor to multiply by concentration of an ionic radical and CO_2 to obtain the elemental equivalent

Conversion	Factor
NH_3 to NH_3-N	0.824
NH_4^+ to NH_4^+-N	0.778
NO_3^- to NO_3^--N	0.226
NO_2^- to NO_2^--N	0.304
PO_4^{3-} to PO_4^{3-}-P	0.326
HPO_4^{2-} to HPO_4^{2-}-P	0.323
$H_2PO_4^-$ to $H_2PO_4^-$-P	0.320
SO_4^{2-} to SO_4^{2-}-S	0.333
H_2S to H_2S-S	0.941
CO_2 to CO_2-C	0.273

The concentrations in elemental form can be converted to the ion concentration by dividing by the factor, e.g., 0.226 mg/L NO_3^- N ÷ 0.226 = 1 mg/L NO_3^-.

NO_3^- and NO_3-N, NO_2^- and NO_2^--N, SO_4^{2-} and SO_4^{2-}-S, etc. Some conversion factors are given in Table A.2.

It also is useful to note that the dimensions for molarity and normality are moles per liter and equivalents per liter, respectively. Thus, multiplying molarity or normality by volume in liters gives moles and equivalents, respectively. The same logic applies for multiplying millimoles or milliequivalents per milliliter by volume in milliliters.

In some calculations, the quantity of a dissolved substance in a particular volume may be sought. It is helpful to remember that 1 mg/L is the same as 1 g/m^3, because there are 1000 L in a cubic meter. Likewise, 1 μg/L is the same as 1 mg/m^3 for the same reason.

Index

© Springer Nature Switzerland AG 2020
C. E. Boyd, *Water Quality*, https://doi.org/10.1007/978-3-030-23335-8

Printed in the United States
By Bookmasters